THE RISE OF STATISTICAL THINKING
1820-1900

THE RISE OF STATISTICAL THINKING

1820-1900

Theodore M. Porter

1986

PRINCETON UNIVERSITY PRESS

Published by Princeton University Press,
41 William Street, Princeton, New Jersey 08540
In the United Kingdom:
Princeton University Press, Guildford, Surrey

Library of Congress Cataloging in Publication Data will be
found on the last printed page of this book

ISBN 0-691-08416-5

This book has been composed in Electra

Clothbound editions of Princeton University Press books
are printed on acid-free paper, and binding materials
are chosen for strength and durability

Printed in the United States of America by
Princeton University Press
Princeton, New Jersey

This book is dedicated to my parents

CONTENTS

ABBREVIATIONS

AOB	*Annuaire de l'observatoire de Bruxelles.*
AQP	Adolphe Quetelet Papers, Bibliothèque nationale de Belgique, Brussels, Belgium.
ATBM	Wilhelm Lexis, *Abhandlungen zur Theorie der Bevölkerungs- und Moralstatistik* (Jena, 1903).
Archive	*Archive for History of Exact Sciences.*
BAAS	*Reports of the Meetings of the British Association for the Advancement of Science.*
BCCS	*Bulletin de la commission centrale de statistique* (of Belgium).
BJHS	*British Journal for the History of Science.*
Cournot	A. A. Cournot, *Exposition de la théorie des chances et des probabilités* (Paris, 1843).
DSB	Charles C. Gillispie, ed., *Dictionary of Scientific Biography* (16 vols., New York, 1970-1980).
ESP	Karl Pearson, *Early Statistical Papers* (Cambridge, 1948).
FGP	Francis Galton Papers, the Library, University College, London.
Galton	Karl Pearson, *The Life, Letters, and Labours of Francis Galton* (3 vols. in 4, Cambridge, 1914-1930).
HSPS	*Historical Studies in the Physical Sciences.*
JAI	*Journal of the Anthropological Institute of Great Britain and Ireland.*
Jbb	*Jahrbücher für Nationalökonomie und Statistik.*
JCMP	James Clerk Maxwell Papers, Cambridge University Library, Cambridge, England.
JHB	*Journal of the History of Biology.*
JRSS	*Journal of the Royal Statistical Society* (since 1887).
	Journal of the Statistical Society (1873-1886).
	Journal of the Statistical Society of London (1838-1872).
MacKenzie	Donald MacKenzie, *Statistics in Britain, 1865-1930: The Social Construction of Scientific Knowledge* (Edinburgh, 1981).

Maxwell	Lewis Campbell and William Garnett, *The Life of James Clerk Maxwell* (London, 1882).
NMAB	*Nouveaux mémoires de l'académie royale des sciences et belles-lettres de Bruxelles* (after 1840, *de Belgique*).
Oeuvres	Pierre Simon de Laplace, *Oeuvres complètes* (14 vols., Paris, 1878-1912).
Phil Mag	*London, Edinburgh and Dublin Philosophical Magazine and Journal of Science.*
Phil Trans	*Philosophical Transactions of the Royal Society of London.*
PPE	Francis Ysidro Edgeworth, *Papers Relating to Political Economy* (3 vols., London, 1925).
PRI	*Proceedings of the Meetings of the Members of the Royal Institution of Great Britain.*
Prob Rev	Lorenz Krüger, Lorraine Daston, Michael Heidelberger, eds., *The Probabilistic Revolution*, vol. 1, *Ideas in History* (Cambridge, Mass., 1987).
PRSL	*Proceedings of the Royal Society of London.*
PS	Ludwig Boltzmann, *Populäre Schriften* (Leipzig, 1905).
SHSP1	Egon S. Pearson and Maurice Kendall, eds., *Studies in the History of Statistics and Probability*, vol. 1 (London, 1970).
SHSP2	Maurice Kendall and R. L. Plackett, eds., *Studies in the History of Statistics and Probability*, vol. 2 (New York, 1977).
Stigler	Stephen M. Stigler, *The History of Statistics: The Measurement of Uncertainty before* 1900 (Cambridge, Mass., 1986).
SP	W. D. Niven, ed., *Scientific Papers of James Clerk Maxwell* (2 vols., Cambridge, Eng., 1890).
TPSC	*Transactions of the Philosophical Society of Cambridge.*
WA	Ludwig Boltzmann, *Wissenschaftliche Abhandlungen*, Fritz Hasenöhrl, ed. (3 vols., Leipzig, 1909).
ZGSW	*Zeitschrift für die gesammte Staatswissenschaft.*

PREFACE

This study is a history of statistical thinking as it developed among social scientists, biologists, and physicists. Their aim was to extend the reach of exact science into the social and biological domains by analyzing large-scale phenomena quantitatively in terms of the collective behavior of numerous individuals. Those individuals were essentially unknowable, either because they were too small to be seen or because they were exceedingly numerous and diverse. Statistics hence served not just as an adjunct to observation, but as the basis of theory, and the growth of statistical thinking accompanied the rise of several new areas of science during the nineteenth century. Most notable among these were demography and social statistics, statistical mechanics, and population genetics. The study of variation within these fields was the direct source of modern mathematical statistics, which in fact grew out of the biological study of heredity around 1900.

This year should mark a real advance in the appreciation and understanding of statistical thinking and its place in the development of modern thought, because five works on the history of probability and statistics are currently in press or have recently appeared in print. The others include books by Lorraine Daston, Ian Hacking, and Stephen Stigler, and a two-volume collection from a 1982-83 research project directed by Lorenz Krüger and Michael Heidelberger at the Zentrum für interdisziplinäre Forschung of the University of Bielefeld, in which Professors Daston, Hacking, and Stigler and I participated. Had I known how much work was going on concerning the history of statistics and probability when I began my dissertation in 1979, I would have become timid and chosen another topic. As it happened, there is no problem. These works are all quite different, reflecting the rich diversity of sources and effects of statistical thinking. I have nevertheless learned a great deal from my fellow probabilists, often in a form too subtle to be properly credited in footnotes. I must be content to offer my deep appreciation to the ZiF, and to my colleagues there during that splendid year.

A number of people have read all or part of this manuscript in one form or another and offered helpful suggestions or criticisms. These include Diane Campbell, I. B. Cohen, Lorraine Daston, Gerald Geison, Charles Gillispie, Ian Hacking, Bruce Hitchner, Robert Horvath, Mark

Kac, Daniel Kevles, Thomas Kuhn, Mary Morgan, JoAnn Morse, James Secord, John Servos, Stephen Stigler, Geoffrey Sutton, and Norton Wise. To them I am most grateful. I wish also to thank my teachers, especially Harold Bacon, Robert Fox, Gerald Geison, Thomas Kuhn, Paul Robinson, Carl Schorske, and John Servos. For professional guidance, and also, as it happens, for much intellectual stimulation, I thank also Daniel J. Kevles (who has recently published a book on a closely related subject), Norton Wise, and Glenn and Rita Ricardo Campbell. By much my greatest scholarly debt is to Charles Gillispie, my graduate advisor, who was extraordinarily generous with his time, and who never relaxed his standards. This book has benefited mainly from his criticism, but his humor and frequent encouragement were also greatly appreciated.

Parts of this book have appeared in different form in *Historical Studies in the Physical Sciences*, *British Journal for the History of Science*, and in Lorraine Daston et al., eds., *The Probabilistic Revolution* (Bradford Books/MIT Press). For assistance in putting the manuscript into its present form I thank Gail Peterson, who typed it, Robin Good, who checked references for me, and M. Roy Harris, for computer translation. I thank also the secretarial staff of the University of Virginia History Department, especially Lottie McCauley and Ella Wood. This work was supported by a Princeton University fellowship and an Andrew Mellon postdoctoral instructorship at the California Institute of Technology. Acknowledgment is due to the Académie royale de Belgique, Cambridge University Library, and the Library, University College London, for permission to quote from the Quetelet, Maxwell, and Galton papers, respectively.

Geoffrey Sutton and JoAnn Morse offered valuable friendship while we were graduate students and since. Many of my ideas about the history of science in general, and much else, are the result of informal discussions with them. It is not easy to know how to thank my parents. Dedicating this book to them is no substitute, but will perhaps be accepted as a token of much more. Diane Campbell has been one of my most helpful critics, and also an ideal companion in the challenging task of making two academic careers within one home.

THEODORE M. PORTER
19 September 1985

THE RISE OF STATISTICAL THINKING
1820-1900

INTRODUCTION

Statistics has become known in the twentieth century as the mathematical tool for analyzing experimental and observational data. Enshrined by public policy as the only reliable basis for judgments as to the efficacy of medical procedures or the safety of chemicals, and adopted by business for such uses as industrial quality control, it is evidently among the products of science whose influence on public and private life has been most pervasive. Statistical analysis has also come to be seen in many scientific disciplines as indispensable for drawing reliable conclusions from empirical results. For some modern fields, such as quantitative genetics, statistical mechanics, and the psychological field of intelligence testing, statistical mathematics is inseparable from actual theory. Not since the invention of the calculus, if ever, has a new field of mathematics found so extensive a domain of application.

The statistical tools used in the modern sciences have virtually all been worked out during the last century. The foundations of mathematical statistics were laid between 1890 and 1930, and the principal families of techniques for analyzing numerical data were established during the same period. The achievement of the founders of mathematical statistics—Pearson, Spearman, Yule, Gosset, Fisher, and others—was formidable, and is justly venerated by their scientific heirs. There is, however, another story to be told, of the background that made this burst of statistical innovation possible. The development of statistics was necessarily preceded by its invention. This was the contribution of the nineteenth century, the culmination of a tradition of statistical thinking that embraced writers from a variety of backgrounds working in areas which were otherwise unconnected.

The invention of statistics was the recognition of a distinct and widely applicable set of procedures based on mathematical probability for studying mass phenomena. Statistics was and continues to be seen as especially valuable for uncovering causal relationships where the individual events are either concealed from view or are highly variable and subject to a host of influences. The identification of statistics as a category of knowledge was first of all a scientific accomplishment, and not a purely mathematical one. To be sure, the central role of probability theory in the history as well as the logic of statistics is plain. But most of the probability mathematics needed by the founders of statistics had been available for almost a century, since the time of Laplace and Gauss. Indeed, practical techniques for the use of such mathematics in the analysis of numerical data were worked out with great sophistication during the early decades of the nineteenth century in the form of error theory, which was used widely in geodesy and observational astronomy. In retrospect, the history of error theory seems to abound in precursors to the principal accomplishments of mathematical statistics. If statistics were just mathematics, the "anticipations" of the error theorists would leave little basis for the claim that Quetelet, Lexis, and Galton were original thinkers in this field.

The identification of precursors, however, is almost always misleading, and it is no less so here. Identical mathematical formulations must yet be viewed as different if they are interpreted differently, especially when their purpose is scientific and not purely mathematical. As Stephen Stigler shows in his new book, the effective use of probabilistic techniques for estimating uncertainty in astronomy and geodesy was by no means sufficient to enable social scientists to apply similar analysis to the problems of their disciplines.[1] Decades after a sophisticated theory of errors had been developed, there remained the great problem of finding points of contact between the mathematical formulas of error theory and the scientific objects of the social and biological sciences, where variation was genuine and important. Once Galton and Pearson had in some measure solved that problem, the formulations of the error theorists could readily be seen to be applicable. In this new context, however, the analysis of error had become something quite different, a method for studying the causes of variation, and not just for measuring it. Only through their successful application to the refractory but rich problems of the social and biological sciences did the probabilistic techniques of error analysis grow into the powerful and flexible method of analysis that we know as mathematical statistics.

The study of variable mass phenomena had different origins from mathematical probability and error theory. It began instead with the development of numerical social science and the formation of what was regarded during the late nineteenth century as the characteristically statistical viewpoint. John Theodore Merz recognized this viewpoint in his 1904 *History of European Thought in the Nineteenth Century*,

[1] See Stigler.

includes a chapter on "the statistical view of nature."[2] Merz's phrase referred not only to the mathematical techniques for analyzing data derived from probability theory, but also, and principally, to the strategy of investigation derived form the numerical science of society and of states. It was this science for which the term "statistics" had been adapted early in the nineteenth century. Its practitioners were initially called "statists," and only late in the nineteenth century did they assume the title of statisticians.

It is no accident that a quantitative, empirical social science of the nineteenth century should have given its name to a branch of applied mathematics, for the contribution of "statists" to the style of reasoning associated with the new mathematics was a fundamental one. To begin, statists familiarized the scientific world and the educated public with the use of aggregate numbers and mean values for studying an inherently variable object. Statistical writers persuaded their contemporaries that systems consisting of numerous autonomous individuals can be studied at a higher level than that of the diverse atomic constituents. They taught them that such systems could be presumed to generate large-scale order and regularity which would be virtually unaffected by the caprice that seemed to prevail in the actions of individuals. Since significant changes in the state of the system would appear only as a consequence of proportionately large causes, a science could be formulated using relative frequencies as its elemental data.

Practicing statists, of course, did not identify their enterprise with any doctrine so abstract as this. Theirs was a world of progress and discontent, of surveys and census figures, into which advanced mathematics and abstruse philosophy rarely intruded. They were, for the most part, reformers and bureaucrats. As nineteenth-century liberals, they were impressed by the power and dynamism of that complex entity, society, and were pleased to find evidence that it exhibited a stability which seemed not to be dependent on the intermittent wisdom of governing authorities. Hence they were delighted by the uniformity from year to year which was found to characterize not only natural events like births and deaths, but also voluntary acts such as marriages and even seemingly senseless and irrational phenomena like crime and suicide. From this was born Adolphe Quetelet's doctrine of "statistical law," which held that these regularities would continue into the future because they

[2] John Theodore Merz, *A History of European Thought in the Nineteenth Century* (1904-1912; 4 vols., New York, 1965), vol. 2, pp. 548-626.

arose necessarily from an underlying stability of the "state of society." Using statistics, it seemed to be possible to uncover general truths about mass phenomena even though the causes of each individual action were unknown and might be wholly inaccessible.

The doctrine that order is to be found in large numbers is the leitmotif of nineteenth-century statistical thinking. The regularity of crime, suicide, and marriage when considered in the mass was invoked repeatedly to justify the application of statistical methods to problems in biology, physics, and economics by writers like Francis Galton, James Clerk Maxwell, Ludwig Boltzmann, Wilhelm Lexis, and F. Y. Edgeworth. Indeed, the use of probability relationships to model real variation in natural phenomena was initially made possible by the recognition of analogies between the objects of these sciences and those of social statistics. We find them not only in veiled allusions; they are openly and explicitly developed in both popular and technical writings.

Thus this book is, in one sense, a history of the influence of ideas developed within social statistics. It is a study of the mathematical expression of what Ernst Mayr calls population thinking,[3] a phrase which points no less clearly than statistics to sources in the human sciences. As in all noteworthy cases of intellectual influence, however, the beneficiaries here were no mere passive recipients of social-scientific dogma. The leading characters in this story were "moral statisticians," economists, kinetic gas theorists, and biometricians. The objects of their work required them to find some way of studying mass phenomena profitably without first having to attain detailed knowledge of the constituent individuals. They were successful precisely because they were able to adapt existing methods and concepts to new objects. In doing so, they contributed as much to the statistical method as to their particular fields.

Just as statistical reasoning was closely associated with the idea of large-scale regularity during the nineteenth century, the history of statistical mathematics before 1890 is, for the most part, a history of the normal or Gaussian distribution. This is the familiar bell-shaped curve, known to nineteenth-century writers as the astronomical error law. Although it had earlier been used in connection with the classical "doctrine of chances," it became closely associated with astronomy as a consequence of its incorporation into the method of least squares for reducing astronomical observations. Since stellar objects have a real po-

[3] Ernst Mayr, *The Growth of Biological Thought: Diversity, Evolution, and Inheritance* (Cambridge, Mass., 1982), pp. 45-47.

sition at a given time, the existence of variation among observations was quite naturally interpreted as the product of error, and the error curve, accordingly, was conceived as descriptive of the imperfections of instruments and of the senses. The practical task of fitting curves of various sorts to astronomical observations, as well as the search for rigorous foundations for the method of least squares, led to much sophisticated mathematical work; but since the object of the exercise was always to manage or estimate error, there was little incentive to study variation for its own sake.

A major transition in thinking about the error law was initiated by the Belgian Adolphe Quetelet. Quetelet journeyed to Paris in 1823 to learn observational astronomy, but while there he was introduced to mathematical probability and was infected by the belief in its universal applicability, as championed by Laplace, Poisson, and Joseph Fourier. The new social science of statistics became for him a branch of the "social physics," patterned closely on celestial physics, for which he wished to lay the foundation. Every possible concept from physics was given a social analogue in this gradualist metaphysic of society, and the error law finally found its place in 1844 as the formula governing deviations from an idealized "average man." Quetelet interpreted the applicability of this law as confirmation that human variability was fundamentally error, but the effect of his discovery was to begin the process by which the error law became a distribution formula, governing variation which was itself seen to have far greater interest than any mere mean value.

The further development of statistical mathematics was, until the end of the nineteenth century, largely the result of work in other natural and social scientific disciplines. Quetelet's belief in the widespread applicability of the error law to variation of all sorts, although not his interpretation of it, won acceptance by the ablest workers on statistical mathematics of the late nineteenth century. James Clerk Maxwell, who learned of Quetelet's use of the error law from an essay review of one of Quetelet's books by John Herschel, proposed that the same formula governed the distribution of molecular velocities in a gas. He, along with Ludwig Boltzmann, made of statistical gas theory one of the great achievements of late nineteenth-century physics and formed an important part of the background to the new quantum theory as well as to Willard Gibbs's statistical mechanics. In social science, Wilhelm Lexis used similar formalism to provide a measure of the stability of statistical series, and Francis Edgeworth showed how error analysis and related techniques might be applied fruitfully to problems in economics such as index numbers. Francis Galton, who was introduced to Quetelet's ideas by the geographer William Spottiswoode, employed the error curve in his study of heredity and, in the end, found that his index of hereditary regression was in fact applicable as a general tool of statistical mathematics to the study of variation in data of all sorts.

At the same time as the mathematics of statistics was being advanced through its application to new objects, the statistical approach began to be seen as distinctive and even as challenging for conventional views of science and of natural law. Quetelet's contemporaries held no uniform view as to the nature of statistical science, but they were all in accord that their discipline consisted of the application of the tried and true method of natural science to a social object. The assertion of what may be called statistical determinism in relation to man and society, however, provoked a backlash of opposition to statistics and a critical analysis of the nature of statistical reasoning. On one level, this reassessment led to wide adoption of the view that statistics could provide little of scientific worth so long as attention was focused on mean values rather than on variation. More abstractly, critics of the idea of statistical law put forward by Quetelet and the historian Henry Thomas Buckle began to argue that the statistical method was inherently an imperfect one, applicable precisely because the remoteness or intrinsic variability of the constituent objects rendered exact deterministic knowledge inaccessible. The elaboration of this view, especially by kinetic theorist and social thinkers, inaugurated what Ian Hacking calls the "erosion of determinism," which has profoundly influenced the scientific world view of the twentieth century.

The modern field of mathematical statistics arose from the diverse applications to which statistical ideas and methods were put during the nineteenth century. It became, under Pearson, Fisher, and others, a mathematical resource for a variety of disciplines for which numerical data could be obtained through experiment or observation. Before 1890, however, and indeed for some decades thereafter, statistical methods and concepts were developed not by mathematicians but by astronomers, social scientists, biologists, and physicists. The development of statistical thinking was a truly interdisciplinary phenomenon for which mathematics had no priority of position; new ideas and approaches arose as a result of the application of techniques borrowed from one or more disciplines to the very different subject matter of another. The great pioneers of statistical thinking were widely read generalists, interested in historical, philosophical, or social issues as well as in their research areas. As Karl Pearson himself pointed out in some lectures he gave on the history of statistics:

> I ought, perhaps to apologise for carrying you so far afield in these lectures. But it is impossible to understand a man's work unless you understand something of his environment. And his environment means the state of affairs social and political of his own age. You might think it possible to write a history of science in the 19th century and not touch theology or politics. I gravely doubt whether you could come down to its actual foundations, without thinking of Clifford and Du Bois Reymond and Huxley from the standpoint of theology and politics. What more removed from those fields than the subject of differential equations? What more removed from morality than the theory of Singular Solutions? Yet you would not grasp the work of De Saint-Venant or Boussinesq unless you realised that they viewed Singular Solutions as the great solution of the problem of Freewill, and I hold a letter of Clerk-Maxwell in which he states that their work on Singular Solutions is epoch making on this very account![4]

Many contributed to a great range of scientific specialties, and all were alert to developments outside their own fields. The general development of a statistical method required effective communication between diverse studies, and this investigation reveals much about the links that have given science a certain measure of unity, making it more than a collection of isolated disciplines.

Statistics has been prominent not only on the margins between disciplines, but also in the nebulous and shifting border region that separates science from nonscience. Statistics has contributed essentially to a considerable expansion of the scientific domain, but it has been for two centuries or more a singularly problematical method of science. Probability was a suspect area of mathematics almost from the beginning, though from a purely technical standpoint its accomplishments were already impressive in the time of Montmort and De Moivre. Influential writers on error theory, such as Augustin Cauchy and James Ivory, ac-

[4] Karl Pearson, *The History of Statistics in the 17th and 18th Centuries Against the Changing Background of Intellectual, Scientific, and Religious Thought*, E. S. Pearson, ed. (London, 1978), p. 360.

cepted the method of least squares but refused to rest content with the probabilistic assumptions in its underpinnings.[5] Laplace himself was inspired by his ambitions for probability to redefine scientific certainty—to limit perfect knowledge to an omniscient but imaginary demon, and to insist that while events in the world are completely determined by preexisting causes, our knowledge of their outcome is necessarily subject to a certain domain of error. The statistical approach also presented serious problems for the kinetic gas theory, which reduced the deterministic laws of thermodynamics to mere regularities. While Maxwell and Boltzmann labored to incorporate the necessary refinements into the kinetic theory, others, such as Planck's student Zermelo, argued that this connection with statistics invalidated atomism altogether.[6]

That probability seemed to imply uncertainty clearly discouraged its use by physicists. From the standpoint of social science, on the other hand, statistical method was synonymous with quantification, and while some were skeptical of the appropriateness of mathematics as a tool of sociology, many more viewed it as the key to exactitude and scientific certainty. Most statistical enthusiasts simply ignored the dependence of statistical reasoning on probability, and those who acknowledged it generally stressed the ties between probability and that most ancient and dignified among the exact sciences, astronomy. The social science of statistics, in the hands of Quetelet and his admirers, constituted a self-conscious attempt to imitate the successful strategy of natural science. Statistical quantification in social science was more commonly seen as exemplary than as problematical, and the aspirations of statistics reveal much about what were taken to be the essential features of science during the nineteenth century.

Finally, the history of statistics sheds light on the relations between abstract science and what are often seen as its applications. In truth, practice was decidedly ahead of theory during the early history of statistics, and "pure" or abstract statistics was the offspring, not the parent, of its applications. The statistical techniques and approaches that were invented by Lexis, Edgeworth, Galton, Pearson, and their successors reflected at once the particular problems to which statistical methods had

[5] James Ivory, "On the Method of Least Squares," *Phil Mag*, 65 (1825), 1-10, 81-88, 161-168. On Cauchy, see Ivo Schneider, "Laplace and the Consequences: The Status of Probability Calculus in the 19th Century," in *Prob Rev*.

[6] Ernst Zermelo, "Über mechanische Erklärungen irreversibler Vorgänge," *Annalen der Physik*, 59 (1896), 793-801.

been applied and the context of ideologies and philosophical attitudes within which statistics had been pursued. Mathematics and physics may, as the Comtean hierarchy of the sciences suggests, have logical primacy over biological and social science, but historically the situation is much more complex and interesting.

There are, to be sure, special reasons why statistical theory should be more powerfully affected by considerations external to its specific subject matter than other areas of mathematics or science. Its main task, after all, has been to provide analytical methods by which practitioners of other disciplines can analyze their numerical data, and the statistician's quest is as much for useful techniques as for timeless truths. Nevertheless, the history of statistics should not be seen in this respect as utterly unique, but as an ideal type for one aspect of the historical process through which modern science has evolved.

Some changes of terminology almost always accompany the emergence of new areas of science, especially when, as in the present case, a significantly new style of thought is involved. A preliminary discussion of some key terms ought therefore to be helpful.

"Statistics" as a plural means to us simply numbers, or more particularly, numbers of things, and there is no acceptable synonym. That usage became standard during the 1830s and 1840s. It seems almost impossible now to talk about such numbers and numerical tables published before that time without using this anachronistic term. That all generations previous to the 1820s managed to get by without it reveals dimly how different was the world they lived in—a world without suicide rates, unemployment figures, and intelligence quotients. To be sure, this prenumerate age was not entirely deprived of statistical tables, but the great explosion of numbers that made the term statistics indispensable occurred during the 1820s and 1830s. The demands it placed on people to classify things so that they could be counted and placed in an appropriate box on some official table, and more generally its impact on the character of the information people need to possess before they feel they understand something, are of the greatest interest and importance.

In the nineteenth century, statistics designated an empirical, usually quantitative, social science. Before that, it was an ill-defined science of states and conditions. The term only came to be applied commonly to a field of applied mathematics in the twentieth century. "Statistics" gained a wide meaning, by which it could be used to refer to mass phenomena of any sort, mainly by analogy with its social object. After the mid-nineteenth century, it became common to investigate collective phenomena using what came to be called the statistical method, the method of reasoning about events in large numbers without being troubled by the intractability of individuals.

The forms of "probability" are less troublesome for the nineteenth century, though probability did not come into its own as a branch of pure mathematics until quite recently. Laplace's substitution of "calculus of probabilities" for the traditional "doctrine of chances" at the end of the eighteenth century was intended to make clear that rational belief or expectation rather than the outcome of games of chance was its proper object. Indeed, this view lingered on, but the mathematics of chance was almost unfailingly referred to as probability or calculus of probability after Laplace, and I see no reason to depart from the modern usage.

"Determinism" was until the mid-nineteenth century a theory of the will—a denial of human freedom—and some dictionaries still give this as its first meaning. Partly as a result of statistical discussion, it assumed during the 1850s and 1860s its modern and more general meaning, by which the future of the world is held to be wholly determined by its present configuration. It differs from fatalism in that it rests on natural laws of cause and effect rather than on some transcendant force. Its opposite, we may note, underwent a similar change. "Indeterminism" now refers to the view that some events in the world are not wholly determined by natural causes, but are, at least to some extent, irreducibly random or stochastic. Indeterminism may be contrasted with probabilism, which implies simply that our knowledge does not permit perfect prediction, though there may be no exceptions to complete causality in the world.

The phrase "law of large numbers" was coined by Poisson in 1835. To him, it referred to the proposition that the frequencies of events must, over the long run, conform to the mean of their probabilities when those probabilities fluctuate randomly around some fixed, underlying value. Virtually everyone else who recited the phrase in the nineteenth century made no distinction between Poisson's theorem and the one in Jakob Bernoulli's 1713 *Ars Conjectandi*, according to which the frequency of events must conform over the long run to the fixed probability governing each trial. It is most convenient here to adopt this undiscriminating usage, which really expresses simply the observed regularity of statistical aggregates. More recently, Poisson's phrase has been used to denote the rule that errors of mean values conform to the normal distribution. I know of nobody who used the phrase in this way before Emile Dormoy in 1874, whose work is mentioned in chapter 8.

There are, incidentally, a number of terms for the normal distribution, but, fortunately, there is little opportunity for confusion. The standard nineteenth-century phrases were "error curve" and "error law"; the eponymous "Gaussian" became common in the late nineteenth century, and "normal law" was used by Pearson in 1894, as was "standard deviation." The nineteenth-century measure of the width of a distribution was the "probable error," the magnitude of error which precisely one-half of the measurements would, over the long run, exceed. These few terms, fortunately, nearly exhaust the technical vocabulary of nineteenth-century statistical thinking, apart from error theory and insurance mathematics. The technical and mathematical content in the following chapters is minimal, and, except for a few pages here and there, should be readily comprehensible even to readers with no mathematical training.

PART ONE

"In this life we want nothing but Facts, sir, nothing but Facts." Thomas Gradgrind

"So many people are employed in situations of trust; so many people, out of so many, will be dishonest. I have heard you talk, a hundred times, of its being a law. How can I help laws?" Tom Gradgrind (Junior)

—CHARLES DICKENS, *Hard Times* (1854)

Humanity is regarded as a sort of volume of German memoirs, of the kind described by Carlyle, and all it wants . . . is an index.—ANONYMOUS, in *Meliora* (1866)

I can scarcely conceive any more wholesome study for a stern unbending Tory than the examination of statistics, for he cannot fail to recognize the grand irrecusable law, as true in politics as in everything else, that movement must always be progressive and never retrogressive.

—L. L. PRICE (ca. 1883)

THE SOCIAL CALCULUS

The modern periodic census was introduced in the most advanced states of Europe and America around the beginning of the nineteenth century, and spread over much of the world in subsequent years. Records of population, health, and related matters, however, had been collected intermittently in a variety of territories since ancient times. Most often, the chief purpose of this statistical activity has been the promotion of bureaucratic efficiency. Without detailed records, centralized administration is almost inconceivable, and numerical tabulation has long been recognized as an especially convenient form for certain kinds of information. Until about 1800, the growing movement to investigate these numbers in the spirit of the new natural philosophy was likewise justified as a strategy for consolidating and rationalizing state power.

The importance of bureaucratic efficiency was by no means forgotten when political arithmetic gave way to the new social science of statistics at the beginning of the nineteenth century. Increasingly, however, statistical writers became persuaded that society was far more than a passive recipient of legislative initiatives. Always dynamic, often recalcitrant, society evidently possessed considerable autonomy, and had to be understood before the aims of the state could be put into effect. Statistics was for the most part a liberal enterprise, pursued by business and professional men who favored a narrower definition of the function of the state even while working to enlist it in some particular reform. The most fervent advocates of systematic bureaucratic expansion still conceded that the state could act successfully only within the constraints defined by the nature of society.

Hence, statistical authors of the scientific persuasion set themselves to uncover the principles that governed society, both in its present condition and, especially, as a historical object. The concept of "statistical law" was first presented to the world around 1830 as an early result of this search. As a social truth it was propagated widely and refined or disputed by decades of writers. Shortly afterwards, statistical regularity came to be seen as the basis for a new understanding of probability, the frequency interpretation, which facilitated its application to real events in nature as well as society. The idea of statistical regularity was thus of signal importance for the mathematical development of statistics.

Chapter One

STATISTICS AS SOCIAL SCIENCE

THE POLITICS OF POLITICAL ARITHMETIC

The systematic study of social numbers in the spirit of natural philosophy was pioneered during the 1660s, and was known for about a century and a half as political arithmetic. Its purpose, when not confined to the calculation of insurance or annuity rates, was the promotion of sound, well-informed state policy. John Graunt observed in his pathbreaking "Observation upon the Bills of Mortality" of 1662:

> That whereas the Art of Governing, and the true Politicks, is how to preserve the Subject in Peace and Plenty; that men study only that part of it which teacheth how to supplant and over-reach one another, and how, not by fair out-running but by tripping up each other's heels, to win the Prize.
>
> Now, the Foundation or Elements of this honest harmless Policy is to understand the Land, and the Hands of the Territory, to be governed according to all their intrinsick and accidental differences.[1]

Graunt's scholarly claims were modest. "I hope," he began, that readers "will not expect from me, not professing Letters, things demonstrated with the same certainty, wherewith Learned men determine in their Schools; but will take it well, that I should offer at a new thing, and could forbear presuming to meddle where any of the Learned Pens have ever touched before, and that I have taken the pains, and been at the charge of setting out those Tables, whereby all men may both correct my Positions, and raise others of their own."[2] Still, he desired to conduct his inquiry philosophically, and his work was dedicated to Robert Mo-

[1] John Graunt, *Observations upon the Bills of Mortality* (1676; 1st ed., 1662), repr. in C. H. Hull, ed., *The Economic Writings of Sir William Petty* (2 vols., Cambridge, Eng., 1899), vol. 2, pp. 395-396. On political arithmetic, see Peter Buck, "Seventeenth-Century Political Arithmetic: Civil Strife and Vital Statistics," *Isis*, 68 (1977), 67-84.

[2] Graunt, *Observations* (n. 1), p. 334.

ray, head of the "Knights and Burgesses" that "sit in the Parliament of Nature,"[3] as well as to the high official, Lord John Roberts, who might be moved to execute some of its recommendations.

William Petty, who invented the phrase "political arithmetic" and is thought by many to have had a hand in the composition of Graunt's work, was in full accord with his friend as to the purpose of these studies. Political arithmetic was, in his view, the application of Baconian principles to the art of government. Bacon, he wrote, had drawn "a judicious Parallel . . . between the Body Natural and Body Politick," and it was evident "that to practice upon the Politick, without knowing the Symmetry, Fabrick, and Proportion of it, is as casual as the practice of Old-women and Empyricks."[4] Petty sought always to bring "puzling and perplext Matters . . . to Terms of Number, Weight and Measure,"[5] so that official policy might be grounded in an understanding of the land and its inhabitants.

Implicit in the use by political arithmeticians of social numbers was the belief that the wealth and strength of the state depended strongly on the number and character of its subjects. Accordingly, the sovereign was frequently enjoined to take measures to protect the lives or health of his subjects. For example, Petty pointed out that considerable expenditures were required to raise a man or woman to maturity, and he advised the king that money spent to combat the plague could bring higher rewards than the most lucrative of investments, since it would preserve part of the vastly greater sum wrapped up in the lives of those who might otherwise perish. Still, it was the standpoint of the sovereign that was almost always assumed, and the members of society typically appeared as objects that could and should be manipulated at will. Petty's philosophy appeared sometimes as authoritarian as that of his old master, Thomas Hobbes. In his *Treatise of Ireland*, Petty proposed that all Irishmen, save a few cowherds, should be forceably transported to England, for since the value of an English life far surpassed that of an Irish one, the wealth of the kingdom would thereby be greatly augmented.[6]

Petty envisioned political arithmetic as embracing a variety of schemes through which number and calculation could be applied to

[3] *Ibid.*, p. 325. Moray was president of the newly created Royal Society of London.

[4] William Petty, "The Political Anatomy of Ireland" (1691), in *Economic Writings* (n. 1), vol. 1, p. 129.

[5] Petty, "A Treatise of Ireland" (1687), in *ibid.*, vol. 2, p. 554.

[6] *Ibid.*, pp. 555-561. See also vol. 1, pp. 108-109.

subjects. The Irish resettlement project, while exceptionally high-handed, was not uncharacteristic methodologically. Another illustration of Petty's computational rationalism is provided by his calculation of the number of clergy requisite for the English people, based solely on the land area. Petty's successors made political arithmetic a more sober discipline, closely tied to the empirical collection of population records and especially to the preparation of accurate life tables for the purpose of calculating insurance and annuity rates.

Even so, political arithmetic continued to inspire vigorous controversy, for eighteenth-century mercantilists generally regarded either population size or its rate of growth as the supreme criterion of a prosperous and well-governed country. Thus Montesquieu employed a dubious comparison of ancient and modern populations to affirm the superiority of ancient morals,[7] while David Hume observed that, after taking account of climate and geography, "it seems natural to expect that wherever there are most happiness and virtue, and the wisest institutions, there will also be most people."[8] Jean-Jacques Rousseau exalted the enumerator as the preeminent judge of policy: "The government under which . . . the citizens do most increase and multiply, is infallibly the best. Similarly, the government under which a people diminishes in number and wastes away, is the worst. Experts in calculation! I leave it to you to count, to measure, to compare."[9]

Obligingly, the experts in calculation struggled to find a reliable method for measuring population size and deciding if it was expanding or contracting. Rousseau's infallible guide, however, proved evasive, for census data were nonexistent, birth registers were imperfect, and it was difficult to decide upon a community for which the ratio of total population to annual births would be representative of an entire state. It was more practical to turn the equation around and cite evidence of changes in prosperity or general happiness as proof that the population must be increasing or declining. Accordingly, the social and political determinants of population growth were widely discussed and debated long before Malthus composed his famous essay.[10]

[7] See Montesquieu, *Persian Letters* (1721; New York: Penguin, 1973), pp. 202ff, and also the first chapter in Book 2 of *De l'esprit des lois* (Paris, 1746).

[8] David Hume, "Of the Populousness of Ancient Nations," in *Essays: Moral, Political and Literary* (1741-42; Oxford, 1964), p. 385.

[9] Rousseau, "The Social Contract," in Ernest Barker, ed., *Social Contract* (Oxford, 1960), 167-307, p. 280.

[10] On the population debate in England, for example, see David V. Glass, *Numbering the People: The Great Demography Controversy* (London, 1978).

Similarly, presumed influences on population were routinely invoked to attack customs, creeds, and circumstances of every description. Cities almost always showed a surplus of deaths over births, obviously attributable to the idleness, luxury, and corruption that they nourished, and were the bane of political arithmeticians. Thomas Short called them "Golgothas, or Places of the Waste and Destruction of Mankind, but seldom of their Increase, and often least Prolific."[11] British and German authors rarely failed to castigate the Catholic Church, whose adherence to priestly celibacy was blamed for the alleged depopulation of popish lands. "This superstitious and dangerous tenet," wrote Robert Wallace, "most justly deserves to be esteemed a doctrine of those devils, who are the seducers and destroyers of mankind, and is very suitable to the views and designs of a church, which has discovered such an enormous ambition, and made such havock of the human race, in order to raise, establish and preserve an usurped and tyrannical power."[12] Alcohol, gaming, promiscuity, and bad air were likewise condemned by the political arithmetician as moralist.

The most ambitious eighteenth-century work on population was Johann Peter Süssmilch's treatise on the "divine order" in these demographic affairs, which went through four editions between 1740 and 1798 and grew to three thick volumes by the final printing. God's first commandment was to "be fruitful and multiply" (Genesis 1), Süssmilch wrote, although he conceded that the human institution most conducive to population increase, marriage, was, "not so much a fruit of religion and of Christianity, as a necessary consequence of nature and sound reason. But is it not good, if Christianity confirms, inculcates, and blesses the command of nature?"[13] Since this "first, fundamental law of earth" was grounded in human welfare and the prosperity of states, every institution, creed, habit, and law could be judged against one universal standard—did it promote or inhibit the growth of population? Süssmilch undertook to discover the conditions which most en-

[11] Thomas Short, *A Complete History of the Increase and Decrease of Mankind* (London, 1767), p. i.

[12] Robert Wallace, *A Dissertation on the Numbers of Mankind in Antient and Modern Times in Which the Superior Populousness of Antiquity is Maintained* (Edinburgh, 1753), p. 87.

[13] Johann Peter Süssmilch, *Die göttliche Ordnung in den Veränderungen des menschlichen Geschlechts aus der Geburt, dem Tode und der Fortpflanzung desselben erwiesen* (2 vols., Berlin, 3rd ed., 1765), vol. 1, p. 446. On Süssmilch see Jacqueline Hecht, "Johann Peter Süssmilch: Point alpha ou omega de la science démographique naive," *Annales de démographie historique*, 1979, 101-134.

couraged population increase by comparing the available information from a variety of countries, both past and present. He made extensive use of parish records and maintained an extensive correspondence. He also relied heavily on the publications of such writers as Antoine Deparcieux of France; Pehr Wilhelm Wargentin of Sweden; Nicolas Struyck, Willem Kersseboom, and Bernard Nieuwentyt of the Netherlands; and British authors Graunt, Petty, Gregory King, and John Arbuthnot.

Süssmilch's theology led to a system of ethical and political maxims founded on the same desideratum of maximizing population. A Protestant pastor, he naturally decried alcohol, gambling, prostitution, urban life, priestly celibacy, and similar ills. He disapproved equally of war, which, he explained, was based simply on princely misperception. It was perfectly plain to this apostle of procreation that the inclination of princes to covet one another's territory arose from the universal desire to gain new subjects for their states. Hence all the unpleasant marching and shooting of war, of which Süssmilch had direct experience, was unnecessary, for the desired object could be attained directly by removing the obstacles to natural population increase. The principle of geometric growth, through which a single couple had produced hundreds of millions of descendants in a few thousand years, insured that the number of loyal subjects would multiply at a rate far surpassing the most optimistic expectations for victory in combat.

Princely success, then, was to be measured by achievement in the domestic sphere, and Süssmilch was ready with voluminous advice for achieving it. He proposed certain measures, such as state support of medical care and systematic inoculation for smallpox, to reduce the death rate, but his campaign was aimed principally at the supply side. Taxes, he remarked, must be kept low, for a heavy burden of taxation robs men of the subsistence necessary to enter married life, and thereby leads to the diminution rather than increase of the true wealth of the state. Distribution of land was crucial, for the unavailability of agricultural territory obliged many couples to delay marriage for years. The Roman empire had been built by thrifty, industrious farmers inhabiting plots of land just large enough to support their families. By imitating the ancient Romans, and eliminating obstacles to the subdivision of tracts of land such as primogeniture, modern states could likewise achieve greatness. Above all, it was essential to defend "Liberty and Property"—Süssmilch used the English words—for these served both to promote marriage and to discourage emigration. Poland and Spain had suffered population decline because they ignored these values; Germany, by upholding them "could become the happiest, most powerful, and richest land."[14]

Süssmilch, like most political arithmeticians, advocated expansion of the government apparatus for collecting population numbers and, more important, for acting on them. Continental political arithmetic was an offspring not only of the scientific ambition of the Enlightenment, but also of enlightened despotism, and even in Great Britain this knowledge was to be put at the service of king and Parliament. To advocate the use of extensive statistical information was to favor centralization and bureaucratization, and hence, as French progressives such as Turgot and Condorcet most earnestly hoped, to bypass conservative and particularistic interests like church and nobility.[15] The ultimate beneficiaries of this rationalization ought to include the mass of private citizens, but its immediate effect was to be the consolidation of state power.

THE NUMBERS OF A DYNAMIC SOCIETY

Political arithmetic was supplanted by statistics in France and Great Britain around the beginning of the nineteenth century. The shift in terminology was accompanied by a subtle mutation of concepts that can be seen as one of the most important in the history of statistical thinking. There was little in the background of the name "statistics" to presage such developments, but a certain measure of unity appears in these changes when they are considered in their context.

Statistics derives from a German term, *Statistik*, first used as a substantive by the Göttingen professor Gottfried Achenwall in 1749. It was perhaps an unfortunate choice, for its etymology was ambiguous and its definition remained a matter of debate for more than a century. Even its proper subject matter was disputed, although most writers of the early nineteenth century agreed that it was intrinsically a science concerned with states, or at least with those matters that ought to be known to the

[14] *Ibid.*, p. 557.

[15] See Keith Baker, *Condorcet: From Natural Philosophy to Social Mathematics* (Chicago, 1975), chaps. 4 and 5. In England, political arithmeticians of the late eighteenth century, led by Richard Price, sought to ally their field with republicanism rather than bureaucratic centralism; see Peter Buck, "People Who Counted: Political Arithmetic in the Eighteenth Century," *Isis*, 73 (1982), 28-45.

"statist." Statistics initially had no more to do with the collection or analysis of numbers than did geography or history. Its task was to describe, and numerical tables were involved only to the extent the author found them available and appropriate to the particular subject matter.

The anglicized form of this German term was introduced by John Sinclair, hub of a network of Presbyterian pastors whose collective labor made possible his 21-volume compilation, the *Statistical Account of Scotland*. Sinclair, who claimed that he had deliberately adopted this foreign expression in order to draw attention to his project, sought to distinguish his enterprise from the German one by noting that whereas the latter dealt with "political strength" and "matters of state," his inquiries were designed to ascertain the "quantum of happiness" enjoyed by the inhabitants of a country, "and the means of its future improvement."[16] It is difficult to establish just when people came to see statistics as a science that was specifically defined in terms of numerical information, for the tradition of statistical surveys and statistical accounts permitted numbers to assume ever greater prominence without forcing any discontinuity. The transition seems to have taken place in Great Britain almost subconsciously, and with a minimum of debate. Bisset Hawkins was probably not too far ahead of his countrymen in 1829 when he defined medical statistics as "the application of numbers to illustrate the natural history of man in health and disease," and pointed to an analogous statistical field in political economy.[17]

Statistique evolved in the same direction, albeit along a less continuous path. Statistics was identified in France with numerical information about society as early as 1820, when Charles Dupin defended the placement of statistics among the *sciences mathématiques* rather than the *sciences morales et politiques* in the Parisian journal *Revue encyclopédique*. Similarly, the great official compilation of population and mortality figures for Paris and the Seine was published under the title *Recherches statistiques* beginning in 1821. The Genevan *Bibliotèque universelle*, however, published purely descriptive contributions under this title until 1828, when it switched abruptly to an exclusively nu-

[16] Sir John Sinclair, ed., *The Statistical Account of Scotland* (21 vols., Edinburgh, 1791-1799), vol 20, Appendix, p. xiii, to which August Ludwig Schlözer responded that Sinclair must never have seen a handbook of statistics if he thinks it concerns only political power; *Theorie der Statistik nebst Ideen über das Studium der Politik überhaupt* (Göttingen, 1804), p. 17. The *Oxford English Dictionary*, my source for the Sinclair quote, shows an earlier use of the term in an English translation of a German work.

[17] Bisset Hawkins, *Elements of Medical Statistics* (London, 1829), p. 2.

merical definition. As late as 1830, statistical articles were scarcely distinguishable from geographical ones—and were classed among the *sciences géographiques*—in the Parisian *Bulletin universel des sciences et de l'industrie*. The durability of the old German definition of statistics, however, is probably less important than the circumstance that by the late 1820s a tradition of numerical social studies had become established in France and Belgium and had laid claim to the title "statistics."

The quantitative science of statistics inherited from its German namesake an exceptionally wide scope, extending from geography and climate to government, economics, agriculture, trade, population, and culture. In its extended sense it included also facts from medicine as well as the natural history of man. Statists conducted surveys of institutions of all sorts and maintained records of trade, industrial progress, labor, poverty, education, sanitation, and crime. Although their science came to be identified most closely with the censuses that Great Britain, France, Prussia, the United States, and other countries began conducting around the beginning of the nineteenth century, its scope was far broader than had been envisioned for political arithmetic, except by William Petty himself.

Political arithmetic, as I have suggested, was associated with centralizing bureaucracy. The information that numbers could provide was vital for controlling the population, and especially for augmenting tax revenue. More fundamentally, however, the ideal of enumeration was one which few other than agents of the crown would seriously have entertained, at least on the Continent, under the Old Regime—and monarchs typically regarded demographic figures as a state secret, too sensitive to publish. Implicitly, at least, statistics tended to equalize subjects. It makes no sense to count people if their common personhood is not seen as somehow more significant than their differences. The Old Regime saw not autonomous persons, but members of estates. They possessed not individual rights, but a maze of privileges, given by history, identified with nature, and inherited through birth. The social world was too intricately differentiated for a mere census to tell much about what really mattered

Such Old Regime hierarchies had already broken down to a considerable extent in eighteenth-century England, and the French Revolution was instrumental in destroying them on the Continent, although social heterogeneity remained a major concern of statisticians, especially in Germany, throughout the nineteenth century. Needless to say, royal absolutism was not always the beneficiary of these changes; perhaps bureaucratic centralism was, but its victory was an ambiguous one. If statistics provided bureaucracies with some of the knowledge that is indispensable to power, they also suggested certain limitations to this power. The limitations in question are not constitutional ones, but constraints that now seemed to exist independently of any particular formal arrangements of government. For the expansion in the scope of numerical investigations was accompanied by an important change in the conception of their object.

That shift can be characterized through a comparison of the most eminent students of population before and after the French Revolution, Süssmilch and Thomas Robert Malthus. Süssmilch, always presupposing that population increase was the highest aim of every leader, devoted his work to showing what the state could do to promote the growth of population. Malthus, who held that a high population density is the principal cause of misery and poor health in a country, stressed instead the constraints imposed by laws of population on the possible forms of governmental and social organization. Population was no longer something pliable, to be manipulated by enlightened leaders, but the product of recalcitrant customs and natural laws which stood outside the domain of mere politics. Government could not dominate society, for it was itself constrained by society.

According to Malthus, the principle of population contradicted the utopian schemes of Condorcet and Godwin by showing "that though human institutions appear to be, and indeed often are, the obvious and obtrusive causes of much mischief to society, they are, in reality, light and superficial, in comparison with those deeper-seated causes of evil, which result from the laws of nature and the passions of mankind."[18] Malthus held society to be a dynamic and potentially unstable force, an incipient source of turmoil that threatened to confront the lover of English freedoms with a choice between revolution and repression. To avert this, Malthus upheld the need for public education to acquaint the people with the true causes of their misery. Political leadership was not unimportant, for a wise government could perhaps chart a safe course through these troubled seas. For this, however, it required a familiarity with the principles of political economy, and also that "cleared insight

[18] Thomas Robert Malthus, *An Essay on the Principle of Population* (2 vols., London, 6th ed., 1826), vol. 2, p. 20.

into the internal structure of human society"[19] which could be derived from statistical investigations.

Malthus's views are characteristic of a set of attitudes that emerged in the wake of the French Revolution and that underlay the statistical movements of the second quarter of the nineteenth century. As Stefan Collini has argued, both the historicism and the new interest in "social science" of the early nineteenth century were manifestations of a new sense among European thinkers and writers that "society" was in reality a more fundamental dimension of existence than either state or government.[20] Statistics, one of the earliest of the would-be social sciences, came to be conceived in France and England as the empirical arm of political economy. Although the laws of economic behavior could be discovered deductively, their very universality rendered them unsuitable for capturing the particular circumstances of a country in any given state of historical development. One of the principal tasks of numerical statistics from the beginning was to fill this gap, to chart the course of economic and social evolution.[21]

The "era of enthusiasm" in statistics was thus inspired by a new sense of the power and dynamism of society. Society was regarded both as a source of progress, revealed by the beginnings of industrialization, and as a cause of instability, typified by the French Revolution and by continuing unrest in Great Britain as well as France. Statistical investigation was not the product of sociological fatalism, however, but of cautious hopefulness for improvement. Statistics reflected a liberal temperament and a search for reform that flourished not during the years of repression following the Congress of Vienna, but the late 1820s and especially the 1830s. The statists sought to bring a measure of expertise to social questions, to replace the contradictory preconceptions of the interested parties by the certainty of careful empirical observation. They believed that the confusion of politics could be replaced by an orderly reign of facts.[22]

The origins, if not the zenith, of the great statistical enthusiasm of the

[19] *Ibid.*, vol. 1, p. 20.

[20] Stefan Collini, "Political Theory and the Science of Society in Victorian Britain," *Historical Journal*, 23 (1980), 203-231, pp. 203-204.

[21] See Victor L. Hilts, "*Aliis Exterendum*, or, the Origins of the Statistical Society of London," *Isis*, 69 (1978), 21-43.

[22] See [William Cooke Taylor], "Objects and Advantages of Statistical Science," *Foreign Quarterly Review*, 16 (1835), 205-229. (For identification of anonymous authors in British reviews I have used Walter Houghton, ed., *Wellesley Index to Victorian Periodicals* [4 vols., Toronto, 1966 et seq.].) The label "era of enthusiasm" was introduced by Harald Westergaard, *Contributions to the History of Statistics* (London, 1932).

early nineteenth century are to be found in France. The passion for information reared its head already during the Consulate and Empire, although the ambitions of the early Bureau de statistique so far exceeded its resources that the extraordinarily detailed census it undertook to carry out through the prefects yielded virtually no usable data.[23] In the end, an impatient Napoleon suppressed the bureau in 1811, as he had abolished the second class of the Institute, moral and political sciences, in 1803. The restored monarchy had little desire to resuscitate it, although the diverse records that were collected and published under Louis XVIII and Charles X assisted the private statistical writers who began to flourish around 1820. Among these records were the accounts of the health of army recruits, which first appeared in 1819; the comprehensive French judicial statistics, beginning in 1827; and the pathbreaking *Recherches statistiques sur la ville de Paris et le département de la Seine*, published in 1821, 1823, 1826, and 1829 under Chabrol and the physicist Joseph Fourier. Finally, the census was reconstituted under the July monarchy and placed under Alexandre Moreau de Jonnès.

The statistical initiative in France was taken chiefly by advocates of public health, particularly by army surgeons released from service at the conclusion of the Napoleonic wars. Initially, these writers focused their attention on the salubrity of orphanages, prisons, and poorhouses, usually with the specific aim of instigating reform of the institution in question. Other public-spirited mutations of the old descriptive statistics followed shortly thereafter. Among the most noteworthy and influential was the use of statistics in the campaign for public education. A magical significance was attached to the proposition that education leads to a reduction of crime, which was conclusively demonstrated again and again through the shameless manipulation or misinterpretation of numerical records. A. Taillandier was able to establish, without any information on literacy rates in the population at large, that "the definitive result of these researches on the instruction of prisoners reveals that 67 out of 100 are able neither to read nor write. What stronger proof could there be that ignorance, like idleness, is the mother of all vices?"[24] In the same vein, the patient forebearance of statisticians from admitting preconceived opinions was revealed in the announcement of a prize for the best

[23] See Marie-Noëlle Bourguet, "Décrire, compter, calculer: The debate over Statistics during the Napoleonic Period," in *Prob Rev*.

[24] A. Taillandier, Review of "Compte général de l'administration de la justice criminelle en France," *Revue encyclopédique*, 40 (1828), 600-612, p. 612.

memoir on the abolition of the death penalty by the Société française de statistique universelle: "By attaining statistical certainty that fewer crimes are committed where the penalty of death has been abolished, the influence that a gentler and more genuinely philosophical legislation exerts on the criminality of human actions can be better appreciated."[25]

Although the occasional appearance of numerical results directly contradicting preconceptions of this sort obliged statistical authors to maintain a certain degree of flexibility in interpreting their figures, the general disposition of these writers was to present their findings as direct and incontrovertible proof of the propositions they seemed to support. Statistics, wrote Alphonse De Candolle, "has become an inexhaustible arsenal of double-edged weapons or arguments applicable to everything, which have silenced many by their numerical form."[26] Less cynically, if more mysteriously, Moreau de Jonnès claimed that the figures of statistics "are like the hieroglyphics of ancient Egypt, where the lessons of history, the precepts of wisdom, and the secrets of the future were concealed in mysterious characters. They reveal the increase in the power of empires, the progress of arts and of civilization, and the ascent or retrogression of the European societies."[27] Statistics "does not have the power to act," he wrote, "but it has the power to reveal, and happily, in our day, this is practically the same thing."[28]

Although an attempt was made in 1830 to form a French statistical society, the result was unsatisfactory, and consequently the leading writers possessed no institution or journal dedicated specifically to statistics. Statisticians were prominent in the classe des sciences morales et politiques of the Academy when it was revived under the July monarchy, but their leading organ was probably the *Annales d'hygiène publique*, and their thought retained a medical cast even in reference to purely social problems. The editors of that journal announced in their 1829 prospectus that public hygiene "can, by its association with philosophy and with legislation, exert a great influence on the march of the human spirit. It should enlighten the moralist and cooperate in the noble task

[25] "Introduction," *Bulletin de la société française de statistique universelle*, vol. 1 (1830-1831), p. 33.

[26] Alphonse De Candolle, "Considérations sur la statistique des délits," *Bibliothèque universelle des sciences, belles-lettres et arts*, 104 (1830), *Littérature*, 159-186, p. 160.

[27] Alexandre Moreau de Jonnès, "Tableau statistique du commerce de la France," *Revue encyclopédique*, 31 (1826), 27-46, p. 27.

[28] Jonnès, *Eléments de statistique* (Paris, 2d ed., 1856), p. 5.

of diminishing the number of social infirmities. Scarcities and crimes are maladies of society that it should work to diminish."[29] Statistics could provide an understanding not only of the prevailing causes of death and disease, but also of crime and revolution, respectively the chronic and epidemic disorders of the human spirit.

Michelle Perrot characterized the French moral statisticians as bourgeois reformers, seeking to control deviant behavior of all sorts with numbers. Similarly, William Coleman has argued that these quantifiers sought "to replace the long reign of opinion, party interest, and political confusion with a secure core of well-established social facts and rigorously deduced truths" based on social science. Although there was some connection between public hygiene and the St. Simonian movement, most of these writers were reluctant to endorse deep and systematic state intervention in private concerns. Perhaps their subservience to classical political economy was, as Coleman intimates, a failure of the diagnostician to countenance the cure—whatever that might have been. More significant for our purposes, however, is that these sociomedical investigations of 1815-1848 represented "the earliest systematic effort to seize the complex interrelations of social structure and change," involving a search for the regularities that underlay the evident disorder of French social development during the early nineteenth century.[30]

The desire to understand contemporary social transformations, and to establish a scientific basis for social policy, was also at the heart of the statistical movement in early Victorian Britain. An official census had been set up in 1801 in response to manpower needs during the Napoleonic wars, but the first four decennial censuses were considered unsatisfactory even at the time, since information was collected primarily through the networks of the established church and because the head, a

[29] "Prospectus," *Annales d'hygiène publique et de médecine légale*, 1 (1829), v-vii. See also Bernard-Pierre Lécuyer, "Médecins et observateurs sociaux: Les *Annales d'hygiène publique et de médecine légale*, 1820–1850," in *Pour une histoire de la statistique (Journées d'étude sur l'histoire de la statistique)*, vol. 1, *Contributions* (Paris, 1977), pp. 445-476; "L'hygiène en France avant Pasteur," unpublished paper; "Démographie, statistique, et hygiène publique sous la monarchie censitaire," *Annales de démographie historique*, 1977, pp. 215-245; René Le Mée, "La statistique démographique officielle de 1815 à 1870 en France," *ibid.*, 1979, pp. 251-279. On the attempted formation of a French statistical society, see letters from Villermé to Quetelet, 1 Jan. 1830 and 21 Jan. 1831 in cahier 2560 *AQP*. The Société française de statistique universelle had little or no support from the leading French statisticians.

[30] William Coleman, *Death is a Social Disease: Public Health and Political Economy in Early Industrial France* (Madison, 1982), p. 275; also Michelle Perrot, "Premières mesures des faits sociaux: les debuts de la statistique criminelle en France (1790-1830)," in *Pour une histoire* (n.29), pp. 125-137; Ian Hacking, "Biopower and the Avalanche of Numbers," *Humanities in Society*, 5 (1982), 279-295.

Tory named John Rickmann, refused to collect occupational records of the sort desired by political economists. The great burst in official, as in private, statistical activity occurred during the 1830s, when a statistical office was set up under G. R. Porter at the Board of Trade (1832), and the General Register Office was created (1837) to collect vital statistics and to supervise a greatly expanded census beginning in 1841. This was the period of the "condition of England question," of heightened fears about social dislocation and the prospect of revolution, but it was also a time when the strategy of conservative repression was largely rejected. The study of social numbers embodied hope as well as fear, and it is not by chance that statistics became "the favourite study of the present age"[31] in Great Britain precisely during those years of upheaval that brought forth the Reform Act of 1832, the Factory Act of 1833, and the new Poor Law of 1834.

The characteristic institution of statistical enthusiasm in Great Britain was the private statistical society. The first important association of British "statists" was the statistical section of the British Association for the Advancement of Science, formed in 1833, just two years after the founding of the parent body. Statistics represented the British Association's foray into social science, a move not universally applauded. The creation of the new section was inspired by the presence at the Cambridge meeting of Malthus and, especially, Quetelet, who discussed his own work on the statistics of crime and suicide. As Lawrence Goldmann has recently shown, the inspiration for its founding came primarily from Richard Jones, who opposed the abstract, deductive style of Ricardian economics and sought to replace it with a historical and empirical approach. Charles Babbage subverted the rules of the B.A.A.S. in order to gain approval for the new section.

The founding members, especially Malthus, Babbage, Richard Jones, W. H. Sykes, and J. E. Drinkwater, formed the original core of the Statistical Society of London—progenitor of the present-day Royal Statistical Society—which was organized in March 1834. During the interim, a different group of reforming physicians and industrialists had formed a statistical association in Manchester, which was the most active and successful of these societies during the first decade of their ex-

[31] [Herman Merivale], "Moral and Intellectual Statistics of France," *Edinburgh Review*, 69 (1839), 49-74, p. 49. On the response of political economists to this political instability, see Maxine Berg, *The Machinery Question and the Making of Political Economy* (Cambridge, Eng., 1980).

istence. A host of similar organizations—perhaps twenty—were formed or publicly contemplated in various provincial cities during the ensuing two decades, but all soon collapsed.[32] Hence British statistics was principally identified during the Victorian period with the statistical societies of London and Manchester, Section F of the British Association, the General Register Office and Board of Trade, and with the masses of data compiled into Blue Books under various parliamentary directives.

The active members of the early statistical societies were far less likely to be seriously interested in natural science or mathematics than to be politically engaged. Most of the leading members of the Manchester organization, such as James Phillips Kay, William Langton, Benjamin Heywood, and Samuel and W. R. Greg, were industrialists, or at least had close family ties to industry. More particularly, they were, as Michael Cullen remarks, "improving" employers, seeking to ameliorate sharp class divisions and to prevent social upheaval through kind treatment of workers, the inculcation of morality, and the provision of education. Although they were philosophically opposed to government meddling, especially in the domain of commerce, they agitated for an active state role in sanitation and, like their French contemporaries, tended to view public education as a panacea for crime and social unrest.

The Statistical Society of London perhaps fits even better than does Manchester Michael Cullen's characterization of statistical work as "a movement of the reforming establishment, Whig to Liberal in politics."[33] Active researchers, however, were distinctly more radical than the great body of political worthies who offered the new organization little more than the prestige of their names and two guineas. The membership skyrocketed, reaching 318 within a year of its founding, but very

[32] See Michael Cullen, *The Statistical Movement in Early Victorian Britain: The Foundations of Empirical Social Research* (Hassocks, 1975); Hilts, "*Aliis exterendum*" (n. 21); Victor L. Hilts, *Statist and Statistician: Three Studies in the History of 19th-Century English Statistical Thought* (New York, 1981); T. S. Ashton, *Economic and Social Investigations in Manchester: A Centenary History of the Manchester Statistical Society* (London, 1934); David Elesh, "The Manchester Statistical Society: A Case Study of Discontinuity in the History of Empirical Social Research," *Journal of the History of the Behavioral Sciences*, 8 (1972), 280-301, 407-417; Philip Abrams, *The Origins of British Sociology 1834–1914* (Chicago, 1968), chap. 3; Jack Morrell and Arnold Thackray, *Gentlemen of Science: Early Years of the British Association for the Advancement of Science* (Oxford, 1981), pp. 291-296; Berg, *Machinery Question* (n. 31), pp. 291-314; Lawrence Goldmann, "The Origins of British 'Social Science': Political Economy, Natural Science and Statistics, 1830-1835," *Historical Journal*, 26 (1983), 587-616.

[33] Cullen, *Statistical Movement* (n. 32), p. 82.

few were willing to undertake the arduous labor necessary to prepare real statistical reports. Moreover, the society was a bit nonplussed by the vague definition and uncertain frontiers of the field it had so enthusiastically entered. The leading natural scientist among its founding members, Charles Babbage, quickly lost interest. William Whewell, who was a close friend and correspondent of Jones, and who had invited Quetelet to the 1833 British Association meetings, was one of the earliest members of the council of the Statistical Society of London, but by August 1834 his portrait of the society's activities was hardly encouraging. He wrote to Quetelet: "You will find that the Statistical Section which sprung up under your auspices at Cambridge is grown into a Statistical Society in London, with many of our noblemen and members of Parliament for its members. Our Committee has had several meetings, but we are still somewhat embarrassed by the extent of our subject. I have no doubt, however, that we shall do something." Six months later, Whewell begged Quetelet to suggest a project for the society he had helped to found, and indicated that the society was attempting to draw up a questionnaire for distribution. He was himself, he confessed, losing interest.[34]

Later in 1835 a questionnaire was at last completed and sent out under the auspices of the society by George Richardson Porter, one of its most active members and also head of the statistical office at the Board of Trade. The "Inquiries on Education in Belgium," sent to Quetelet, included under the heading "Results of Education" a set of questions which suggest the degree of sophistication in social science attained by this fledgling organization:

1. What has been the effect of the extension of education on the habits of the People? Have they become more orderly, abstemious, contented, or the reverse?
2. What is the proportion of crime under the two heads of offenses against property and against the person, to population and property in the different Provinces?
3. What is the proportion of crimes to education? Are the educated found to be more exempt than the uneducated, or the reverse?
4. What description of crime prevails most in the educated provinces, crimes against property or against the person?

[34] See letters of 4 Aug. 1834 and 3 Feb. 1835, Whewell to Quetelet, cahier 2644, *AQP*. The letter is reprinted in Isaac Todhunter, *William Whewell* (2 vols., London, 1876), p. 185.

5. What proportion of criminals, especially in the grosser class of crimes could read and write, in the returns of 1833, or 1834?
6. What is the political effect of the diffusion of education, as evinced by the increase of newspapers, pamphlets, etc. etc. etc.?
7. What is the number of Books published during the last year and how classified?[35]

This was hardly the type of questionnaire that could be answered without bias, and Porter divulged a few of his own in a subsequent missive. The level of crime, he wrote, "is disgraceful to us as a nation and proceeds from the deplorable ignorance in which the great bulk of the people are kept." The cause, he explained, was the established church, whose power and wealth would be reduced if the people were instructed. Its leaders accordingly had sought to keep education in their own hands and to render the knowledge imparted "as little hurtful to themselves as possible by adulterating it with their own sectarian prejudices."[36]

Porter's tirades against Anglicanism are by no means defining attributes of the statistical movement as a whole, but they do indicate something of the extent to which statistical facts were sought to bolster particular social and political programs. Cullen argues that statistical research was aimed primarily at vindicating industrial progress by laying the blame for social ills on other causes, including alcohol, moral degeneracy, and the growth of cities.[37] More charitably, one can at least see that the statistical enthusiasts held strong commitments. Ignorance and filth were seen as responsible for the prevalence of disease, the rampant growth of crime, and the threat of domestic turmoil among the working classes. Statistical investigation, it was presumed, would provide empirical support for the reforms that were obviously needed. Thus the Royal Cornwall Polytechnic Society issued the following statement in 1837:

> If "the proper study of mankind be man," the value of statistical information can no longer be doubted. It stimulates the benevolence and gives aim and effect to the energies, of the philanthropist; it fur-

[35] Porter to Quetelet, 28 May 1835, cahier 2041, *AQP*.

[36] *Ibid.*, 4 June 1838.

[37] Cullen, *Statistical Movement* (n. 32), p. 144 and *passim*. Bernard-Pierre Lécuyer indicates that the same was often true in France; see "Les maladies professionnelles dans les *Annales d'hygiène publique et de médecine légale*, ou une première approche de l'usure au travail," *Le Mouvement Social*, no. 124 (1983), 45-69.

> nishes the legislator with materials on which to found remedial measures for social derangement, and plans for increasing the mass of social happiness; and though its conclusions often consist in a bare numerical statement of aggregate results, yet they come home with all the authority of stubborn facts, and often tell more than the most elaborate moral appeal. From information thus furnished, it cannot be questioned that the public attention has been fastened, with an intensity never before given to the subject, upon the physical and moral degradation of the poorer classes in the metropolis and many of our large towns. The appeal thus made has been nobly responded to; the dry facts have been interpreted; and means have been adopted for carrying the blessings of education, order, and virtue into those dark recesses where ignorance, vice, and misrule appeared to have fortified themselves in impenetrable obscurity.[38]

Notwithstanding their obsession with filth and criminality, and their dedication to public health and education, British statists were not content to be simply activists and reformers. Theirs was to be the empirical science of society, yielding the certainty of science and deserving the respect appertaining thereto. The philosophical justification they gave for the scientific character of their activity was problematical, however, and the London society received much notoriety for it. The statistical section of the British Association had been suspect from the beginning because it was feared that investigation into the affairs of man would engender controversy and might even lead to the breakup of the larger organization.[39] Partly as a defensive move, and partly to reassure interested political leaders that their support of statistics would not embarrass them, the statists adopted the position that they were concerned exclusively with facts. The council of the London Statistical Society announced the following general principle:

> The Science of Statistics differs from Political Economy, because, although it has the same end in view, it does not discuss causes nor

[38] "Extract from the Annual Report of the Royal Cornwall Polytechnic Society for the Year 1837," *JRSS*, 1 (1838), 190.

[39] See, for, example, the report of Adam Sedgwick's address of 1834 in "Proceedings of the British Association," *The Edinburgh New Philosophical Journal*, 17 (1834), 369-374, p. 372, which was recognized as a criticism of statistics, and perhaps also phrenology. See also "Phrenology and the British Association," *The Phrenological Journal and Miscellany*, 9 (1834-1836), 120-126; cf. Roger Cooter, *The Cultural Meaning of Popular Science: Phrenology and the Organization of Consent in Nineteenth-Century Britain* (Cambridge, Eng., 1984), esp. chap. 3.

> reason upon probable effects; it seeks only to collect, arrange, and compare, that class of facts which alone can form the basis of correct conclusions with respect to social and political government. . . .
>
> Like other sciences, that of Statistics seeks to deduce from well-established facts certain general principles which interest and affect mankind; it uses the same instruments of comparison, calculation, and deduction: but its peculiarity is that it proceeds wholly by the accumulation and comparison of facts, and does not admit of any kind of speculation.[40]

The motto of the new society was *Aliis exterendum*—"to be threshed out by others"—and the council declared that "the first and most essential rule" of the conduct of the society was "to exclude all opinions."[41] As late as 1861, William Farr wrote to Florence Nightingale: "We do not want impressions. We want facts. . . . Again I must repeat my objections to intermingling Causation with Statistics. . . . The statistician has nothing to do with causation; he is almost certain in the present state of knowledge to err. . . . You complain that your report would be dry. The dryer the better. Statistics should be the dryest of all reading."[42]

The "most essential rule" of statistics was, of course, ignored in practice. Statistical investigation—the study of school attendance records and of death rates in relation to sanitary conditions, or the conduct of surveys of factories and neighborhoods to learn about diet—required more time and energy than busy people were willing to invest unless they had a real interest in the outcome. Doubtless the search for a "science of government, the principles of which would be discovered outside the realm of partisan dissention and arise from the accumulation of simple, irrefutable social facts," as John Eyler characterizes it,[43] was futile from the beginning. Nevertheless, the facts helped to inform official policy and the disavowal of speculation perhaps added weight to the conclusions of statists in the view of the public. "It is of the essence of statistics," wrote the political philosopher George Cornewall Lewis, "that its object is scientific, not practical; that it is intended to represent the

[40] "Introduction," *JRSS*, 1 (1838), pp. 1, 3.

[41] Cullen, *Statistical Movement* (n. 32), p. 85. See also Hilts, "*Aliis Exterendum*" (n. 21).

[42] See Marion Diamond and Mervyn Stone, "Nightingale on Quetelet," *JRSS*, A, 144 (1981), 66-79, 176-213, 332-351, p. 70.

[43] John M. Eyler, *Victorian Social Medicine. The Ideas and Methods of William Farr* (Baltimore, 1979), p. 16.

truth of facts, not to subserve some immediate purpose of administration or legislation." At the same time, he held statistics indispensable for the practical statesman: "It is by comparing the numbers of the subject under consideration . . . that his practical judgement ought to be formed." Without numbers, legislation is ill-informed or haphazard.[44]

Although an American census was mandated by the Constitution and carried out decenially beginning in 1790, a decade before any other country except Sweden had instituted such a periodical tabulation, American statistical work seems to have had little influence on European. The tone of early American statistical writing is scarcely distinguishable from that of Britain. Already in 1811 we find James Mease proclaiming the virtues of pure numerical facts, and others suggesting that when these facts were fully known, differences of opinion must disappear. In the United States, as in Britain, it became popular during the 1830s and 1840s to produce great compilations showing the impressive progress of the nation, and implying that lands less fully blessed with freedom, including the slave South as well as most of the Old World, were destined to fall hopelessly behind if they did not reform. Battles between North and South were fought with numbers long before they were fought with soldiers, most instructively, perhaps, in the form of a debate inspired by some errors in the 1840 census over the question of whether insanity was in fact ten times more prevalent among free northern blacks than among southern slaves. Finally, crime, intemperance, and the other familiar objects of moral statistics were as much at issue among American statists as in Britain and France.[45]

Nowhere else was statistics pursued with quite the level of enthusiasm as in Britain, and numerical statistics developed more slowly outside the politically and economically most advanced states of western Europe. To be sure, disciplines of that name flourished in Italy and, especially, Germany, during the early decades of the nineteenth century, but this was the old university statistics. While it can hardly be faulted for failing to anticipate what statistics would become in other countries some years later, this variety of statistics lacked a clearly defined scope, purpose, or method. German statisticians debated these matters endlessly and failed

[44] George Cornewall Lewis, *A Treatise on the Methods of Observation and Reasoning in Politics* (2 vols., London, 1852), vol. 1, pp. 133-134.

[45] See Patricia Cline Cohen, *A Calculating People: The Spread of Numeracy in Early America* (Chicago, 1982), chaps. 5 and 6. On statistics in America, see also James Cassedy, *Demography in Early America: Beginnings of the Statistical Mind* (Cambridge, Mass., 1969); Cassedy, *American Medicine and Statistical Thinking, 1800-1860* (Cambridge, Mass., 1984).

to reach agreement on anything. Statistics was variously seen as the descriptive science of states, or the study of the state or condition (*Zustand*) of any object at a given time—both interpretations being consistent with the Latin root *status*, though only the former could be read into the German *Staat*.

Government statistical agencies had been set up in four German-language states prior to 1848—Prussia, Austria, Bavaria, and Württemberg. The great surge in official statistics, however, came in the aftermath of the revolution, during the early years of German industrialization. Statistical bureaus were already operating in virtually every German state a decade before the unification brought a greater measure of standardization to these things. At the same time, numerical statistics became increasingly influential among social and political scientists at the universities, and it appears that here, as in France and Britain, the sudden growth of interest in statistics can be attributed to an increase of anxiety about social change and instability.

Numerical statistics entered the German universities as a consequence of the successful effort to redefine the discipline in terms compatible with the work of census bureaus and with current usage in French and English. The use of numbers had occasionally been championed by university statisticians such as Schlözer, who thought them especially suitable for rendering descriptions concise and systematic, and, during the 1840s, by Christoph Bernoulli, descended from the great Swiss scientific family, who advocated a new science of *Populationistik*.[46] Generally, however, numbers were regarded during the early nineteenth century as secondary and even superficial. Such was the opinion of the Tübingen professor Johannes Fallati, leading academic statistician in Germany during the 1840s.[47] Fallati, however, believed numerical statistics to be essential for practical administration and reform, and he championed the development of statistical organizations on the model of the societies that had sprung up in London, Manchester, Bristol, and Ulster to collect, organize, and disseminate this useful if shallow variety of statistical information.[48] In 1847 a statistical

[46] Schlözer, *Theorie* (n. 16), p. 20; Christoph Bernoulli, *Handbuch der Populationistik, oder der Völker- und Menschenkunde nach statistischer Erhebnissen* (Ulm, 1841), and "Vorwort," *Schweizerisches Archiv für Statistik und Nationalökonomie*, 1 (1827).

[47] Johannes Fallati, *Einleitung in die Wissenschaft der Statistik* (Tübingen, 1843), pp. 104-106.

[48] Fallati, *Die statistischen Vereine der Englaender* (Tübingen, 1840); "Gedanken über Mittel und Wege zu Hebung der praktischen Statistik mit besonderer Rücksicht auf Deutschland," ZGSW, 3 (1847), 496-557.

society of public-spirited persons from all of Germany was set up by Freiherrn von Reden, a Hanoverian bureaucrat who had for some time been involved in seeking solutions to the problems of German laborers and craftsmen. His Verein für deutsche Statistik[49] did not survive the events of 1848, but the cause it embodied did, for the concerns that had inspired it were only magnified. When Carl Knies published a study defending the numerical method of statistics early in 1848, he did so on the basis of its alleged superiority for investigating the condition of society, and maintained that the descriptive method of statistics was more appropriate for studying the state.[50]

The numerical social science of statistics never became as popular a movement in Germany as it had in Britain and France, perhaps because the field remained under the control of government bureaucrats and university professors. It did, however, effectively supplant descriptive statistics by 1860, and a remarkable volume of statistical writing for general readers and philosophical thinkers as well as political and social scientists and government administrators was produced during the remainder of the nineteenth century. Statistics was pursued in close connection with historical economics, especially by the "academic socialists" of the Verein für Sozialpolitik, and was a principal tool in the effort to understand and solve the social problems associated with German industrialization.

[49] See F. W. Reden, ed., *Zeitschrift des Vereins für deutsche Statistik* (2 vols., 1848-1849); also J. v. W., "Die Errichtung statistischer Büreaus und statistischer Privatvereine," *Deutsche Vierteljahrsschrift*, 9 (1846), no. 3, 95-128.

[50] C.G.A. Knies, *Die Statistik als selbständige Wissenschaft* (Kassel, 1850), p. 23. See also Eberhard A. Jonák, *Theorie der Statistik in Grundzügen* (Vienna, 1856); Leopold Neumann, "Aphoristische Betrachtungen über Statistik, ihre Behandlung und ihre neueren Leistungen," *Oesterreichische Vierteljahresschrift für Rechts- und Staatswissenschaft*, 3 (1859), 87-114.

Chapter Two

THE LAWS THAT GOVERN CHAOS

The council of the Statistical Society of London was perhaps a bit extreme when it argued that brute facts were at once the distinctive feature of modern science and the foundation of the social discipline statistics, but the standard nineteenth-century apology for statistical science was based on the utter reliability of its results. Nineteenth-century statisticians revelled in the existence of government records such as census data, which liberated them from the need for surmise and conjecture—both standard equipment for political arithmetic in the previous century. The special merit of statistics was its insistence on accurate and exhaustive enumeration, its exclusion of guesses and approximations.

This science of pure facts was of doubtful standing even from the beginning, however. R. John Robertson, reviewing the first volume of the *Journal of the Statistical Society of London* for the Benthamite *Westminster Review*, mocked the pretense of the council that opinion was to be excluded from this science. Opinion, he observed, is just a Latin word for thought: "It is not needful in the present day to discourage thinkers, they are not too numerous."[1] Robertson argued that facts can be interpreted only in the light of theories, or indeed that theories are precisely "facts as viewed by the most powerful minds,"[2] and that statists had wasted much labor trying to prove silly or even incoherent claims because they had failed to analyze concepts and establish clear definitions. "No mere record and arrangement of facts can constitute a science,"[3] he wrote, and numerical statistics ought to be understood as a method—"a mode of arranging and stating facts which belong to various sciences."[4]

[1] [R. John Robertson], "Transactions of the Statistical Society of London, vol. 1, part 1," *Westminster Review*, 29 (1838), 45-72, p. 49.

[2] *Ibid.*, p. 48.

[3] *Ibid.*, pp. 68-69.

[4] *Ibid.*, p. 70. The exclusion of opinions was not generally well received; see also Thomas Chalmers' essay, which I have seen only in German translation, *Die kirchliche Armenpflege*, Otto von Gerlach, trans. (Berlin, 1847), chap. 12; "A Grumbler" [Anon], "Statistics," *Fraser's Magazine*, 52 (1855), 91-95.

That the vaunted science of statistics was merely a method and the statistical society composed of "hewers and drawers to those engaged on any edifice of physical science" was wholly unacceptable to the Victorian statists. "Statistics, by their very name, are defined to be the observation necessary to the social or moral sciences, to the sciences of the *statist*, to whom the statesman and the legislator must resort for the principles on which to legislate and govern."[5] Statists were by no means in agreement, however, that the exclusion of opinions was forever to be the defining attribute of their science. More common was the view of J. E. Portlock, organizer of the short-lived statistical society of Ulster, who believed that statistics represented the empirical stage of a social science that had previously been based on mere conjecture and would soon yield reliable laws of the sort attained through the same procedure by "astronomy, zoology, botany, chemistry, and geology."[6] Portlock, however, never characterized in any detail the nature of this universal scientific procedure, and in practice the task of the numerical science of statistics remained the collection and presentation of tabular information on subjects of political or social interest.

QUETELET AND THE NUMERICAL REGULARITIES OF SOCIETY

Adolphe Quetelet was among the few nineteenth-century statisticians who pursued a numerical social science of laws, not just of facts. Like Comte and St. Simon, he believed in a dynamic social entity endowed with properties and tendencies that would not be significantly altered or impeded by the whimsical acts of political leaders. Quetelet even pirated Comte's phrase, *physique sociale*, as the title of his new science. His proposed method, however, departed radically from the positivist way, for whereas Comte insisted on the distinctiveness of each level on the hierarchy of the sciences and eschewed the use of mathematics or even of numbers in the physiological and social disciplines,[7] Quetelet main-

[5] "Sixth Annual Report of the Council of the Statistical Society of London," *JRSS*, 3 (1840), 1-13, pp. 1-2.

[6] J. E. Portlock, "An Address Explanatory of the Objects and Advantages of Statistical Enquiries," *JRSS*, 1 (1838), 316-317, p. 317.

[7] See Auguste Comte, "Plan des travaux scientifiques nécessaires pour réorganiser la société" (1822), *Opuscules de philosophie sociale, 1819-1828* (Paris, 1883), pp. 159-163, 172; *Cours de philosophie positive* (6 vols., Paris, 1830-1842), vol. 2 (1835), p. 371; vol. 4 (1839), pp. 7, 511-516.

tained that a single method was appropriate for every science and that the social physicist could do no better than to imitate the celestial one.

Quetelet came to statistics from astronomy, and his commitment to the use of mathematics in the social sciences distinguished his approach from that of his reform-oriented statistical contemporaries. At the same time, his active interest in social policy and even in certain concrete reforms as well as the wide scope of his ambitions for statistics separated his work from that of the various astronomers and mathematicians of his day who wrote demographic models and computed life tables. Quetelet was almost unique in the early nineteenth century in combining the characteristic concerns of the statistical movement with the technical tools of astronomers and probabilists. His contribution to statistical thought and to the mathematics of statistical analysis was a characteristic if not unsurprising product of his syncretic approach. Social physics was an elaborate metaphor that integrated Quetelet's genuine concern for the advancement of scientific knowledge with his desire to turn science to the promotion of sound government and social improvement. Mathematics would bring order to the apparent social chaos.

In many ways, Quetelet's work in statistics reflected the experiences of a whole career. Born in 1796, soon after the conquest and annexation of Austria's Belgian provinces by France, Quetelet was educated in a French *lyçee* and retained close cultural as well as scientific ties to France throughout his life. Although he was more devoted to art and literature than to mathematics during his youth, he did not fail to notice the exalted role assigned to science under the Napoleonic empire.

> It was known that the sciences were honored, that their power had never been greater: it was known that the man whose military glory resounded then throughout Europe took to heart the goal of reflecting onto them a part of the prestige that surrounded him, and that he had raised the most illustrious savants to the dignity of princes and of the highest functionaries of the empire. Such munificence, which honored him who exercised it as much as the savants who were its object, maintained that source of illustrious men that had arisen in the midst of the revolutionary exaltation.[8]

After the defeat of France, Quetelet became a teacher of mathematics in Ghent at a *collège* set up by the government of the newly united Neth-

[8] Adolphe Quetelet, *Histoire des sciences mathématiques chez les Belges* (Brussels, 1864), p. 312.

erlands. Soon afterwards he wrote a doctoral dissertation under the mathematician and old revolutionary J. G. Garnier, who regaled his pupil with tales of life among the great savants of Paris. A few years later the ambitious Quetelet gained tentative approval from the government for the construction of an observatory in Brussels, where he had taken a new position and where he was already scheming to build a scientific empire. As a result, he was authorized to make a pilgrimage to the great scientific metropolis in order to learn enough astronomy to equip and run the proposed observatory. In Paris he was taken in by Bouvard at the Royal Observatory and given the necessary instruction in the instruments and methods of scientific observation. Among these was the method of least squares, by then used routinely to reduce astronomical observations, which in the Laplacian tradition was closely associated with the general field of mathematical probability and even with population and mortality studies. By the time he returned to Brussels, Quetelet had become infected with an enthusiasm for statistics, the social science of the observatory.

Quetelet stressed the similitude of statistics with astronomy throughout his career, and he always attributed his acquisition of a "taste for statistics," specifically to this venture to Paris of 1823. He identified his statistical teachers as "the great French school of mathematics and social science," noting especially Fourier, Laplace, Lacroix and Poisson, and he claimed to have "had the good fortune of enjoying the lessons" of the "two great masters" Laplace and Fourier.[9] Although he did correspond with Fourier, who once sent him a letter containing a quotable line about the need for statisticians to be well-versed in mathematics if their science was to progress, it seems unlikely that he actually received formal instruction from these men. Nevertheless, it is clear that he studied their written work carefully, that he admired them greatly, and that he identified his own goals with their ideas and with the remarkable achievements of contemporary French science generally.

Having familiarized himself with the relations of statistics to the most advanced mathematics of probability theory under the guidance of these august personages, Quetelet was little impressed by the ordinary statistical enthusiasts of his generation. He consistently denied that his science was the province of physicians and amateur reformers, insisting in-

[9] See Frank Hankins, *Adolphe Quetelet as Statistician*, in Columbia University Studies in History, Economics and Public Law, 31, no. 4 (New York, 1908), 19-20; also Quetelet, "Des lois concernant le développement de l'homme," *AOB*, 38 (1871), 205-206.

stead that the true foundation for statistics had been established by mathematicians and astronomers. "It is not to doctors that we owe the first tables of mortality, they were calculated by the celebrated astronomer Halley," he wrote. Indeed the astronomer's love of natural order provided the foundation for statistical science: "The laws that concern man, and those that govern social development, have always had a special attraction for the philosopher, and perhaps most especially for those who have directed their attention to the system of the universe. Accustomed to considering the laws of the material world, and struck with the admirable harmony that reigns there, they can not be persuaded that similar laws do not exist in the animate world."[10] Statistics was a science like other sciences, to be presented in scientific periodicals such as his own *Correspondance mathématique et physique.*[11]

The focus of Quetelet's earliest work in statistics was the study of periodic phenomena. This was already evident in his first statistical memoir, presented to the Brussels Academy in 1825, although his announced purpose there was the improvement of Belgian insurance tables. Arguing that analogy with the regularities of plant and animal life "entitles us to believe that the influences of these laws extend even to the human species," Quetelet proposed to examine the relationship of births and deaths to time of year in order to determine "whether it is possible to ascertain in this regard some law of nature."[12] As it turned out, the natural laws of mortality and natality proved to be statistical generalizations of the form $y = a + b \sin x$, where y was the number of births or deaths, x the time of year, suitably normalized, and a and b arbitrary constants, to be determined empirically. Such formulas, which Quetelet called statistical laws, were typical results of his early investigations. They were based partly on curve-fitting, partly on analogy, and were not interpreted concretely or grounded in physical models.

The motif of periodic phenomena supplied also the connection between social and natural science in Quetelet's early work. Quetelet maintained that tides, weather, flowering of plants, terrestrial magnetism, and events in the life of man constituted a unified set of phenomena, which could be studied by a single method. These studies were to

[10] See Quetelet's "Notice scientifique" for his book *Sur l'homme*, *AOB*, 7 (1840), 230.

[11] See Quetelet, "Avertissement et observations sur les recherches statistiques insérées dans ce recueil," *Correspondance mathématique et physique*, 4 (1829), 77-82.

[12] Quetelet, "Mémoire sur les lois des naissances et da la mortalité à Bruxelles," *NMAB*, 2 (1826), 493-512, p. 496.

be based on exhaustive measurement, the quantitative form of natural history that Susan Faye Cannon has called Humboldtian science.[13] "Instead of words facts are wanted," Quetelet wrote, "and sage observations instead of vague hypotheses and unfounded systems. . . . This manner of only proceeding scientifically characterizes the nineteenth century, which is destined to occupy one of the highest places in the annals of the human spirit."[14] The diversity and complexity of these phenomena—the volume of facts required—was not a disadvantage but an asset, for Quetelet was an energetic organizer and indefatigable correspondent. Already in the early 1820s he had decided that the newly reconstituted Brussels Academy could labor most effectively as a group, working precisely on the class of objects forming the heart of Quetelet's research project, whose variety was so great that it would never yield to the individual scientist.[15] Quetelet was also a leader of the international network of quantitative natural historians that rose to prominence during the 1830s.

The young Quetelet was highly impressed by the rationalistic French program for applying mathematical probability to such problems as the reliability of testimonies, the correctness of judicial decisions, and the validity of electoral results. He proclaimed the universality of the rule of numbers—*Mundum regunt numeri*—and argued that "all that can be numerically expressed" falls under the jurisdiction of probability theory.[16] Placing statistics under the domain of mathematical probability was his highest aim. Quetelet was disdainful of the mass of statistical reformers who flourished during the early nineteenth century. He inveighed against the "strange abuses" of statistics perpetrated by dilettantes and scientific illiterates who wished only to gain support for preconceived ideas, and argued that the contemporary infatuation with statistics had "rather retarded than accelerated its advance."[17]

Perhaps these complaints were justified, but his confidence in the redemptive power of probability theory was at best prophetic. The avail-

[13] Susan Faye Cannon, "Humboldtian Science," in *Science in Culture: The Early Victorian Period* (New York, 1978), chap. 3.

[14] Quetelet, "Recherches statistiques sur le Royaume des Pays-Bas," *NMAB*, 5 (1829), separate pagination, p. ii.

[15] Quetelet, *Sciences mathématiques et physiques chez les Belges au commencement du XIXe siècle* (Brussels, 1866), p. 9; *Histoire* (n. 8), pp. 375-376.

[16] Quetelet, *Popular Instructions on the Calculation of Probabilities* (1828), Richard Beamish, trans. (London, 1839), p. 108.

[17] Quetelet, "Sur l'appréciation des documents statistiques, et en particulier sur l'appréciation des moyennes," *BCCS*, 2 (1844), reprint, p. 1.

able techniques of error analysis were of little use in analyzing official aggregate tables based on complete enumeration, and this was the standard material used by nineteenth-century statisticians. If probability theory was to contribute to the statistical work of Quetelet and his contemporaries, it was the genius to create new techniques, not the skill to employ existing ones, that was required. In fact, Quetelet almost never used mathematics in his empirical work on statistics, and he developed no new evaluative tools that might have been put to use by others. Although he incessantly exhorted his colleagues to furnish error estimates along with mean or composite numbers, he never did so in his own writings. The most sophisticated test Quetelet ever applied in published work to assess the reliability of his numbers was to divide the data at random into two or three groups and then compare their respective means.

Lacking, as he did, the genius to formulate a usable mathematical procedure for analyzing statistical information, Quetelet had recourse to a strategy especially suitable for deeply ambitious but fundamentally sensible persons of moderate ability. In his practical work, he compiled and arranged statistical data as best he could in order to learn something of the composition and perhaps the causes of aggregate phenomena like natality, mortality, marriage, crime, and suicide. He gave tables of these events according to age, sex, profession, and place of residence, and discussed the implications of his numbers in a coherent and temperate manner. At the same time, virtually independent of this practical work, he developed an extravagant system of metaphors and similes linking the social domain to the theories and even the mathematics of physics and astronomy. This was the much-vaunted science of social physics. It embodied Quetelet's bid to become the Newton of statistics, and not merely—as his friend Villermé kept offering—its nineteenth-century Süssmilch.[18]

If social physics is to be conceived in part as a testament to the confidence and ambition of the astronomer, it must also be recognized as a paean to social order in the spirit of gradualist liberalism. The prominence of antirevolutionary metaphor in its basic structure can perhaps be partly explained in terms of Quetelet's experience as a scientific empire builder in unstable times. The September 1830 Belgian revolution, which secured the independence of the southern provinces from the united Netherlands, seemed also to wreak havoc with Quetelet's scien-

[18] See letters of 31 Aug. 1835 and 30 April 1837 in cahier 2560, *AQP*.

tific plans. Some of his protégés left their posts to join the military, and many scientific positions at universities, colleges, and museums were eliminated.[19] His cherished observatory, by then nearly complete after years of hard work and planning, was occupied by the radical Liège volunteers to assist in the defense of Brussels; "shots were fired wildly through the windows, blood spilled in many places," and the structure was then "converted into a fortress" and "surrounded with ditches and ramparts."[20] In the end, the observatory and Quetelet's position as royal astronomer were salvaged—after serious discussion of using the observatory as a magazine—but the "intellectual movement" which he felt to have been taking hold, mostly as the result of his own efforts, was reduced to nought.[21] If the *éloges* Quetelet wrote as permanent secretary of the Brussels Academy are to be believed, virtually no Belgian savants passed through the revolution without serious damage to their careers.[22] He later pronounced it a universal truth that the abrupt changes characteristic of political revolutions invariably bring about a loss of the delicate "living force" of science.

Mécanique sociale, the direct ancestor of *physique sociale* and social correlate of *mécanique celeste*, was originally announced in a paper read by Quetelet in March 1831. This, Quetelet's earliest paper after the September revolution, introduced the first of what soon became a battery of metaphors linking the social order to planetary astronomy. Prominent among these was the distinction between the constant forces of nature and perturbational forces, generated by the conscious decisions of man. Although Quetelet maintained that in social mechanics, as in astronomy, the effect of the perturbational force must at first be disregarded, he did not delay posing what he evidently viewed as the critical question regarding perturbations of the social system. Laplace had claimed to refute the old view that planetary movements were unstable, and required Providential intercession to restore their natural orbits. "Can the forces

[19] Quetelet was evidently upset by his own loss of income as a result of the termination of public lectures at the Brussels Museum; see Villermé to Quetelet, 7 Dec. 1834, cahier 2560, AQP.

[20] Quetelet, "Lettre à M. le Bourgmestre, 15 dec. 1831," *AOB*, 1 (1834), 285; letter to Bouvard, 5 Nov. 1830, quoted in Joseph Lottin, *Quetelet: Statisticien et Sociologue* (Louvain, 1912), p. 52; also Ed. Mailly, "Essai sur la vie et les ouvrages de Lambert-Adolphe-Jacques Quetelet," *Annuaire de l'Académie royale des sciences, des lettres, et des beaux arts de Belgique*, 41 (1875), 109-279, pp. 186-187.

[21] See Quetelet, "Aperçu de l'Etat actuel des Sciences Mathématiques chez les Belges," *BAAS* (1839), p. 58; *Histoire* (n. 8), p. 370.

[22] See Quetelet, *Sciences mathématiques* (n. 15), esp. chapter on Quetelet's friend Dandelin.

of man," Quetelet asked, "compromise the stability of the social system."[23]

Notwithstanding the evident implications of this metaphor, perturbational forces were no mere secondary effects on the course of the social system. These were the moral forces of man, the living forces of progress, providing the impetus to overcome the dead hand of nature which threatened to render society "stationary and incapable of amelioration."[24] Quetelet invested in these forces all the ambivalence evoked by the idea of rapid social change in a cautious, nineteenth-century liberal. He wrote:

> These forces that characterize man are the living forces of his nature, but . . . does there exist something analogous to the principle of conservation of living force in nature? What, moreover, is their destination? Can they influence the march of a system or imperil its existence? Or, indeed, are they, like the internal forces of a system, entirely unable to modify its trajectory or the conditions of its stability? Analogy would lead one to believe that in the social state one may anticipate finding in general all the principles of conservation that one observes in natural phenomena.[25]

Although social physics was intended to promote social improvement by serving as a guide to the legislator, it served most prominently as an allegory of stability and lawlike certainty. That Quetelet's new science was inspired by his experience during an age of revolution is suggested most concretely by the preface to a particularly extravagant essay on the "Analogies between physical laws and moral laws" which he published in a book of 1848. He wrote: "In a moment when passions were acutely excited by political events, I sought, in order to distract myself, to establish analogies between the principles of modern mechanics and what was taking place in front of me."[26] More generally, he exalted lawfulness in every domain, holding it quite literally as tantamount to godliness. This appears clearly in his standard defense against the charge of materialism and fatalism:

[23] Quetelet, "Recherches sur la loi de la croissance de l'homme," *NMAB*, 7 (1832), separate pagination, p. 7.

[24] Quetelet, "Recherches sur le poids de l'homme aux différens ages," *NMAB*, 7 (1832), separate pagination, p. 11.

[25] Quetelet, "Loi de croissance" (n. 23), p. 2.

[26] Quetelet, *Du système social et des lois qui le régissent* (Paris, 1848), p. 104.

> After having viewed the course that has been followed by the sciences in regard to worlds, can we not attempt to follow them with regard to men; would it not be absurd to believe that among all the things that occur in accordance with such admirable laws, the human species alone rests mindlessly abandoned to itself, and that it possesses no principle of conservation? We do not fear to say that such a supposition would be more injurious to divinity than the research itself that we propose to carry out.[27]

Quetelet's fascination with the possibility of subjecting ostensibly uncontrolled social phenomena to scientific order was at the heart of his dedication to the concept of statistical laws. Characteristically, he gave this idea its fullest development in reference to such events as crime and suicide, the immoral materials for that "moral statistics" which was central to the early statistical movements in Britain and France as well as the Low Countries. In 1827 the French government began publishing records of criminal activity in its *Compte général de l'administration de la justice criminelle*. They revealed, to the amazement of numerous readers, that criminal activity varied little from year to year. Quetelet reported in 1829 his shock at the "frightening regularity with which the same crimes are reproduced."[28] André-Michel Guerry, the Parisian lawyer and astute analyst of social numbers generally placed alongside Quetelet as a founder of the science of "moral statistics,"[29] was also astonished that crime could display such regularity. "If we consider now the infinite number of circumstances that can cause the commission of crime, . . . we will not know how to conceive that in the end result, their conjunction leads to such constant effects."[30]

That a writer like Quetelet, well informed about mathematical probability and conversant with the leading works of political arithmetic, could have been startled by the regularity of these numbers seems itself to be a bit surprising. There was, after all, a rich tradition of admiration for the constancy and stability of population statistics dating back to Arbuthnot's amazing discovery that the ratio of male to female births in

[27] Quetelet, "Recherches sur le penchant au crime aux différens ages," *NMAB*, 7 (1832), separate pagination, p. 4.

[28] Quetelet, "Recherches statistiques" (n. 14), p. 28.

[29] In fact, this phrase, like the concerns that underlay it, had already been current in France for at least several years. See, for example, Charles Dupin, "Effets de l'enseignement populaire sur les prosperités de la France," reviewed by Aubert de Vitry in *Bulletin universel des sciences et de l'industrie*, 6th sec., *Bulletin des sciences géographiques*, 8 (1826), 329-336.

[30] A. M. Guerry, *Essai sur la statistique morale de la France* (Paris, 1833), p. 11.

London exceeded unity by a small proportion every year, without exception. Jakob Bernoulli's demonstration that probabilistic events of any sort can be expected to give rise to stable ratios if repeated often enough was intended as an explanation for regularities such as this, and it was repeated in most subsequent general works on mathematical probability, including that of Laplace. Quetelet's reaction makes it clear that the extension of this principle to events so irrational, disorderly, and antisocial as crime had not been anticipated.

In fact, demographic regularities were generally viewed as exemplars of natural theology, or at least of the harmonies of nature, even by writers who accepted and understood Bernoulli's argument. The mathematician De Moivre, like John Arbuthnot and William Derham, held up the regularity of births, marriages, and even of deaths by age as compelling evidence of the wisdom and benevolence of the Creator. Süssmilch pointed to the birth differential and to the mortality rates of boys and girls to establish that the totality of these processes led to a perfect balance of the sexes by the time the age for marriage was reached, thus facilitating the great goal of all human activity: maximal population increase. He rejoiced also in the stability of death tables, noting that it yielded "extraordinarily great advantages for the state and other arrangements in the life of man, for it contains the basis and points to the rules, according to which annuities and tontines may be arranged and determined."[31] It also guarantees that young men and women will exist in equal numbers at the age of marriage. No blind accident, he remarked, "could place goal and means in so beautiful a union together."[32]

Süssmilch's table attracted the interest of Immanuel Kant, who was deeply impressed by these instances of the emergence of large-scale regularity from local chaos. Kant used this truth of political arithmetic to illustrate how, despite the undirected behavior of every individual, the philosopher might hope to "discover a natural purpose in this idiotic course of things human. . . . a history of a definite plan for creatures who have no plan of their own." Teleology could prevail in history, just as order emerges in political arithmetic:

> Whatever concept one may hold, from a metaphysical point of view, concerning the freedom of the will, certainly its appearances,

[31] J. P. Süssmilch, *Die göttliche Ordnung in den Veränderungen des menschlichen Geschlechts* (2 vols., Berlin, 3d ed., 1765), vol. 2, pp. 366-367.
[32] *Ibid.*, p. 289.

> which are human actions, like every other natural event are determined by universal laws. However obscure their causes, history, which is concerned with narrating these appearances, permits us to hope that if we attend to the play of freedom of the human will in the large, we may be able to discern a regular movement in it, and that what seems complex and chaotic in the single individual may be seen from the standpoint of the human race as a whole to be a steady and progressive though slow evolution of its original endowment. Since the free will of man has obvious influence upon marriages, births, and deaths, they seem to be subject to no rule by which the number of them could be reckoned in advance. Yet the annual tables of them in the major countries prove that they occur according to laws as stable as [those of] the unstable weather, which we likewise cannot determine in advance, but which, in the large, maintain the growth of plants, the flow of rivers, and other natural events in an unbroken, uniform course. Individuals and even whole peoples think little on this. Each, according to his own inclination follows his own purpose, often in opposition to others; yet each individual and people, as if following some guiding thread, go toward a natural but to each of them unknown goal; all work toward furthering it, even if they would set little store by it if they did know it.[33]

Until the nineteenth century, then, statistical regularity was generally seen as pertaining to the natural history of man, and as indicating divine wisdom and planning. The first well-publicized instance of statistical order which could not be plausibly interpreted in this way was Laplace's announcement in the *Philosophical Essay on Probabilities* that the number of dead letters in the Paris postal system was constant from year to year. The uniformity of murder, theft, and suicide was even more difficult to explain in natural-theological terms. Quetelet was, in some sense, able to do so, but only by embracing a cosmology at once physicalist and theological that made mass regularity the expected outcome of natural processes in every domain.

Quetelet interpreted the regularity of crime as proof that statistical laws may be true when applied to groups even though they are false in relation to any particular individual. Beyond that, he implied that the

[33] Immanuel Kant, *On History* (1784), Lewis White Beck, ed. (Indianapolis, 1963), pp. 11-12.

obliteration of the particular by the general was responsible for the very preservation of society. The prevalence of mass regularity, he wrote, "teaches that the action of man is restrained in such a circle, that the great laws of nature are forever exempted from his influence; it also shows that laws of conservation can exist in the moral world, just as they are found in the physical world."[34] As a consequence of the fundamental truths revealed by the uniformity of statistics, society could be seen to be an entity in its own right, independent of the whims and idiosyncracies of its constituent individuals.

Indeed, statistical regularity seemed to Quetelet to provide the key to social science. He cast the "law of large numbers," Poisson's form of Bernoulli's theorem, as the fundamental axiom of social physics. It confirms, after all, that general effects in society are always produced by general causes, since chance, or accidental causes, can have no influence on events when they are considered collectively. In particular, social facts can never be generated by the arbitrary, unmotivated, spontaneous, or otherwise inexplicable acts of the human free will. However, "capricious" and "disordered" may be the action of the human will, "whatever concerns the human species, considered *en masse*, belongs to the domain of physical facts; the greater the number of individuals, the more the individual will is submerged beneath the series of general facts which depend on the general causes according to which society exists and is conserved."[35]

Quetelet's most celebrated construct, *l'homme moyen*, or the average man, was similarly dependent on his belief in the regularity of statistical events. Quetelet maintained that this abstract being, defined in terms of the average of all human attributes in a given country, could be treated as the "type" of the nation, the representative of a society in social science comparable to the center of gravity in physics. After all, deviations from the average must necessarily cancel themselves out whenever a great number of instances is considered. Hence for the average man "all things will occur in conformity with the mean results obtained for a society. If one seeks to establish, in some way, the basis of a social physics, it is he whom one should consider, without disturbing oneself with particular cases or anomalies, and without studying whether some given individual can undergo a greater or lesser development in one of his fac-

[34] Quetelet, "De l'influence de libre arbitre de l'homme sur les faits sociaux," *BCCS*, 3 (1847), 135-155, p. 136.

[35] Quetelet, "Penchant au crime" (n. 27), p. 80.

ulties."[36] Quetelet allowed his average man a temporal dimension as well as physical and moral ones, thus giving him a mean rate of growth and moral development over his average life.

The calculation of *l'homme moyen physique* was no problem for Quetelet, since it merely involved collecting measurements of height, weight, and the dimensions of the various limbs and organs. *L'homme moyen moral*, on the other hand, offered certain obstacles, for human individuals do not present themselves to the scientist endowed with measurable quantities of courage, criminality, or affection. In this respect, the average man was a far more tractable problem than the concrete individual. In principle, wrote Quetelet, the courage or criminality of a real person could be established if that person were placed in a great number of experimental situations, and a record kept of the number of courageous or criminal acts elicited. This would be interesting, but it was wholly unnecessary for social physics. Instead, the physicist need only arrange that courageous and criminal acts be recorded throughout society, as the latter already were, and then the average man could be assigned a "penchant for crime" equal to the number of criminal acts committed divided by the population. In this way, a set of discrete acts by distinct individuals was transformed into a continuous magnitude, the penchant, which was an attribute of the average man.[37]

It seems wrong to argue, as Paul Lazarsfeld did, that Quetelet's method for assigning numbers to social attributes was independent of his idea of the average man.[38] His "penchants" were, after all, traits of *l'homme moyen*, and it is far from clear that the assignment of numerical values from aggregate statistics to this abstract being involves a genuine "contribution to quantification in sociology" except insofar as the concept of the average man proved to be a useful tool. Quetelet thought this a valuable exercise largely because *l'homme moyen* could be used as an analogue to certain concepts in celestial physics. In addition, he found it appealing to distribute responsibility for crime over the whole community. To speak of the tendency to crime within a given age class was

[36] Quetelet, *Sur l'homme et le développement de ses facultés, ou essai de physique sociale* (1835; 2 vols., Brussels, 1836), vol. 1, pp. 21-22.

[37] Quetelet, "Penchant au crime" (n. 27) and "Sur la statistique morale et les principes qui doivent en former la base," *NMAB*, 21 (1848). Quetelet was not the first to seek statistical indices of moral attributes. The prominent Italian statistician Melchiorre Gioja, *Filosofia della statistica* (1826; Mendrisio, 1839), p. 570, wrote: "The most certain measure of immorality is, as everyone recognizes, the ratio of the number of crimes to the number of inhabitants."

[38] Paul Lazarsfeld, "Notes on the History of Quantification in Sociology—Trends, Sources, and Problems," *Isis*, 52 (1961), 164-181.

more compatible with Quetelet's reformist impulse and his sociological aims than to discuss crime in terms of the wicked acts of a specifiable group of malignant persons.

Crime, then, was a social phenomenon, whose frequency, like that of marriage, should be attributed "not to the will of individuals, but to the customs of that concrete being that we call the people, and that we regard as endowed with its own will and customs, from which it is difficult to make it depart."[39] The cause of crime, he explained, may be identified precisely as the "state of society." Since Quetelet's often prolix discussions of crime contain only the vaguest hint of what aspects of the social condition were responsible for it (he was better on the causes of fluctuations in the marriage rate), his analysis hardly seems a landmark in the history of criminology. His object, however, was not so much to propose an immediate cure for crime as to establish the need for statistical analysis as part of a campaign to remove its causes gradually.[40] Crime, he wrote, is like a "budget that is paid with frightening regularity," or "a tribute that man acquits with greater regularity than that which he owes to nature or the state treasury." Indeed, "every social condition presupposes a certain number and a certain arrangement of offenses as a necessary condition of its organization. This observation, which may appear discouraging at first, becomes, on the contrary, consoling, when it is examined closely, because it shows the possibility of improving men, by modifying their institutions, their customs, the state of their enlightenment." Really, he concluded, "it only presents us an extension of a law already well known to all philosophers who have occupied themselves with society in relation to physics: that so long as the same causes subsist, one may expect the continuation of the same effects."[41]

In retrospect, Quetelet's importance as a social scientist seems modest, for his social physics was and remained completely unworkable. His great goal was to measure the changes experienced by *l'homme moyen* over time in order to ascertain the general law of social development. That is, by plotting the trajectory of the average man through history, he hoped to discover the forces acting on the "social body" and hence to predict its future course. Although he continued to speak of this as the

[39] Quetelet, "Libre arbitre" (n. 34), p. 142.

[40] See Quetelet, "Sur le poids de l'homme aux différens ages," *Annales d'hygiène publique et de médecine légale*, 10 (1833), 5-27.

[41] Quetelet, *Sur l'homme* (n. 36), vol. 1, pp. 8-11.

realization of widespread hopes for an exact social science throughout his life, not a single trajectory calculation is to be found in all of Quetelet's works. Quetelet the practicing statistician was always more sensible than Quetelet the social physicist.

Quetelet's lasting contribution to science, though related to his social ideas, was more abstract. He succeeded in persuading some illustrious successors of the advantage that could be gained in certain cases by turning attention away from the concrete causes of individual phenomena and concentrating instead on the statistical information presented by the larger whole. Quetelet implied that the viability of this approach was not uniquely tied to society, but was an immediate and general corollary of the universal truth that constant causes must yield constant effects. His metaphorical science of social physics became an important source of analogies for scientists in other fields, for it seemed to reveal plainly that statistical laws can prevail for a mass even when the constituent individuals are too numerous or too inscrutable for their actions to be understood in any detail.

LIBERAL POLITICS AND STATISTICAL LAWS

Notwithstanding Friedrich Engels' disdainful remarks on "utopian socialism," few if any of the pioneer social scientists of the early nineteenth century believed that the great reorganization of society foreshadowed by their theories was to be a pure triumph of their own inventive genius. Comte and the St. Simonians, for instance, argued explicitly that social and scientific developments had both made possible their ideas and prepared the way for their actualization. They conceived themselves as discovering, not inventing, the social condition of the future.

Quetelet, unlike the St. Simonians, was no dialectician; there were no "critical" epochs required in his version of the historical process. Progress, he believed, was smooth and continuous, and he saw no advantage in a radical change of social organization. The course of the social body through history was along a path that was inherently progressive, towards the eminently desirable goal of effacing *l'homme physique* by *l'homme intelectuelle.*[42] To get there it is most efficient and least trou-

[42] *Ibid.*, vol. 2, p. 285.

blesome to follow a straight line; Quetelet was a "rigorous partisan of the principle of least action that the Supreme Being follows in all things."[43]

The cause of this secular progress, according to Quetelet, was the living "moral force" of science and learning, which tended always to increase. Since this force was conserved whenever it acted continuously, social change could not be averted by even the most skillful reactionary government. Any attempt to block the course of progress could only result in an accumulation of grievances, a deep well of repressed forces that must sooner or later escape in a violent explosion. This would lead to a wastage of some of the moral force, accompanied by much misery. Even revolutions, fortunately, could do no more than inflict a temporary setback. When Quetelet wrote to Prince Albert of Britain in June 1848 lamenting the "veritable moral cholera, which spreads its ravages over the whole of Europe," he was able to find solace in the laws of social physics: "Scourges strike the moral as well as the physical structure of humanity and however destructive their effects may be, it is at least consoling to think that they cannot in any way alter the external laws that guide us. Their action is transitory and time will soon have healed the wounds of the social body; but it is not the same with individuals."[44]

Given that society was governed naturally by statistical laws, its political government was constrained to an ancillary role. The wise legislator would not try to impose his will on the social system, but would seek first to determine the direction and magnitude of secular social evolution—that is, of the average man (constancy was a characteristic of statistical laws only in the short term). Once the contribution of the constant causes was known, that of the accidental causes could be found through a process of vector subtraction. It would then be possible to restore a just state of equilibrium, or to minimize the perturbations, by opposing this resultant with an equal and oppositely directed force. "The whole art of governing," he proclaimed, "resides in estimating the nature and direction of this resultant. It is necessary to know perfectly the forces and tendencies of the factions that ordinarily divide a state, in order to judge the most appropriate means for combating and paralyzing them."[45] Certainly Quetelet's social physics did not eliminate the need for an active state, but neither did it give the state much control over the

[43] Quetelet, *Du système social* (n. 26), p. 110.

[44] Quoted in Harriet H. Shoen, "Prince Albert and the Application of Statistics to Problems of Government," *Osiris*, 5 (1938), 286-287.

[45] Quetelet, *Du système social* (n. 26), p. 289.

autonomous domain of society. Statistics supported a liberal, mildly bureaucratic politics, even as it exalted the practical function of the social scientist and assured him that society was on course.

The notion of statistical law achieved its fullest expression in Great Britain during the 1850s, the decade when laissez-faire liberalism reached its intellectual apogee. Liberals and even progressive conservatives became increasingly sympathetic to the idea that the role of government could and would be sharply reduced, while radicals propagated the claim that the historic role of government had been to protect regressive interests from the forces of change, to impede the natural course of progress, and to obstruct the general tendency towards prosperity and freedom. Social science began at this time to flourish in the rationalistic form of conjectural history,[46] which, in its British guise, regarded historical development as intrinsic to society and largely beyond the influence of kings and legislators. The recent successful campaign to repeal the Corn Laws suggested that the enlightenment of society would at last compel government to retreat into its proper sphere.

Expressions of amazement and admiration at the impressive regularity uncovered through statistical research can be found in a variety of mid-century writings. Lord Stanley, a leader of the Conservative party, claimed as an "axiom" of social science the maxim that "the moral and physical condition of the human race" was governed by constant statistical laws.[47] The economist Nassau Senior noted the remarkable result that "the human will obeys laws nearly as certain as those which regulate matter."[48] Robert Chambers observed in the *Vestiges of Creation*: "Man is seen to be an enigma only as an individual, in mass, he is a mathematical problem."[49] Henry Holland proposed that the "law of averages" had "acquired of late a wonderful extension and generality of use; attaining results, from the progressive multiplication of facts, which are ever more nearly approaching to the fixedness and certainty of mathematical formulas."[50] Even Charles Dickens was sufficiently impressed by a recent report of the Registrar General on marriage to print the fol-

[46] See J. W. Burrow, *Evolution and Society: A Study in Victorian Social Theory* (London, 1970), and J.D.Y. Peel, *Herbert Spencer: The Evolution of a Sociologist* (New York, 1971).

[47] Lord Stanley, "Opening Address . . . as President of Section F . . . ," *JRSS*, 19 (1856), 305-310, p. 305.

[48] Nassau W. Senior, "Opening Address . . . ," *JRSS*, 23 (1860), 357-362, p. 359.

[49] Robert Chambers, *Vestiges of Creation* (New York, 1846), pp. 333-334, quoted in C. C. Gillispie, *Genesis and Geology* (New York, 1959), p. 157.

[50] [Henry Holland], "Human Longevity," *Edinburgh Review*, 105 (1857), 46-77, pp. 54-55.

lowing reflections by Frederick Knight Hunt in *Household Words*: "Not content with making lightning run messages, chemistry polish boots, and steam deliver parcels and passengers, the savants are superseding the astrologers of old days, and the gipsies and wise women of modern ones, by finding out and revealing the hitherto hidden laws which rule that charming mystery of mysteries—that lode star of young maidens and gay bachelors—matrimony."[51] In the same vein, William Farr, effective head of the General Register Office, modestly remarked:

> It would formerly have been considered a rash prediction in a matter so uncertain as human life to pretend to assert that 9000 of the children born in 1841 would be alive in 1921; such an announcement would have been received with as much incredulity as Halley's prediction of the return of a comet, after the lapse of 77 years. What knew Halley of the vast realms of aether in which that comet disappeared? Upon what grounds did he dare to expect its re-appearance from the distant regions of the heavens? Halley believed in the constancy of the laws of nature; hence he ventured from an observation of parts of the comet's course to calculate the time in which the whole would be described; and it will shortly be proved that the experience of a century has verified quite as remarkable predictions of the duration of human generations.[52]

The political viewpoint that underlay much of this talk is more readily apparent in some remarks made by William Newmarch. When the International Statistical Congress met in London in 1860, Newmarch greeted the assembled native and foreign dignitaries on behalf of the Statistical Society of London with the following observations, which, it is recorded, were received with great applause.

> On the part of the Statistical Society, which is an entirely voluntary association of individuals, not connected with the State, and, I think I may say, not in the smallest degree desiring to be connected with the State, we pride ourselves upon our entire independence, we feel that if we are to maintain our ground in this country of free

[51] [F. K. Hunt], "A Few Facts about Matrimony," in Charles Dickens, ed., *Household Words*, 1 (1850), p. 374 (for authorship, see Anne Lohrli, comp., *Household Words* [Toronto, 1973], p. 319).

[52] William Farr [written under the name of the titulary head of the G.R.O., George Graham], "Report," in *Fifth Annual Report of the Registrar General of Births, Deaths, and Marriages* (1843), p. 21.

> and open thought, . . . where every man, we firmly believe, obtains the reward which his merit deserves, [the society should not seek government support]. . . . [W]e certainly act upon this maxim, that if our society cannot support itself by its own intrinsic merits, and by its own usefulness, we are quite ready to close it tomorrow.[53]

Newmarch interpreted the laws of statistics in the same spirit. Writing in the *Economist* newspaper, he explained how governments had begun to learn that legislation could only accomplish its aims if it was formulated in accordance with the natural principles of these things. Official support for statistics, he proposed, embodied a desperate attempt to understand the condition of society, deriving from

> nothing less than the necessity under which all Governments are rapidly finding themselves placed, of understanding as clearly and fully as possible the composition of the social forces which, so far, Governments have been assumed to control, but which now, most men agree, really control Governments. The world has got rid of a good many intermediate agencies, all of them supposed originally to be masters, where in truth, they were even less than servants. The rain and the sun have long passed from under the administration of magicians and fortune-tellers; religion has mostly reduced its pontiffs and priests into simple ministers with very circumscribed functions; commerce has cast aside legislative protection as a reed of the rottenest fibre; and now, men are gradually finding out that all attempts at making or administering laws which do not rest upon an accurate view of the social circumstances of the case, are neither more nor less than imposture in one of its most gigantic and perilous forms. . . .
>
> Crime is no longer to be repressed by mere severity; Education is no longer within the control of the maxims which preceded printing,—Law is found to be a science perhaps the most difficult of any—Justice means more than tricks and plausibilities of procedures;—Taxation, Commerce, Trade, Wages, Prices, Police, Competition, Possession of land,—every topic from the greatest to the least which the old legislators dealt with according to . . . ca-

[53] William Newmarch, Opening Address, in *Report of the Proceedings of the Fourth Session of the International Statistical Congress*, 1860 (London, 1861), p. 116.

price . . . have all been found to have laws of their own, complete and irrefragable.[54]

The most determined and influential effort to harness the laws of history in opposition to government meddling was Henry Thomas Buckle's ambitious but uncompleted project, the *History of Civilization in England.* Buckle was the child of a prosperous Cambridge-educated merchant father and a devout Yorkshire Calvinist mother, and if his reputation had remained what it was during the nineteenth century, the childhood of this lifelong bachelor would doubtless have been memorialized by a psychobiographer. As it is, we know that he was a spoiled child, raised with little direction or discipline, and that he became radicalized in politics as in religion around the time of his grand tour of the Continent, in 1841. The influence of the maternal religion, however, can perhaps still be seen in the deterministic cast of mind revealed by his early enthusiasm for phrenology and by the explicit arguments in his *History.* The introductory section of his work—the second volume of which, incidentally, was dedicated to the memory of his mother—interpreted belief in strict scientific causality as derived from religious fatalism.[55]

Buckle was an archetypical autodidact, conversant with if not master of a variety of fields and authors. His study originated as a complete history of the world, constructed on entirely new principles, but he soon found that a radical new interpretation required new research and that one lifetime would not suffice to finish such a project. He decided, reluctantly, to confine his ambitions and to write instead a history of England. The wisdom of this choice he justified in terms of Quetelet's metaphor of social physics. England, he explained, had been the least affected of all countries by destructive perturbations, and its progress reflected most clearly the natural evolution of society. All other societies had somehow been displaced from the preordained path of progressive liberalization. This appeared clearly in his general introduction and in the diverse studies of social pathology which were to fill the first three volumes of the general history, thereby permitting him to salvage the results of his first ten years of research. France, it seemed, had departed

[54] [William Newmarch], "Some Observations on the Present Position of Statistical Inquiry with Suggestions for Improving the Organization and Efficiency of the International Statistical Congress," *JRSS*, 23 (1860), 362-369, pp. 362-363.

[55] There is no good modern work on Buckle's career. On his life, see A. H. Huth, *The Life and Writings of Henry Thomas Buckle* (New York, 1880).

from the true way because its people were excessively dependent on the state, while Scotland and Spain revealed in their histories the distortions caused by the excessive power of religious institutions. The perturbations of Germany and the United States, involving respectively an excess in the concentration and in the diffusion of knowledge, were to make up the third volume of the introductory material, but Buckle died before he could write it. Liberal England, that model of healthy development and the ostensible object of the whole exercise, was discussed only incidentally in the volumes that were actually completed.

Buckle's avowed goal was to elevate history to the rank of the other sciences. Although history "deserves the highest faculties of science," he wrote, the public had become accustomed to such mediocrity among historians that "any author who from indolence of thought, or from natural incapacity, is unfit to deal with the highest branches of knowledge, has only to pass some years in reading a certain number of books, and then he is qualified to be an historian."[56] Buckle denounced the presentation of history as a chronicle of kings and battles and insisted on the need to look beyond the surface confusion of particular events in order to establish the simple, general, underlying principles. Such investigation would demonstrate, he proposed, that the substance of history was not politics, but society—not court intrigues, declarations of war, and issuance of church decrees, but the slow, continuous advancement and diffusion of knowledge. Government activity might seem whimsical and unpredictable, but the regular course of social development confirmed that "in the moral world, as in the physical world, nothing is anomalous; nothing is unnatural; nothing is strange. All is order, symmetry, and law."[57]

Buckle's fondness for an intellectual and social approach to history reflected his political viewpoint. His book carried on a sustained polemic against corporate entities such as governments and religious institutions. Church and court were mere superstructures, lacking any autonomous principles of development, whose only contribution to history had been to hold society to the institutional forms of previous ages. Buckle was consistently cynical of the motives that determined the activities of public institutions. He wrote of Scottish preachers, for example, that "like every corporation, which has ever existed, whether spiritual or

[56] Henry Thomas Buckle, *History of Civilization in England* (2 vols., 1857-1861; New York, 1913), vol. 1, p. 3; see also p. 671.

[57] *Ibid.*, vol. 2, p. 25.

temporal, their supreme and paramount principle was to maintain their own power."[58] But it mattered little whether rulers were generous or self-interested. Charles III, he observed, had the best interest of his people in mind when he sought to free them from Spain's superstitious and bigoted clergy. In the absence of enlightenment, however, such efforts were "worse than futile" for they served only to strengthen the people's sympathy for their clerical oppressors.[59]

On occasion, Buckle thought, unwise policy had led to lamentable consequences. By blocking the development of Protestantism, the leaders of France, Spain, and Italy had locked their nations into a system of beliefs that could only confuse and retard their development, producing disequilibrium and stimulating social turmoil. In the long run, however, institutions were powerless, and Buckle scoffed at the "folly of lawgivers" to think that their enactments could affect any phenomenon—such as suicide—that was governed by general laws.[60] Legislation, always derivative in its development, was at best a measure, and never a cause, of progress. "No great political improvement, no great reform, either legislative or executive, has ever been originated in any country by its rulers,"[61] Buckle wrote. Moreover, "every great reform which has been effected has consisted, not in doing something new, but in undoing something old."[62] "Lawgivers are nearly always the obstructors of society, instead of its helpers."[63]

Statistics was more critical to the justification of Buckle's project than to its execution, and social numbers were not cited in the substance of his historical discussion. Buckle invoked the results of statistics as compelling evidence for the existence of social laws, rather than as instances of the general principles governing history; indeed, his rhetoric of perfect statistical regularity was not easy to square with an approach that presupposed the inherent dynamism of society. Nevertheless, the success of statistical science provided, in Buckle's view, a crucial lesson to the historian. Isolated events always seem unpredictable and confusing, and the historian must not lose himself in their intricacies. Instead, he must take a broad view and seek general principles. Although social phenomena have proven intractable to metaphysicians, who seek through

[58] *Ibid.*, p. 193.
[59] *Ibid.*, p. 91.
[60] *Ibid.*, vol. 1, p. 19n.
[61] *Ibid.*, p. 198.
[62] *Ibid.*, pp. 199-200.
[63] *Ibid.*, vol. 2, p. 244.

introspection the rules that operate in individual minds, they prove themselves wholly accessible to the historian who has learned the essential lesson of statistics. Buckle cited the proven regularities of statistical social science in order to persuade his readers that mankind was no exception to the universal reign of natural order. The instances he seized upon were the very ones that had most impressed Quetelet, and were taken directly from the work of the Belgian statistician—murder, suicide, the misaddressing of letters. As Quetelet himself had marvelled, not only do murder and suicide occur at a constant rate in any given society, but so do the categories of those committed with guns, knives, or poison, and by asphyxiation, hanging, or drowning. However diverse and irrational the behavior of individuals may seem, the collective regularity of human behavior proves that each act is a necessary consequence of unvarying social laws.

Indeed, in his insistence on the universality of law, Buckle even went beyond Quetelet. The latter, ever faithful to the principle of moderation, aimed to promote the influence of science on legislation, and he had no desire to entangle statistics in sectarian controversy. To this end, he worked out a compromise position, a reconciliation of statistical law with free will that would exonerate statistics from the stigma of materialism and fatalism without giving up the possibility of a rigorous social science patterned after celestial mechanics. The heart of his argument was a distinction between the laws of society and those of individuals. Society, he held, is the product of a social contract, into which individuals have invested a portion of their freedom. Even the laws of society are not so rigid that they cannot be altered through the activity of the legislator—who seems somehow to stand outside the universe of law. Individuals participate in the laws of society in the sense that they are all accidental variants of the average man, and hence possess, as a first approximation, identical penchants for crime, suicide, marriage, and so on. Still, the "moral causes that leave their mark on social phenomena are . . . inherent in the nation, not in individuals,"[64] and the individuals have been accorded a limited domain for the exercise of their free will. The combined effect of a multitude of free decisions is always to neutralize them. Accordingly, the will, though free, can have only a trifling effect on society as a whole. Quetelet relegated it to the domain of accidental causes, and insisted that the constant causes which were

[64] Quetelet, "Statistique morale" (n. 37), p. 6. See also "De l'influence" (n. 34), and *Du système social* (n. 26), pp. 71-73.

alone of interest to social physics were quite uninfluenced by human freedom.

Buckle accepted this distinction between levels of human existence, but denied that freedom pertained to individuals any more than it did to societies. The autonomy of the will is a mere fiction, appropriate for cultures dependent for their maintenance on hunting and gathering or subject to numerous tornadoes and earthquakes, but not for a prosperous and educated modern society that was increasingly taking charge of its own fate. In Buckle's view, the possibility of scientific history would be contradicted if some "mysterious and providential" force, such as divine authority or unfettered human will, could act on society. The fundamental prerequisite of scientific history was the absolute universality of law, the denial of "chance or supernatural influence."[65] The regularities of statistics, he thought, were compelling not just as evidence of the lawfulness of social evolution, but also of individual actions. If human behavior were subject to free will, we could expect to see no relation between the quantity of vicious behavior and the state of society. Indeed, nothing would seem more disorderly or incomprehensible than transgressions against the moral order. Yet crimes occur in the same numbers year after year, obeying ostensible social laws that left no place for divine guidance of history, metaphysical free will, or individual moral responsibility. Suicide, for example,

> is merely the product of the general condition of society. . . . The individual felon only carries into effect what is a necessary consequence of preceding circumstances. In a given state of society, a certain number of persons must put an end to their own life. This is the general law; and the special question as to who shall commit the crime depends of course upon special laws; which, however, in their total action, must obey the large social law to which they are all subordinate. . . . [T]he offences of men are the result not so much of the vices of the individual offender as of the state of society into which that individual is thrown.[66]

The existence of statistical laws also demonstrated the impotency of government to sustain the malicious influence that it had always worked on society. In the long run, Buckle maintained, the progressive laws of social development must prevail even in nations where their action has

[65] Buckle, *History* (n. 56), vol. 1, p. 6.
[66] *Ibid.*, pp. 20-22.

been partially arrested. He expressed admiration and astonishment that deviations from the "moral laws" revealed by statistics were so small, despite the complicated condition of the social world. "From the circumstance that the discrepancies are so trifling," he remarked, "we may form some idea of the prodigious energy of those vast social laws, which, though constantly interrupted, seem to triumph over every obstacle, and which, when examined by the aid of large numbers, scarcely undergo any sensible perturbation."[67]

Buckle was possibly the most enthusiastic and beyond doubt the most influential popularizer of Quetelet's ideas on statistical regularity, but similar remarks may be found in publications ranging from the *Journal of the Institute of Actuaries* to the *Revue des deux mondes.*[68] Quetelet's principal books were translated into English and German, while Buckle's two completed volumes went through edition after edition in English, French, German, Russian, and various other languages. Quetelet and Buckle were read, moreover, by a generation of Europeans who were imbued with a sense of society as a fundamental and preeminently historical entity that was capable of having its own laws. Statistics acquainted these readers with a new form in which natural laws could reveal themselves, one whose distinctive features were noted by writer after writer. The individuals are so numerous, and subject to so complex an array of circumstances, that it is impossible to foresee with any reliability their future behavior. Yet whenever a large number of individuals is considered at once, "the influence of contingencies seems to disappear before that of general laws."[69]

The principle of statistical regularity, or of the stability of the mean, was invoked incidentally in a variety of works on social and political subjects during the nineteenth century. Charles Morgan, following Quetelet, explained how legislation need not take account of numerous individual natures, but could instead be founded upon a knowledge of the

[67] *Ibid.*, p. 23.

[68] Samuel Brown, "On the Uniform Action of the Human Will, as exhibited by its Mean Results in Social Statistics," *Assurance Magazine and Journal of the Institute of Actuaries*, 2 (1852), 341-351; Louis Etienne, "Le positivisme dans l'histoire," *Revue des deux mondes*, 74 (1868), 375-408.

[69] [Herman Merivale], "Moral and Intellectual Statistics of France," *Edinburgh Review*, 69 (1839), 49-74, p. 51. Statistical regularity, or "the law of large numbers," was also invoked by the laissez-faire economist Frédéric Bastiat, who argued that the spontaneous order of a free economy, revealed in the existence of stable wage and interest rates, is wholly analogous to the operation of an insurance company. See Bastiat, *Economic Harmonies* (1851), W. Hayden Bayers, trans. (Princeton, 1964), chap. 14, pp. 361-406.

"abstract being 'man' " which can be attained through statistics.[70] George Cornewall Lewis proposed that the law of averages could be employed to determine the "prevailing character" of a government, the "medium state between opposite extremes." He wrote: "We may be unable to predicate any invariable and universal tendency of a form of government, just as we may be unable to say that all men live a certain number of years. But as we can say of men, that the average duration of their life is a certain number of years—so we may say of a form of government, that it has a certain prevailing average character."[71] John Stuart Mill mentioned statistics as characteristic of his "Inverse Deductive Method," and began in 1862 to cite Buckle in his *Logic* for the principle that the "very events which in their own nature appear most capricious and uncertain, and which in any individual case no attainable degree of knowledge would enable us to foresee, occur, when considerable numbers are taken into the account, with a degree of regularity approaching to mathematical."[72] He speculated on the number of years it would take to average out the effect on history of a Napoleon or Alexander, and argued that statesmen may rely on probable statements regarding multitudes, for "what is true approximately of all individuals is true absolutely of all masses."[73]

Similarly Karl Marx explained how Quetelet's doctrine of the average man could be used to define a uniform standard of labor, and hence to furnish an exact and defensible interpretation of the labor theory of value.[74] William Whewell illustrated the "method of means" in science

[70] Charles Morgan, Review of Quetelet's *On Man* in *Athenaeum* (1835), 593-595, 611-613, 658-661, p. 593.

[71] G. C. Lewis, *A Treatise on the Methods of Observation and Reasoning in Politics* (2 vols., London, 1852), vol. 2, pp. 84-85.

[72] John Stuart Mill, *A System of Logic: Ratiocinative and Inductive*, variorum edition in *Collected Works*, vols. 7-8, J. M. Robson, ed. (Toronto, 1973), p. 932. On the other hand, Mill asserted in his *Principles of Political Economy with Some of their Applications to Social Philosophy* (1848), W. J. Ashley, ed. (London, 1923), p. 704, on the authority of Thomas Tooke, "that even so long a period as half a century may include a much greater proportion of abundant and a smaller of deficient seasons than is properly due to it. A mere average, therefore, might lead to conclusions only the more misleading, for their deceptive semblance of accuracy. There would be less danger of error in taking the average of only a small number of years, and correcting it by a conjectural allowance for the character of the seasons, than in trusting to a longer average without any such corrections."

[73] *Ibid.*, p. 603.

[74] Karl Marx, *Capital* (3 vols., New York, 1967), vol. 1, p. 323; see also his remarks on average prices in vol. 2, pp. 860-861. Marx also cited Edmund Burke, about whom he otherwise had little favorable to say, on the point of average labor. Robert Owen, who was not cited here, but whose work Marx knew and praised (with qualification), made precisely the same point, also in defense of a strict labor theory of value, in *Report to the County of Lanark* (Glasgow, 1821), p. 7.

by citing the regularity of dead letters.[75] M. L. Wolowski, who held that "the law of large numbers governs the moral world, as it governs the physical," argued that statistical laws provided essential empirical confirmation for economic deductions.[76] Mark Pattison, in his review of Buckle, asserted that "social history can only be composed upon statistical data," for in "dealing with the individual human being, everything is uncertainty; it is only of man in the aggregate that results can be calculated with accuracy."[77] Perhaps the most enthusiastic statistician of all was Florence Nightingale, who was enthralled by the laws Quetelet discovered and who sought to integrate statistics into administration not only by collecting and popularizing pertinent numbers, but also by endowing statistical research or setting up a tripos at Oxford based on Quetelet's *Physique sociale.*[78]

The leaders of the new social and behavioral sciences of the late nineteenth century were generally more attracted to biological, historical, and anthropological ideas than statistical ones, but statistics was not absent from their programs. The young Wilhelm Wundt dismissed Buckle with a long, critical footnote in his programmatic book of 1862, but repeated many of Buckle's pronouncements and expressed hope that use of numbers might enable him to penetrate to the unconscious. Accordingly, he called for a natural history of human society based on the laws of statistics, arguing: "It can be stated without exaggeration that more psychology can be learned from statistical averages than from all philosophies, except Aristotle."[79] Wundt later became a critic of statistical law, as he shifted to a less naturalistic, more holistic position. Although the mature Wundt thought the existence of historical laws inconsistent with the primacy of psychology among the mental sciences, he at least promoted error analysis in experimental psychology.

75 William Whewell, *The Philosophy of the Inductive Sciences* (2 vols., London, 2d ed., 1847), vol. 2, p. 405.

76 M. L. Wolowski, *Etudes d'économie politique et de statistique* (Paris, 1848), pp. 397, 401.

77 [Mark Pattison], "History of Civilization in England," *Westminster Review*, 12 (1857), 375-399, p. 392.

78 Marion Diamond and Mervyn Stone, "Nightingale on Quetelet," *JRSS*, A, 144 (1981), 66-79, 176-213, 332-351; see also I. Bernard Cohen, "Florence Nightingale," *Scientific American*, 250 (1984), March, 128-137.

79 Wilhelm Wundt, *Beiträge zur Theorie der Sinneswahrnehmung* (Leipzig, 1862), p. xxv. See Carl F. Graumann, "Experiment, Statistics, History: Wundt's First Program of Psychology," in W. G. Bringman, R. D. Tweney, eds., *Wundt Studies, A Centennial Collection* (Toronto, 1980), pp. 33-41, or, better, Solomon Diamond, "Buckle, Wundt, and Psychology's Use of History," *Isis*, 75 (1984), 143-152.

Herbert Spencer's sociology was first of all evolutionary, and was exposited principally in the form of biological analogies. As part of his campaign to discourage government meddling, Spencer expressed disdain for those parts of statistics that seemed to yield easy answers, such as sanitary statistics, and he was particularly unenthusiastic about the evident tendency of the medical profession to organize itself "after the fashion of the clergy"[80] and take over the domain of public health. Educational statistics, the results of which remained monumentally inconclusive, suited his temperament better. Spencer, however, was by no means an obscurantist, and though he thought the main contribution of social science would be to educate people on the matter of unintended consequences, he was firmly persuaded that its claim to science was just. To this end he cited the results of statistics, even though he made no effort to found theory on them. He argued for the beneficial effects of natural selection with an invocation of Quetelet, contending that the death of many is redeemed by the survival "in the average of cases" of one "perfect specimen."[81] Though he claimed to be little impressed by Buckle, he defended the possibility of social science against Buckle's critics Charles Kingsley and J. A. Froude with statistical arguments. Granting that "the results of individual will are incalculable," he insisted "Mr. Froude himself so far believes in the doctrine of averages as to hold that legislative interdicts . . . will restrain the great majority of men in ways which can be predicted."[82] Social science may not be as exact as astronomy, but "if there is some precision, there is some science."[83]

Emile Durkheim holds the rare distinction among early sociologists of actually having written a statistical work, his study of suicide. This was, of course, the most statistical of topics, having engaged the serious attention of many major European statisticians and legions of minor ones since the time of Quetelet. Durkheim in fact cited a fair number of them—Quetelet, Adolph Wagner, M. W. Drobisch, Georg Mayr, and Alexander von Oettingen. He also attempted a critique of statistical knowledge, though the result was among the most confused of all such efforts. His use of numbers to draw sociological conclusions, on the

[80] Herbert Spencer, *Social Statics, Abridged and Revised* (1st ed., 1850, 1892; Osnabrück, 1966), p. 199.

[81] *Ibid.*, p. 233; also p. 203.

[82] Spencer, *The Study of Sociology* (9th ed., 1880; Osnabrück, 1966), p. 46. On Buckle, see Spencer, *An Autobiography* (1904; 2 vols., Osnabrück, 1966), vol. 2, p. 4.

[83] Spencer, *Study of Sociology* (n. 82), p. 39.

other hand, was more sophisticated than virtually any of his predecessors. The relevant point here, however, is that even at the end of the nineteenth century the regularities of statistics bore impressive testimony to the power and reality of the social domain. In the *Rules of Sociological Method* (1895), Durkheim identified the "average type" with the normal, as opposed to the pathological, and with the "group mind (*l'âme collective*)," arguing that statistics furnished the means to isolate those "currents of opinion" that "impel certain groups either to more marriages, or to more suicides, or to a higher or lower birthrate, etc."[84] In *Suicide* (1897) he used statistics to illustrate how "the individual is dominated by a moral reality greater than himself," one so pressing that Durkheim felt obliged to reconcile "liberty with the determinism revealed by the statistical data."[85] For this, he argued in the manner of Quetelet that the social force "does not determine one individual rather than another," but only "exacts a definite number of certain kinds of actions."[86] Following his German predecessors, Durkheim preferred to interpret his figures as expressing a "collective impulse" that could not be reduced to the mean of individual dispositions.[87] The power of these social tendencies, however, was nowhere more compellingly presented to view than in the collected acts of numerous individuals driven to suicide or marriage not by the particular circumstances that they cite to rationalize their deeds, but by deep reasons "unknown to consciousness" arising in the collective social being.[88]

The evident success of statistics as an approach to social science was not interpreted by contemporaries as vindication of a metaphysic which regarded the laws governing certain domains as only probable. On the contrary, statistical laws were deliberately formulated to extend the certainty of sciences like astronomy and mechanics to knowledge of phenomena which hitherto had resisted exact scientific investigation. As will appear in part 3, statistics carried sufficient prestige at the time of Buckle that his arguments were able to set off a considerable debate over

[84] Emile Durkheim, *The Rules of Sociological Method* (1895), George E. G. Catlin, ed., S. A. Solovay and J. H. Mueller, trans. (Glencoe, N.Y., 1964), pp. 8, 56.

[85] Emile Durkheim, *Suicide: A Study in Sociology* (1897), J. A. Spaulding and George Simpson, trans. (New York, 1951), pp. 38-39.

[86] *Ibid.*, p. 325.

[87] *Ibid.*, p. 306.

[88] *Ibid.*, p. 297. On these matters, see also Stephen Turner, "The Search for a Methodology of Social Science: Durkheim, Weber, and the Nineteenth-Century Problem of Cause, Probability, and Action," in *Boston Studies in the Philosophy of Science*, vol. 92, 1985. This work was not yet available to me.

the possible incompatibility of statistical laws of moral behavior with traditional notions of free will. To be sure, the statist recognized that his adoption of a numerical approach constituted abandonment of the search for laws of individual behavior, but this was not taken to imply that chance had any objective role there. At worst, the action of individuals might be like a coin toss—too complex and unstable to foretell reliably, but nonetheless wholly determined by conditioning influences and natural laws. Indeed, it was not universally recognized that the exact number of crimes or suicides in a society was dependent even trivially on the whims of individual decisions. Buckle seemed to believe, and was certainly read by some to imply, that the root cause of crime was a fixed sum of immorality arising from the ignorance and degradation in society, which must be exhausted every year. If a particular individual did not yield to temptation, others would be impelled with yet greater force until the annual quota of crime had been reached.

During the second half of the nineteenth century, confidence in the value and reliability of statistical laws reached the point that the social science of statistics could become a model for certain areas of the physical and biological sciences. Analogies and similes of social science were used repeatedly to justify the application of what came to be known as statistical reasoning or the statistical method to problems of thermodynamics, heredity, and price fluctuations. In the end, this form of knowledge would indeed become associated with uncertainty and even indeterminism, but the initial use of statistics in relation to subjects which nobody had any reason to think behaved whimsically or irregularly required first of all that statistics be deemed capable of producing knowledge no less certain or lawlike than the rules by which nature itself was governed. Statistical principles bore sufficient prestige to be applied to economics, to biology, and even to the most dignified of sciences, physics, where they were used to work out the implications of the kinetic gas theory. Maxwell, Boltzmann, Galton, and Edgeworth interpreted Buckle's statistical observations as justification for the presumption that any phenomenon composed of numerous independent events could be expected to exhibit impressive regularity in the mass. The greatest confusion at one level is not only consistent with but implies remarkable stability at another, which manifests itself in the form of statistical laws.

Chapter Three

FROM NATURE'S URN TO THE INSURANCE OFFICE

During the eighteenth century, as Lorraine Daston has shown, probability was customarily interpreted as the calculus of reasonableness for a world of imperfect knowledge. Enlightenment thinkers applied the mathematics of chance to an implausibly rich variety of issues. They used it to demonstrate the rationality of smallpox inoculation, to show how degrees of belief should be apportioned among testimonies of various sorts, and even to establish or preclude the wisdom of belief in biblical miracles. Condorcet, followed by Laplace and Poisson, furnished calculations of the probability that juries of a given form and size would reach a just verdict in any given case. These, however, were not the only calculations presented to the world as evidence of the power of what De Moivre called the "doctrine of chances." Probabilists also stressed the applicability of their subject to actuarial and demographic matters. Probability calculations based on mortality records had been used increasingly to set rates for life insurance and annuity purchases—though, interestingly, not for maritime and fire insurance—since Edmond Halley published the first life table in 1693. Mathematicians all over Europe, but especially in the great commercial states, the Netherlands and Great Britain, applied their skill to political arithmetic during the eighteenth century. Some of Laplace's most important contributions arose from his work on population estimates and other demographic problems.[1]

Writings by eighteenth-century probabilists on insurance were plentiful, but they evidently had little influence on the interpretations given to probability. The mathematical structure of probability theory was de-

[1] See Lorraine Daston, "The Reasonable Calculus: Classical Probability Theory, 1650-1840" (Ph.D. Dissertation, Harvard University, 1979); also Charles C. Gillispie, "Laplace," in *DSB*, vol. 15; Gillispie, "Probability and Politics: Laplace, Condorcet, and Turgot," *Proceedings of the American Philosophical Society*, 116 (1972), 1-20; Gillispie, *Science and Polity in France at the End of the Old Regime* (Princeton, 1980), chap. 1.

rived from its earliest object, games of chance. Probabilities of single events were supposed to be known *a priori*, and calculation was used to reach formulas for compound or multiple events of various sorts. Gaming seemed too vile and contemptible, however, to serve as the real interpretation for so profound a study, and the eighteenth-century frame of mind did not permit a domain of mathematics to exist apart from its interpretation.[2] Not chance events, but degrees of belief, were taken to be the fundamental object of probability theory. The inescapable games of chance, the casting of dice and drawing from urns, were explained away as heuristic tools, analogies to the apportionment of belief. "It is remarkable," wrote Laplace, "that a science, which commenced with the consideration of games of chance, should be elevated to the rank of the most important subjects of human knowledge."[3]

Beyond doubt the most famous of Laplace's ideas, in his own time as in ours, was the imaginary being whose perfect vision and unlimited powers of calculation would enable it to know both past and future with complete certainty. Laplace was not seeking to explain the ways of God. He had need of this hypothesis, the "infinite intelligence," to make clear his conception of probability and to explain why he assigned it so prominent a place in human thought. Chance meant for Laplace not the irreducibly random, but the fortuitous production of patterns through the interaction of a multitude of independent causes. If all the events of nature could be perceived simultaneously, and if our techniques of calculation were sufficiently advanced, we would have no more need of probability than the infinite intelligence. Finite humanity, however, had no prospect of attaining omniscience. Probability was thus essential for estimating, and above all for reducing, the level of human error. Its application to nature was predicated on a certain level of knowledge, but its proper object was human ignorance.

Even as philosophy, the classical theory of probability did not stand wholly apart from the objective domain of frequencies. When pressed to give an ultimate justification for basing belief on probability calculations, Condorcet had recourse to Jakob Bernoulli's theorem about the conformity of an indefinitely long sequence of similar events to the underlying probabilities. Laplace, whose reputation as a probabilist rested largely on the techniques he developed for extending the theory to great

[2] Daston, "Reasonable Calculus" (n. 1), p. 78.

[3] Pierre Simon de Laplace, *Philosophical Essay on Probabilities*, F. W. Truscott and F. L. Emory, trans. (New York, 2d ed., 1917), p. 195.

numbers of events, introduced the term *possibilité* for the actual ratios of events presented over the long run by nature. He was convinced that these ratios were generally stable, and he developed analytical tools for determining the degree of convergence of subjective *probabilités* to objective *possibilités*.[4] Classical probability writers saw no contradiction between the view that probability statements represent the uncertainty of the rational mind about a particular event and the claim that probability estimates should reflect long-run distributions of events in nature or society. Most subscribed, after all, to a relatively passive form of associationist psychology, which made belief almost a direct reflection of experience. Probability assessments measured the grooves etched by impressions over time in the *tabula rasa* of the mind.[5]

Nevertheless, there was much in the classical treatment of probability that could not be justified in terms of experience or even of expectation of stable frequencies. Laplace, following a tradition that originated with Jakob Bernoulli, assumed that the mind naturally distributes belief equally among complementary alternatives about which it knows nothing. This "principle of indifference," as it came to be called, was invoked as required to justify an assumption of equipossibility in relation to such chance events as a single toss of an untested coin. It also figured prominently in various calculations of a more daring and ambitious character, such as the method of inference developed by Laplace and termed the "probability of causes." One of the classic instances of this genre was his determination of the probability that a fixed cause, rather than chance, was responsible for the revolution of all the planets in the same direction, and roughly the same plane, around the sun. The probability that the planets would so align themselves if distributed at random had been given in 1734 by Daniel Bernoulli, and could be computed straightforwardly, but the inverse probability of the existence of a fixed cause depended on certain assumptions. One of these was an *a priori* probability for the existence of a fixed cause. Laplace, applying the principle of indifference, chose one half.

All told, however, Laplace and Condorcet were far more interested in the applications of probability than in its epistemological foundations.

[4] See Condorcet, *Essai sur l'application de l'analyse à la probabilité des décisions* (Paris, 1785), in Keith M. Baker, ed., *Condorcet: Selected Writings* (Indianapolis, 1976), p. 39; Laplace, "Mémoire sur les Approximations qui sont fonctions de très grands nombres" (1785), *Oeuvres*, vol. 10, p. 310.

[5] Daston, "Reasonable Calculus" (n. 1), pp. 174-218.

One of the first thinkers to examine at length the pertinent philosophical issues was Augustus De Morgan, whose work on the logic of probability helped to establish it as a worthy philosophical problem. De Morgan was the most influential of a group of members of the Royal Astronomical Society who sought to do for probability what John Herschel, William Whewell, Charles Babbage, George Peacock, and George Airy, all originally at Cambridge, had begun to do for the algebraic methods of Laplace's *Mécanique céleste* and Fourier's essay on heat.[6] It was evident to De Morgan, Herschel, Thomas Galloway, John Lubbock, and Francis Baily that England had fallen far behind France in probability theory, and that it would be a great achievement simply to introduce their countrymen to the profound mathematical reasoning of Laplace's *Théorie analytique des probabilités*. Unlike the algebraists, they were less concerned to establish an autonomous mathematical research tradition than to learn those parts of probability that were needed by science and commerce. In particular, they were interested in the probabilistic method of least squares—which, by 1830, was indispensable for observational astronomers—and also in the mathematics of insurance. Interestingly, the principal founder of the Astronomical Society, Francis Baily, had originally made his fortune in the insurance industry, and at least a few of these men derived income from positions on the boards of insurance institutions at different times.[7] As mathematical probabilists, they made no claim to important advances over Laplace and Poisson. They sought instead to make Laplace accessible, and their writings can be found in the *Encyclopaedia Metropolitana*, the *Encyclopaedia Britannica*, Lardner's *Cabinet Cyclopaedia*, the *Penny Cyclopaedia*, and the Library of Useful Knowledge.

De Morgan was no frequentist. On the contrary, he adopted an uncompromisingly subjectivist framework for the logic of probability. He was no less enthusiastic than Laplace and Condorcet about the probabilistic study of testimonies and judicial decisions, and his own demonstration that a mathematical Robinson Crusoe who had hitherto ob-

[6] See Susan Faye Cannon, "The Cambridge Network," in *Science and Culture: The Early Victorian Period* (New York, 1978), chap. 2, and Maurice Crosland and Crosbie Smith, "The Transmission of Physics from France to Britain: 1800-1840," *HSPS*, 9 (1978), 1-61.

[7] See Francis Baily, *The Doctrine of Life-Annuities and Assurances Analytically Investigated and Practically Explained* (2 vols., London, 1813); John Herschel, "Memoir of Francis Baily," in *Journal of a Tour in Unsettled Parts of North America in 1796 and 1797*, Augustus De Morgan, ed. (London, 1856). The election of De Morgan and Galloway to positions with insurance offices is mentioned in Sophia Elizabeth De Morgan, *Memoir of Augustus De Morgan* (London, 1882), pp. 60-61, 110, 181, 279-280, 363.

served seven ships, all bearing flags, should properly conclude with a probability of 8/9 that the next vessel to appear would also bear a flag became notorious.[8] De Morgan also upheld subjective probability as a logic of tolerance which, if understood, would serve as an antidote to religious bigotry. Although he was not wealthy, he resigned from University College, London, in protest against religious interference with faculty appointments, and he insisted that all knowledge except pure mathematics and logic was probabilistic—that is, uncertain. He wrote in his book of 1838:

> Probability is the feeling of the mind, not the inherent property of a set of circumstances. . . . [R]eal probabilities may be different to different persons. The abomination called intolerance, in most cases in which it is accompanied by sincerity, arises from inability to see this distinction. A believes one opinion, B another, C has no opinion at all. One of them, say A, proceeds either to burn B or C, or to hang them, or imprison them . . . ; and the pretext is, that B and C are morally inexcusable for not believing what is true. Now substituting for what *is* true that which A believes to be true, he either cannot or will not see that it depends upon the constitution of the minds of B and C what shall be the result of discussion upon them.[9]

For this and other reasons, De Morgan maintained that no measure of probability can ever be objective. "I throw away objective probability altogether," he wrote, "and consider the word as meaning the state of the mind with respect to an assertion, a coming event, or any other matter on which absolute knowledge does not exist." Probability, he maintained, was indistinguishable from psychology, and a perfect measure of probability would depend on weighing the force of sundry impressions on the human mind.[10] This, however, was not to endorse pure irrationalism, or to give up all hope of a mathematical theory, for the judging mind was not impervious to influence from the real world. De Morgan refuted d'Alembert's famous objection to the assumption of equiprobability—which for De Morgan was absolutely central—by noting: "If

[8] See John V. Strong, "The Infinite Ballot Box of Nature: De Morgan, Boole and Jevons on Probability and the Logic of Induction," in *PSA: Proceedings of the Philosophy of Science Association* (1976), vol. 1, 197-211, pp. 200-201.

[9] Augustus De Morgan, *An Essay on Probabilities and on Their Application to Life Contingencies and Insurance Offices* (London, 1838), pp. 7-8.

[10] De Morgan, *Formal Logic* (1847; London, 1926), p. 199.

any individual really feel himself certain, in spite of authority and principle as here laid down, that the preceding cases [in tosses of a coin—namely H, TH, TT] are equally probable, he is fully justified in adopting 2/3 instead of 3/4 [for the probability of 'heads at least once' in two tosses]." However, let him spend an evening betting against an opponent on that supposition, and he would rapidly come around to a more orthodox view.[11]

The circumstance that experience was permitted to correct deviant notions about probability suggests that even De Morgan allowed something in probability which was not entirely subjective. Indeed, De Morgan was an insurance enthusiast, exalting it as "the most enlightened and benevolent form which the projects of self-interest ever took. It is, in fact, in a limited sense and a practicable method, the agreement of a community to consider the goods of its individual members as common."[12] He was especially attached to friendly societies, whose benefits could "raise working men to an unknown degree of independence as the poor laws are removed."[13] Insurance, the focus of De Morgan's only full book of probability, required attention to its objective aspect. He explained: "The word probability has two different senses, the collision of which is a grand source of confusion: it is used to refer both to the state of the mind, and to the external dispositions which are to regulate the long run of events: to our strength of prediction, and also to the capacity of circumstances to fulfil our prediction."[14] De Morgan thought that the word probability should be reserved for degrees of belief, and presupposed no capacity to make reliable predictions. The objective ratios generated by nature and society he preferred to designate with the term "facility." He made frequent use of this statistical idea, and he stressed that probability calculations could take on a measure of objective reliability when applied to long series of events. "Uniformity is the law of large masses compared with each other," he wrote, citing Quetelet.[15] Similarly, his friend Thomas Galloway marvelled that the "constant approximation to fixed ratios, which is proved by all experience, in the recur-

[11] De Morgan, "Theory of Probabilities," *Encyclopaedia Metropolitana*, vol. 2 (London, 1845), 393-490, p. 401; on equiprobability, see De Morgan, "Probability," in *The Penny Cyclopaedia of the Society for the Diffusion of Useful Knowledge*, vol. 19 (London, 1841), 24-30, p. 27.

[12] De Morgan, *Essay* (n. 9), p. xv.

[13] *Ibid.*, p. 295.

[14] De Morgan, "On the Theory of Errors of Observation," *TPSC*, 10 (1864), 407-427, p. 407.

[15] De Morgan, *Essay* (n. 9), p. 119.

rence of events of the same kind, enables us to apply the calculus of probabilities to many of the most interesting questions connected with our social and political institutions; and to determine the average result of a series of coming events with as much precision as if their chances were determinate, and known *a priori*."[16]

In France, as in Britain, continued adherence to a fundamentally subjectivist understanding of probability was accompanied during the early nineteenth century by an attenuation of interest in this aspect of it. S. F. Lacroix, though enthusiastic about Condorcet's work on testimonies and judicial probabilities, denied his claim that probability was exclusively an expression of imperfect knowledge. Whether a judgment is announced before or after the drawing from an urn, he explained, the long-run tendency will be for the results to conform to the ratios of balls within the urn, just as the most complex social phenomena manifest a remarkable order when considered in large numbers. He wrote: "It is truly upon this connection that all legitimate applications of the calculus of probability are based."[17] Siméon-Denis Poisson upheld the entire spectrum of applications of probability set forth by his master, Laplace, against challenges by Poinsot and others, and he cast his main work on the subject in the form of an essay on the probabilities of judicial decisions. Much of his work, however, was devoted to an extension of Bernoulli's law of large numbers to cases where no single, fixed, underlying probability exists, and the part on judicial decisions involved a statistical analysis of actual court records. The law of large numbers, he proposed, is "the foundation of all applications of the calculus of probabilities," applicable equally to the physical and the moral sciences.[18]

The main break with the classical interpretation of probability occurred during the early 1840s. In 1842 and 1843, four writers from three countries independently proposed interpretations of probability that were fundamentally frequentist in character. So perfect a temporal coincidence can only be regarded as the outcome of chance, but the general timing of this change is not entirely surprising. Applications like the

[16] Thomas Galloway, *A Treatise on Probability* (Edinburgh, 1839; reprinted from *Encyclopaedia Britannica*, 4th ed.), p. 4.

[17] S. F. Lacroix, *Traité élémentaire du calcul des probabilités* (Paris, 3d ed., 1833), p. 69.

[18] S. D. Poisson, *Recherches sur la probabilité des jugements* (Paris, 1837), p. 12. See also his "Recherches sur la probabilité des jugements, principalement en matière criminal," *Comptes rendus hebdomadaires des séances de l'Académie des Sciences*, 1 (1835), 473-494; "Note sur la loi des grands nombres," *ibid.*, 2 (1836), 377-382; "Note sur le calcul des probabilités," *ibid.*, pp. 395-400, with critique by Poinsot.

probability of causes had come increasingly under attack during the early nineteenth century, and were looked upon with a level of skepticism bordering on outright hostility by all four authors. At the same time, the growing prominence of the social science of statistics offered a new justification for the mathematics of gaming that would eventually liberate probabilists from the need to invoke, in defense of their craft, applications like the probability of causes or judicial decisions that had come to seem implausible and even dangerous. Richard Leslie Ellis, John Stuart Mill, A. A. Cournot, and Jakob Friedrick Fries offered reasons for a frequentist interpretation of probability that were as diverse as their national origins. Yet they all chose to explain probability in terms of the regularities produced by chance phenomena, exemplified by the results of statistics.

One line of thought leading to a frequentist view of probability was based on distaste for the ostensible arbitrariness of most *a posteriori*, or inverse, probabilities. An inverse probability is one involving the formula of Thomas Bayes, rediscovered independently by Laplace and applied by him in a variety of ways, among them his determination of the probability that a fixed cause had produced the observed alignment of the solar system. In the more mundane terms of games of chance, a direct probability might be exemplified by the chance that a fair coin tossed ten times would yield eight heads and two tails; the corresponding inverse problem would give the probability that the coin is unfair (to any specified extent), supposing that eight heads and two tails appeared in the first ten tosses. All the prominent frequentists of the nineteenth century distrusted *a posteriori* calculations, especially when they purported to yield truths about the real world. The most extensive arguments were made, however, by a group of British writers.

This critique was initiated in 1842 by R. L. Ellis, who argued bluntly that the "estimates furnished by what is called the theory *à posteriori* of the force of inductive results are illusory."[19] Ellis pointed to inconsistencies in the examples of this genre given by previous authors. De Morgan's lost seaman might have seen seven ships sail by with red flags, in which case he would have been obliged to assign the same probability to the proposition that the next ship would display a red flag as that it would fly any flag at all. By this reasoning, green flags would seem to be excluded entirely. Ellis also argued that the concept of "next event" was

[19] R. L. Ellis, "On the Foundations of the Theory of Probabilities," *TPSC*, 8 (1849), 1-6, p. 6.

nebulous, and wondered if the passage of a skiff would suffice. This ambiguity, he maintained, also vitiated Laplace's argument about the planetary orbits; there was no valid *a priori* reason for excluding from his calculation the comets, which revolve in a variety of planes and in both directions around the sun.

A few years later, a controversy dealing with the same general issue arose in the context of another astronomical problem. In his *Outlines of Astronomy*, John Herschel presented a calculation deriving from a 1767 paper by Michell in which he used the inverse method to determine the probability that the large number of optically adjacent stars observed in the sky implied that some at least were true doubles. The Edinburgh professor of natural philosophy, J. D. Forbes, challenged this argument in 1850. While conceding the qualitative plausibility of the evidence, Forbes maintained that no numerical probability could be justified, for any calculation of probabilities regarding a configuration that had already been observed involves a hypothesis especially framed to accord with the data, and hence to yield a high probability of a fixed cause. Probabilities, he wrote, "have no *absolute* signification with reference to an event which *has* occurred, such as the distribution of stars on the celestial sphere. . . . They represent only the state of *expectation* of the mind of a person before the event has occurred, or having occurred before he is informed of the result."[20] In addition, Forbes argued that the alternative hypothesis—distribution of stars over the celestial globe at random—was inadequately defined, and he was uncertain that it could even be given a precise definition.

These papers by Ellis and Forbes were answered by Herschel and W. F. Donkin,[21] and soon afterwards George Boole entered the fray. Boole, who had already published an outline of his ideas on symbolic logic, readily discerned the applicability of his methods to this controversy over the foundations of probability. His proposed resolution of the conflict, first published in a series of papers for the *Philosophical Magazine* beginning in 1851,[22] was generalized in his important work of 1854, *The*

[20] J. D. Forbes, "On the Alleged Evidence for a Physical Connexion between Stars Forming Binary or Multiple Groups, Deduced from the Doctrine of Chances," *Phil Mag* [3], 37 (1850), 401-427, p. 406. See also Barry Gower, "Astronomy and Probability: Forbes versus Michell on the Distribution of the Stars," *Annals of Science*, 39 (1982), 145-160.

[21] [John Herschel], "Quetelet on Probabilities," *Edinburgh Review*, 92 (1850), 1-57, p. 37; W. F. Donkin, "On Certain Questions Relating to the Theory of Probabilities," *Phil Mag* [4], 1 (1851), 353-368, 458-466; 2 (1851), 55-60.

[22] These papers are reprinted in George Boole, *Studies in Logic and Probability* (London, 1952).

Laws of Thought. Boole explained that the conversion of a direct into an inverse probability always involved two undetermined parameters. In the case of the "probability of causes," it was always necessary to assign some value to the *a priori* probability that a fixed cause exists, and to the probability that this unspecified cause would suffice to produce the observed effect. Such assignment must always be arbitrary, and therefore the eventual solution necessarily involves some questionable assumptions. By tacitly assigning values to these parameters—one half to the first and one to the second—Laplace and De Morgan had given their analyses of causation a misleading appearance of precision. To assume these values on the basis "of the equal distribution of our knowledge, or rather of our ignorance," was, in Boole's view, quite arbitrary, and amounted to nothing more than "the assigning to different states of things of which we know nothing, and upon the very ground that we know nothing, equal degrees of probability."[23]

R. L. Ellis inferred from his objections to the ideas of Laplace and De Morgan that something was fundamentally wrong with the prevailing interpretation of probability. Positing "the inconsistency of the theory of probabilities with any other than a sensational philosophy," he proposed for the first time that probability statements were properly applicable to series of events rather than to mental uncertainty about individual trials. Hence he held that the convergence of long-run frequencies to their respective probabilities must be regarded as tautologous: "I have been unable to sever the judgment that one event is more likely to happen than another, or that it is to be expected in preference to it, from the belief that on the long run it will occur more frequently." Objecting, not quite accurately, that the Napoleonic functionary Laplace had wished to make even the most elementary truths appear in abstruse scientific garb as the result of calculation, Ellis proposed that the idea of probability is spontaneously produced by the continuous endeavor of the active mind "to introduce order and regularity among the objects of its perceptions." Our deep-seated trust in the constancy of nature assures us that stable averages will gradually be generated as "the action of fortuitous causes disappears." Probability must be associated with observation, not ignorance, of phenomena and allied with notions of order and statistical regularity, not chance.[24]

[23] George Boole, *An Investigation of the Laws of Thought on Which are Founded the Mathematical Theories of Logic and Probabilities* (1854; New York reprint, 1958), p. 370.

[24] Ellis, "Foundations" (n. 19), pp. 1-3. See also Ellis's letter to J. D. Forbes, 3 Sept. 1850,

Boole arrived also at a frequentist interpretation, but from idealism rather than Ellis's "sensational philosophy." He wrote, in the tradition of Scottish Common Sense, that logic and probability "set before us what, in the two domains of demonstrative and of probable knowledge, are the essential standards of truth and correctness,—standards not derived from without, but deeply founded in the constitution of the human faculties."[25] Yet he denied probability could be defined in terms of anything so capricious as "the strength of that expectation, viewed as an emotion of the mind." Regularity of masses is a property of nature, he observed:

> The rules which we employ in life-assurance, and in the other statistical applications of the theory of probabilities, are altogether independent of the *mental* phenomena of expectation. They are founded upon the assumption that the future will bear a resemblance to the past; that under the same circumstances the same event will tend to recur with a definite numerical frequency; not upon any attempt to submit to calculation the strength of human hopes and fears.
>
> Now experience actually testifies that events of a given species do, under given circumstances, tend to recur with definite frequency, whether their true causes be known to us or unknown. Of course this tendency is, in general, only manifested when the area of observations is sufficiently large.[26]

Thus, although Boole was a "conceptualist" rather than a "materialist" in John Venn's terminology,[27] he based his understanding of probability on insurance and social statistics rather than on a model of uncertain knowledge derived from the classical urn. He conceded that in certain games of chance the probabilities could be determined directly from knowledge of the composition and balance of coins and dice, but even for these he insisted that a measure of probability "derived from the observation of repeated instances of the success or failure of events" must

in J. C. Shairp, P. G. Tait, and A. Adams-Reilly, *Life and Letters of James David Forbes, F.R.S.* (London, 1873), p. 480: "*Avec des chiffres on peut tout démontrer*, ought to be the motto of most of the philosophical applications of the theory of probabilities—which in its own nature and according to the plain view of it, is only a development of the theory of combinations. To attempt to constitute it into the philosophy of science, is, in effect, to destroy the philosophy of science altogether."

[25] Boole, *Laws of Thought* (n. 23), p. 2.

[26] *Ibid.*, pp. 244-245.

[27] John Venn, *The Logic of Chance* (3d ed., 1888; New York reprint, 1962), pp. ix-x.

be regarded as no less fundamental. To other chance phenomena, including all those of legitimate practical interest, the gambler's *a priori* reasoning was wholly inapplicable. Boole maintained that those rare and useless instances for which results may be obtained by the direct operation of the binomial distribution on primitive probabilities had dominated the attention of mathematicians for too long. The theory of probability was properly conceived as "the mode in which the expected frequency of occurrence of a particular event is dependent upon the known frequency of occurrence of any other events."[28]

Logic, long central to the Scottish university curriculum and increasingly influential in England after Richard Whately restored it at Oxford around 1820,[29] provided also the framework for John Stuart Mill's critique of classical probability. Mill, seeking a *via media* between his father's rationalism and Macaulay's conservative Baconianism, denied that even mathematics was possible independent of experience. He could not countenance the separation of knowledge from the means of attaining it, and he argued that all reasoning proceeds from particulars to particulars. Intermediate generalizations were useful, but they had no privileged status above the concrete observations upon which they were based. Laplace's probability of causes, with its argument from ignorance, looked to Mill like the kind of *a priori* reasoning which supported, on the one hand, Bentham's inflexible deductions and, on the other, the intuitionist's exaltation of unreasoning emotion and tradition.

The first edition of Mill's *Logic*, published in 1843, contained one of the harshest denuniciations of classical probability written in the nineteenth century. He argued, with specific reference to Laplace, that those who preferred to make elaborate calculations rather than to seek improvement of the data had fostered "misapplications of the calculus of probabilities which have made it the real opprobrium of mathematics."[30] Mill dismissed the assignment of probabilities based on indifference as "entirely wrong," and deemed it a "strange pretension . . . that by a system of operations upon numbers, our ignorance can be coined into science."[31] Without an adequate base of empirical information, "to attempt to calculate chances is to convert mere ignorance into danger-

[28] Boole, *Laws of Thought* (n. 23), p. 13.

[29] See Whately's textbook, *Elements of Logic* (London, 2d ed., 1827).

[30] John Stuart Mill, *A System of Logic: Ratiocinative and Inductive,* variorum edition in J. M. Robson, ed., *Collected Works*, vols. 7-8 (Toronto, 1973), p. 538. This quote appeared in all editions.

[31] *Ibid.*, pp. 1140, 1142 (1843 edition only).

ous error by clothing it in the garb of knowledge."[32] Knowledge, whether probable or certain, was yet knowledge, and as such could only be acquired in one way. "Conclusions respecting the probability of a fact rest not upon a different, but upon the very same basis, as conclusions respecting its certainty; namely, not our ignorance, but our knowledge: knowledge obtained by experience of the proportion between the cases in which the fact occurs and those in which it does not occur."[33]

Mill's objections to the Laplacian view of probability were moderated for the second edition of 1846. During the interim, John Herschel had written him a long letter arguing that subjective probability provided a satisfactory measure of the expectation we are justified in forming with respect to an event, and that empirical information was required only for calculating probable ratios of multiple events in the real world.[34] Mill, whose footing was never too sure in actual questions of science and mathematics, acceded to this argument without really accepting it. Although he took out his denunciation of the indifference principle, he let stand the claim that improved information was more useful than the most sophisticated mathematical analysis. Every edition contained a remark that the probabilities arising from the toss of a die are better known from experience than from mere conjecture based on indifference, although Herschel's letter would seem to dictate that the two types of probability are not fully commensurable.[35]

Indeed, Mill remained suspicious of any application of probability to a single case, and preferred always to rely on the regularity of long-term frequencies. He argued that insurance tables, however useful to the companies that compile them, have scarcely any value for individuals, since each will certainly be either above or below average. He joined the chorus of contemporary statists who made the point that science and certainty emerge on a large scale whenever numerous unsystematic causes act at the local level. Beginning with the edition of 1862, he cited Buckle for the principle that "[t]he very events which in their own nature appear most capricious and uncertain, and which in any individual case no attainable degree of knowledge would enable us to foresee, occur, when considerable numbers are taken into the account, with a degree of regularity approaching to mathematical."[36]

[32] *Ibid.*, p. 545 (1843 and 1846 editions).
[33] *Ibid.*, p. 1142 (1843 edition only).
[34] See John V. Strong, "John Stuart Mill, John Herschel, and the 'Probability of Causes,' " in *PSA: Proceedings of the Philosophy of Science Association* (1978), vol. 1, pp. 31-41.
[35] Mill, *Logic* (n. 30), p. 540.
[36] *Ibid.*, p. 932. See also pp. 518, 597, 847.

A. A. Cournot, now known primarily for his work in economics, was a bit more circumspect in criticizing the classical interpretation of probability, but the tendency of his analysis was largely the same as the British frequentists. Applications of probability to testimonies, judicial decisions, and the causes of natural phenomena had been denounced in France by writers of a positivist bent like Destutt de Tracy, Poinsot, and Auguste Comte. Cournot handled these subjects more delicately, allowing that calculation of subjective probabilities was not wholly invalid but denying that it had any use. The ambiguities of *a posteriori* probability, he proposed, would "be rectified once the distinction is brought out between probabilities that have an objective existence, and give the measure of the actual possibility of things, and subjective probabilities, which relate partly to our knowledge and partly to our ignorance."[37] Neither sort of probability was of much interest unless based on wide experience:

> The mathematical theory of chance would offer no more than the charm of speculation if it could only teach us the number of chances for and against an isolated event which will never recur, or which recurs only under very unusual circumstances; as the following will show, however, it assumes a very great importance even for practice when tests of the same risk can be repeated a great number of times under similar conditions.[38]

Cournot set himself two objectives in his book on probability: to provide a clear exposition of the mathematics of chance, and to clear up the philosophical errors that seemed to prevail in so many works. He did not present probability as chiefly of mathematical interest, but as an indispensable tool of the observational sciences, statistics, and practical economic studies. Cournot regretted the customary association of this field with games of chance, though he also thought that the events of social statistics were too complicated for probability to be applied fruitfully until after its methods had been better developed. Astronomy seemed to him a more useful study for perfecting the techniques of numerical analysis by probabilities.[39] Cournot's general goal was to promote the emergence of statistics as a method, and it was this that he saw as the fruition of mathematical probability. The measurement of belief based on the principle of indifference seemed much less promising, and it is

[37] Cournot, p. 155. See also p. iv.
[38] *Ibid.*, p. 46.
[39] *Ibid.*, pp. i, 261-262.

hardly surprising that he emphasized so strongly the aspect of probability connected with stable frequencies. "The calculus of probabilities," he wrote, "has real importance only inasmuch as it is applied to numbers sufficiently large that one must have recourse to approximation formulas."[40]

In Germany, the frequentist viewpoint was introduced in 1842 by a prominent Kantian philosopher, Jakob Friedrich Fries. Like the British, Fries objected strongly to the probability of testimonies and causes, and indeed to "a large part of the doctrine of probability *a posteriori*," for "that which cannot be calculated should not be subjected to pseudo-calculations."[41] The object of his book was to secure the foundations of probability and to define its relation to political arithmetic, insurance, and observational astronomy, and his interpretation was based primarily on confidence in stable statistical frequencies. His argument was not grounded in an exaltation of experience, however, but in a defense of the higher human faculties and an antipathy to the sensualism which he imputed to Enlightenment philosophy and to the French approach to mathematics.

Fries, though competent if not original in mathematics, wanted to dispense with most of the combinatorial formalism of the pure calculus of probability. Since the notion of equipossibility was merely a mathematical abstraction, with no bearing on the investigation of nature, the standard deductive approach was applicable to little more than games of chance. Moreover, the French mathematicians, and especially the revolutionary Condorcet, had ignored the rational faculties that are necessary for all judgments; they had confused proper assessments of probability with the mere repetition of similar cases, or the blind influence of habit. The true object of mathematical probability, he explained, is the discernment of general laws whose action is obscured by special laws. Hence the arbitrary assumptions and sensualistic procedures of *a posteriori* calculation could best be discarded, and the rational investigation of the causes revealed by statistical regularities put in their place.

Fries, then, looked to political arithmetic and insurance as the model for scientific studies involving probability. Like Ellis, he objected to the claim that statistical regularity could be explained in terms of Bernoulli's principle or Poisson's law of large numbers. This was to put the cart be-

[40] *Ibid.*, p. ii.

[41] Jakob Friedrich Fries, *Versuch einer Kritik der Principien der Wahrscheinlichkeitsrechnung* (Brunswick, 1842), in *Sämtliche Schriften*, vol. 14 (Aalen, 1974), pp. v-vi.

fore the horse. The uniformity of statistics was to be explained in terms of the fundamental truth that constant causes always yield proportionate effects. Statistical regularity does not emerge from the chaos of chance; order exists prior to events in the form of natural laws. Hence the immediate object of probability as applied to nature was not the individual event, but the mean value, which alone could be used to uncover the laws of nature. Mathematical probability, wrote Fries, "has absolutely no significance for the predetermination of individual events. Such significance can only emerge if we make an estimate for a sufficiently extensive series of events of a certain sort."[42]

It should not be supposed that the frequency interpretation immediately—or ever—eclipsed completely the subjective interpretation of probability. Something like it did, however, become the common-sense view during the late nineteenth century. The most prominent French and German writers on the subject were not won over to the view that probability was intrinsically objective in character, but neither were they much interested in the allocation of uncertainty among possible outcomes of a single trial, and none seem to have upheld the calculation of probabilities for causes and testimonies. The German Johannes von Kries published a subjective interpretation of probability in 1886, but he argued that the rational estimation of real probabilities almost invariably required an extensive statistical experience. He dismissed equipossibility as a mathematical artifice, which generally was not completely valid even for games of chance, and he thought mathematical probability should be stripped of its pretension of being a general logic of uncertainty.[43] In France, Joseph Bertrand argued that the principle of indifference was not fully coherent, and that "at random" could sometimes be interpreted in different ways to yield utterly disparate results. He further maintained that probability could only be useful in relation to large numbers of similar events, and that its application to man and nature was justified exclusively by statistical regularity. "Le hasard corrige le hasard," he wrote—"Chance corrects itself. General experience reveals the justice of this maxim even to those who fail to recognize its rigor. The word *rigor* is no exaggeration. The results of the free action of chance can be predicted with certainty, without impeding in the least its caprices."[44]

[42] *Ibid.*, p. 23. See also pp. 135-136, 144.

[43] Johannes von Kries, *Die Principien der Wahrscheinlichkeitsrechnung: eine logische Untersuchung* (Freiburg, 1886).

[44] Joseph Bertrand, *Calcul des probabilités* (Paris, 1888), pp. 68-69; also pp. 4-5.

In Great Britain, the situation was more complicated. De Morgan and Herschel continued to defend Laplace's view. The Oxford mathematician W. F. Donkin depicted "a perfect mathematical identity between the fundamental laws of the equilibrium of belief as to the position of a point in space and those of the equilibrium of a perfect elastic fluid . . . [O]ur whole belief expands itself uniformly over infinite space, just as a finite mass of perfectly elastic fluid, if perfectly free, would expand itself throughout all space and have a uniform density."[45] He even derived the astronomical error function from a model based on the analogy between the force of testimony and physical attraction. Stanley Jevons's *Principles of Science* was evidently uninfluenced by the argument of the frequentists, and even Francis Edgeworth adhered to a subjectivist interpretation, albeit a more nuanced one.

The most influential nineteenth-century work on the philosophy of probability, however, was resolutely frequentist, and was in fact the book that gave the frequency interpretation its first full exposition. John Venn argued that quantitative belief cannot be justified with respect to individual nonrepeatable events, and insisted that the ideas of probability depend on a postulate of ultimate statistical regularity. To determine the probability of an event, he argued, requires that the event be placed in a series. The probability value then applies to the series, and not to the individual occurrence, for since a single event can be placed in different series, it might otherwise be assigned several contradictory probabilities. Venn offered the example of the consumptive Englishman in Madeira, a species too rare to have its own life tables. An insurance company, which is concerned only with averages, can classify this man either among consumptive Englishmen or among English residents of Madeira, but no unique probability of death can be assigned to the individual.

Venn maintained that the ultimate basis for probability calculations is experience, and its justification the "notorious fact" that a "wide and somewhat vague kind of regularity that we have called Uniformity" characterizes a vast range of phenomena.[46] He deemed it unfortunate that the convenience for mathematical purposes of *a priori* assumptions, and the association of probability with gambling problems, had "produced the impression in many quarters that they are the proper typical

[45] W. F. Donkin, "On an Analogy relating to the Theory of Probabilities and on the Principle of the Method of Least Squares," *Quarterly Journal of Pure and Applied Mathematics*, 1 (1857), 152-162, p. 153.

[46] Venn, *Logic of Chance* (n. 27), p. 240.

instances to illustrate the theory of chance. Whereas, had the science been concerned with those kinds of events only which in practice are commonly made subjects of insurance, probably no other view would ever have been taken than that it was based upon direct appeal to experience."[47]

[47] *Ibid.*, p. 76.

PART TWO

While revolutions are taking place with a frightening speed, empires are collapsing, and passions ravaging the world, there are savants and philosophers who are following attentively the progress of events, analyzing them and striving to subject them to general laws very nearly as constant as those of astronomy and physics. To prove that man is placed under the empire of fixed laws which direct his will without obstructing his free agency, such is the goal of these . . . works. —DUC DE CARAMAN (1849)

The masses seem to me worthy of notice in only three respects: first as blurred copies of great men, produced on bad paper with worn plates, further as resistance to the great, and finally as the tools of the great; beyond that, may the devil and statistics take them.

FRIEDRICH NIETZSCHE (1874)

There are three kinds of lies: lies, damned lies, and statistics. MARK TWAIN (attributed to Disraeli)

THE SUPREME LAW OF UNREASON

Although the most sophisticated statistical mathematics of the early nineteenth century grew out of the use of error theory in astronomy and related fields, the most fruitful applications of statistical reasoning proved to be elsewhere. So long as probability functions were held to represent the imperfections of measurement and of observation, there was little reason to do more with the variation that emerged than to find ways to estimate it and to eliminate as much of its effect as possible. The source of modern statistics is to be found less in error analysis than in the use of probability as a modeling tool to capture and analyze real variation in nature and society.

Significantly, the use of probability to analyze genuine variation originated in the social science of statistics. In the mid-nineteenth century, statistics seemed the most promising field of application for the calculus of chance. Charles Gouraud evinced especially high expectations for the statistical work of Quetelet in his 1848 history of probability.[1] This shift of interest within probability from error to variation is already to be found, in a general way, in the frequency interpretation. Much more central to the development of a mathematical statistics, however, was Quetelet's insight concerning the distribution of traits in human populations. His idea presents a specific connection of the greatest importance between social and mathematical statistics. It both illustrates and confirms the more general argument of this book concerning the role of social science in the development of statistical thinking.

The object of Quetelet's attention was the probabilistic error function, what Galton called the "supreme law of Unreason." The history of this curve, now known as the Gaussian or normal distribution, is practically coextensive with the history of statistical mathematics during the nineteenth century, and its reinterpretation as a law of genuine variation, rather than of mere error, was the central achievement of nineteenth-century statistical thought. That reinterpretation took place only gradually, through what in retrospect looks like a process of creative misunderstanding. Once again the lead was taken by social thought, and

[1] Charles Gouraud, *Histoire du calcul des probabilités depuis ses origines jusqu'à nos jours* (Paris, 1848), pp. 12, 139.

in this matter its influence on the natural sciences is demonstrable. The main line of development proceeded from Laplace to Quetelet to Maxwell and Galton, and from the error of mean values in demography as well as astronomy to the deviations from an idealized average man to the distribution of molecular velocities in a gas and the inheritance of biological variation in a family. Ultimately, even the analysis of error was transformed by this line of statistical thinking.

Chapter Four

THE ERRORS OF ART AND NATURE

The exponential function $\frac{1}{\sqrt{2\pi}} e^{-x^2/2}$, later to become known variously as the astronomer's error law, the normal distribution, the Gaussian density function, or simply the bell-shaped curve, was introduced to probability theory by Abraham De Moivre. Like most early probability mathematics, it first arose in the context of games of chance; it appeared as the limit of the binomial distribution. Because of its usefulness in combination and permutation problems, the binomial had become the heart of the doctrine of chances, but while it could be used to solve, in principle, a great range of problems, a superhuman computational effort would have been required to find, for example, the probability that 1,000 tosses of a fair coin would yield 480 heads and 520 tails. In 1730, James Stirling, with help from De Moivre, derived an exponential approximation for factorials. De Moivre then showed in a paper of 1733, reprinted in 1738 in the second edition of his *Doctrine of Chances*, that the exponential error function gave a very good approximation to the distribution of possible outcomes for problems like the result of 1,000 coin tosses. Now, for the first time, it was practicable to apply probability theory to indefinitely large numbers of independent events. De Moivre was highly enthusiastic about his new technique, but at the same time thought it represented the hardest problem in the whole Doctrine of Chances.[1]

De Moivre's discovery received little attention until Laplace began writing on probability during the 1770s. Laplace saw this method of approximation as invaluable, for he hoped to use it in conjunction with the new technique of *a posteriori* probability—which he developed independently of Bayes around 1774[2]—to predict distributions of events in the future, or to infer the existence of real causes, from a record of past

[1] On De Moivre, see Ivo Schneider, "Der Mathematiker Abraham De Moivre," *Archive*, 5 (1968-69), 177-317.

[2] See Laplace's classic memoir, "Mémoire sur la probabilité des causes par les événements" (1774), in *Oeuvres*, vol. 8, pp. 27-65.

occurrences. In a paper published in 1781, Laplace put this exponential approximation to work to show that the observed surplus of male over female births in Paris every year without exception required no providential intercession, since, in view of the average male to female ratio and the total number of births, the probability that female births would exceed male in any given year was only 1/259. Using the same methods he confirmed with probability 410457/410458 that the male birth ratio in London was genuinely greater than that in Paris, and calculated the number of citizens who must be counted to assure a certain standard of reliability in a population figure derived from a complete register of births and an estimated ratio of annual births to population.[3] The applicability of the mathematics of chance to demographic problems reaffirmed Laplace's view that probability theory was generally usable as a means for narrowing the range of uncertainty, and he began speaking of De Moivre's exponential function as the law of facility of errors. He hailed the exponential integral as a tool of the probability of causes, and proposed that investigators should use error analysis to demonstrate that an observed effect really exists before concocting arguments to explain it.[4]

Laplace was able to generalize De Moivre's mathematics, making it unnecessary to assume that individual events distribute themselves symmetrically about the underlying mean like fair coin tosses. He also developed the general method of error analysis, and applied it to several problems in a wide variety of fields. Perhaps he was too easily convinced of the quality and independence of the numbers with which he worked, as when he calculated odds of one million to one that the mass of Jupiter fell within bounds subsequently proved to be mistaken by George Airy.[5] Still, he exhibited more skill and resourcefulness in the use of error analysis than would be seen again for more than half a century after his death.

Laplace defined for subsequent generations both the class of problems to which error analysis could be fruitfully applied and the method for

[3] See Laplace, "Mémoire sur les probabilités" (1781), in *Oeuvres*, vol. 9, 383-485, pp. 437, 466; "Sur les naissances, les mariages, et les morts à Paris depuis 1771 jusqu'en 1784, et dans toute l'étendue de la France pendant les années 1781 et 1782" (1786), in *Oeuvres*, vol. 11, 35-46. Daniel Bernoulli had applied De Moivre's result to the problem of birth ratios in 1770-1771, in a work that Laplace doubtless knew. See O. B. Sheynin, "Daniel Bernoulli on the Normal Law," in *SHSP2*, pp. 199-202.

[4] Laplace, "Mémoire sur les approximations qui sont fonctions de très grands nombres, Suite" (1786), in *Oeuvres*, vol. 10, 295-338, p. 319.

[5] See Cournot, pp. 242-243.

doing so. Quetelet, among others, found a particularly valuable exemplar in Laplace's calculation of the probability that atmospheric pressure was subject to a regular cause of diurnal variation. Since the level of the barometer was also subject to a host of irregular causes of variation, implying that the "errors" exceeded in magnitude the effect itself, there was no alternative but to compile a large number of observations and apply the techniques of probability. "As events multiply themselves, their respective probabilities become more and more developed," wrote Laplace, and indeed the mean value for morning and afternoon barometric readings revealed a small difference. Only after an infinite number of observations would the discrepancy point with certainty to a fixed cause, so it was necessary to use error analysis to calculate the probability with which the effect was indicated. The probability that chance alone would produce a mean effect as large as that observed could be expressed in terms of an integral of the error function. That probability proved to be minuscule, and accordingly, Laplace inferred that the effect indicated the existence of a constant cause.[6]

The incorporation of the same function into observational astronomy at the beginning of the nineteenth century guaranteed that it would become known to virtually all physical scientists and many others as well. Astronomers had been debating for several decades the best way to reduce great numbers of observations to a single value or curve, and to estimate the accuracy of this final result based on some hypothesis as to the occurrence of single errors. Thomas Simpson, in 1755, and then Lagrange furnished expressions for the probability of a given magnitude of error in a mean value, and Laplace also wrote on this subject during the eighteenth century.[7] Then in 1805 Legendre announced a general method for reducing multiple observations of an object, such as a star or planet. It involved choosing the curve of some specified type that minimized the sums of the squares of deviations of the individual measurements from it—a procedure which, if the object were stationary, was no different from taking the mean. This, the so-called "method of least squares," was presented by Legendre without probabilistic justification.

Four years later, Carl Friedrich Gauss announced that he had been using this method since 1794. At the same time, he presented a

[6] Laplace, *Théorie analytique des probabilités* (Paris, 3d ed., 1820), in *Oeuvres*, vol. 7, pp. 280, 355-358.

[7] See Stigler; also Isaac Todhunter, *A History of the Mathematical Theory of Probability* (New York, 1949), chap. 15.

formal derivation of it—or rather, worked out the conditions under which it was valid. He concluded that this method yielded the most probable value for the actual position or path of the object in question if it was supposed that errors from the true value were distributed according to the already-familiar error curve of De Moivre and Laplace. Gauss's derivation, like those that followed it, remained controversial, for Gauss offered no justification for his premise, the applicability of the error law, beyond its necessity if the conclusion were to be upheld. His argument was also found exceptionally obscure and difficult, but nevertheless it was enormously influential. In 1810 Laplace provided an alternative derivation of least squares, giving at the same time his familiar argument on the limit of the binomial to establish that errors of mean results in astronomy, like those in population studies, ought to be distributed according to the error law.[8]

Nineteenth-century astronomers and mathematicians produced an enormous volume of literature on the method of least squares, some merely to instruct readers in the techniques for using it, but much that questioned the foundations of the method or sought to clarify its relation to the error law. By the late 1830s a consensus seemed to be established that the error curve applied not only to the errors of calculated mean values, but also to the distribution of individual errors. In 1837 G.H.L. Hagen argued that every particular measurement was a composite phenomenon, and that measurements were subject to the error law for the same reason that the binomial distribution converged to it. The next year, his teacher, F. W. Bessel, specified eleven different classes of random errors that occurred in every telescopic observation, including expansion and contraction of the telescope, errors of the observer, irregularities in the atmosphere, and so on. This plurality of sources of deviation, he explained, accounted for the dominion of the error curve over their results.[9]

[8] See Stigler; also articles on Gauss (by Churchill Eisenhart) and Laplace (by Stephen Stigler) in William H. Kruskal and Judith M. Tanur, eds., *International Encyclopedia of Statistics* (2 vols., New York, 1978), vol. 1, pp. 378-386, 493-499; R. L. Plackett, "The Discovery of the Method of Least Squares," in *SHSP2*, pp. 230-251.

[9] See G.H.L. Hagen, *Grundzüge der Wahrscheinlichkeits-Rechnung* (Berlin, 1837); F. W. Bessel, "Untersuchungen über die Wahrscheinlichkeit der Beobachtungsfehler" (1838), in *Abhandlungen*, vol. 2 (Leipzig, 1875), pp. 372-391; also Eberhard Knobloch, "Zur Grundlagenproblematik der Fehlertheorie," Menso Folkerts, Lindgren, eds., *Festschrift für Helmuth Gericke* (Stuttgart, 1985), pp. 561-590. The curve had already been associated with atmospheric retractions by Christian Kramp, *Analyse des refractions astronomiques et terrestres* (Strasbourg, 1799).

One more extension of the growing domain of the error law was made by Joseph Fourier. Fourier became involved in social statistics early in his career as the result of his administrative as well as scientific activities. Having served as permanent secretary of Napoleon's Institute of Egypt, he was appointed prefect of Isère upon his return to France—in which capacity he supervised for his *département* the great compilation planned by the Bureau de statistique. Although his participation in this unsuccessful project was evidently halfhearted, it was a statistical position that he secured to support himself under the restored monarchy. He became director of the Bureau de statistique of the Département de la Seine under his friend and former pupil at the Ecole polytechnique, the Count de Chabrol, prefect of the Seine. In that capacity, his involvement with the collection and presentation of statistical tables was intimate. The publications of his office were exemplary, and were indispensable for the first generation of statistical enthusiasts in France.[10]

Fourier published five papers on statistics and demography after his appointment to the statistical bureau in Paris, including four weighty introductions to the official statistical compilations. The first of these papers, a lengthy memoir on what he called the analytical theory of insurance, used Daniel Bernoulli's logarithmic function for moral expectation to show how both parties to an insurance contract can derive benefit from it. The next two dealt with population, one a pioneering work in mathematical demography, showing how life expectancies and other quantities could be calculated from age and mortality tables for a population not in equilibrium, the second an empirical study of the population of Paris since the end of the seventeenth century. These were followed by two memoirs on the value or precision of mean results derived from numerous observations.

Fourier was highly impressed by the regularity of statistics, and he maintained as "the foundation of most statistical researches" the proposition "that the indefinite repetition of events that appear to us as fortuitous must make everything irregular in them disappear; in a series of an immense number of facts, only the constant and necessary relations, determined by the nature of things, endure."[11] At the same time, he emphasized that mean values alone were not enough for the science of sta-

[10] See John Herivel, *Joseph Fourier: The Man and the Physicist* (Oxford, 1975), esp. pp. 113, 118.

[11] Joseph Fourier, "Notions générales sur la population," in *Recherches statistiques sur la ville de Paris et le Département de la Seine*, vol. 1 (Paris, 1821; 2d ed., 1833), pp. xxxviii-xxxix.

tistics, and that it was necessary also to find the limits within which the mean is confined.[12] To this end he presented with great simplicity and clarity the technique for evaluating the probability of an error of any given magnitude by computing a standard error of the mean and then checking it against certain probabilities from an integral table for the error curve. His "standard of certainty" in hypothesis testing was a bit more stringent than the current five percent; he thought that a chance of one part in twenty thousand, corresponding to about four times our standard error of the mean, was suitable.[13]

All this had been worked out fully, though never presented very clearly, by Laplace. Fourier also used the error curve, however, as the formula for an actual physical distribution. He did so in his landmark contribution to physics, the *Traité analytique de la chaleur,* where he showed that if all the caloric (heat) in an infinite, one-dimensional conductor were concentrated initially at a single point, and then allowed to disperse freely, the heat at any subsequent time would be distributed in accordance with the error function. Fourier's analysis relied on exact differential equations rather than the calculus of probability, and he asserted proudly that his mathematics was independent of any particular theory of the nature of heat. Still, he expressed favor for a physical model to which Laplace's binomial derivation of the error curve could have been applied. "Of all modes of presenting to ourselves the action of heat," he proposed, "that which seems simplest and most conformable to observation, consists in comparing this action to that of light. Molecules separated from one another reciprocally communicate, across empty space, their rays of heat, just as shining bodies transmit their light." That is, the dispersion of heat could most plausibly be regarded as taking place in tiny, independent increments, as the rays migrated along an intricate course from molecule to molecule. "There is

[12] Fourier, "Mémoire sur la population de la ville de Paris dépuis la fin du XVIIe siècle," in *ibid.*, vol. 2 (1823), p. xx.

[13] Fourier, "Mémoire sur les résultats moyens déduits d'un grand nombre d'observations," in *ibid.*, vol. 3 (1826). Fourier's "degree of approximation of the calculated mean to the true mean,"

$$g = \sqrt{\frac{2}{m}(B - A^2)} \quad = \quad \frac{1}{m}\sqrt{2[(a - A)^2 + (b - A)^2 + \ldots]}$$

where a, b, ... were the individual measures, n the number of observations, A their mean, and B the mean of their squares, is $\sqrt{2}$ times larger than the current "standard error of the mean," and was inserted into an error curve of the form

$$P = \frac{2}{\sqrt{\pi}} \int_0^t e^{-x^2} dx,$$

where t was measured in units of g. See the table in Kramp, *Analyse* (n. 9).

no direct action except between material points extremely near, and it is for this reason that the expression for the flow has the form which we assign to it. The flow thus results from an infinite multitude of actions whose effects are added."[14]

In addition, Fourier was vastly impressed by the extraordinary diversity of fields in which the error law had found application. Indeed, he maintained that the establishment of a mathematical identity among these disparate phenomena pointed to some underlying unity that hitherto had been concealed. Fourier's faith in the universality of the error curve was almost mystical:

> We ought to remark, in concluding this extract, that the principal element of the analysis of probabilities is an exponential integral that has presented itself in several very different mathematical theories. Geometers have considered this function in an abstract manner, and as an element of general analysis. . . . This same function is connected with general physics. It is required in order to characterize the motion of light through gaseous milieux. We have discovered in recent years that it also represents the diffusion of heat in the interior of solid substances. Finally, it determines the probability of errors of the measures and mean results of numerous observations; it reappears in the questions of insurance and in all difficult applications of the science of probabilities.
>
> Thus, mathematical analysis unites the most diverse effects, and discovers in them common properties. Its object has nothing of the contingent, nothing of the fortuitous. Imprinted in all of nature, it is a preexistent element of universal order. This science has necessary relations with all physical causes, and with most of the combinations of spirit. It preserves only those methods that can simultaneously become more extensive and more simple; and its true advances always recur to two fundamental points: public utility, and the study of nature.[15]

All the early applications of the error law to which Fourier alluded, and which have been mentioned here, could be understood in terms of a binomial converging to an exponential, as in De Moivre's original der-

[14] Fourier, *The Analytical Theory of Heat* (1822), Alexander Freeman, trans. (New York, 1945), pp. 32, 460-461.

[15] Fourier, "Extrait d'un mémoire sur la théorie analytique des assurances," *Annales de chimie et de physique* [2], 10 (1819), 177-189, pp. 188-189.

ivation. All but Fourier's law of heat, which was never explicitly tied to mathematical probability except by analogy, were compatible with the classical interpretation of probability. Just as probability was a measure of uncertainty, this exponential function governed the chances of error. It was not really an attribute of nature, but only a measure of human ignorance—of the imperfection of measurement techniques or the inaccuracy of inference from phenomena that occur in finite numbers to their underlying causes. Moreover, the mathematical operations used in conjunction with it had a single purpose, to reduce the error to the narrowest bounds possible. With Quetelet, all that began to change, and a wider conception of statistical mathematics became possible.

QUETELET: ERROR AND VARIATION

When Quetelet announced in 1844 that the astronomer's error law applied also to the distribution of human features such as height and girth, he did more than add one more set of objects to the domain of this probability function; he also began to break down its exclusive association with error. Quetelet intended nothing of the sort, however. His extension of the range of the error law seemed to him an epochal achievement precisely because it vindicated the presuppositions of social physics by proving that the concepts and formalism of astronomy were fully adequate to capture the essential properties of that hitherto mysterious entity, man. This surprising testimony to the power of the method of celestial mechanics bolstered Quetelet's longstanding claim that the social sciences could do no better than to imitate the physical. It revealed with exceptional clarity the need for statisticians to familiarize themselves with advanced mathematics if their science was ever to realize its potential.

It is fully characteristic of Quetelet's scientific style that he interpreted his proudest discovery not as an indication that the mathematics of error had been conceived too narrowly, but as clear evidence that human variation could be understood in the same terms as errors of observation. Indeed, this interpretation was not simply the *ex post facto* rationalization of an empirical discovery. Quetelet had been very close to identifying variation and error ever since the birth of social physics, and the appeal of the error law arose as much from the implication that mathematical order extended even to deviations from mean values—the embodiment of terrestrial imperfection—as from the potential it held out for unifying the study of man with that of the stars. There was, it seems, a preordained harmony between Quetelet's metaphors and the standard Laplacian understanding of the error law.

One way to view Quetelet's average man was simply as an instrument of social analysis. This being "would be of only mediocre interest," Quetelet punned, "if I had not recognized that the average man also enjoys some very remarkable properties in relation to the social system, and which are such as seem to me to open a vast field to a new order of social researches."[16] But Quetelet went immediately beyond this. As Maurice Halbwachs observed in 1913, Quetelet was inspired by the philosophy of Victor Cousin to develop the analogy between the mathematical idea of the average man and the moral idea of the *juste milieu*.[17] The *juste milieu* was a symbol of Louis Philippe's July monarchy, but it could equally well have been applied to post-1830 Belgium, whose first king, Leopold I, was married to Louis Philippe's daughter. It referred simultaneously to the supremacy of the bourgeois class and to the constitutional monarchy, perched precariously between divine-right kingship and radical democracy. At the same time, it embodied an ideal of moderation, especially in political behavior, to which Quetelet already was strongly attached.

The philosophical emblem of these post-1830 governments was Victor Cousin's eclectic philosophy, which Quetelet greatly admired. Cousin defended eclecticism on the ground that no particular philosophy could suffice for *l'humanité tout entière*, and also because it seemed to him the embodiment of the great political achievements of his age. It reflected, as Cousin proclaimed in an encomium on Charles X's restoration of the charter, a synthesis of the old regime's exclusive monarchical principle with the exclusive democratic principle of the great revolution.[18] Cousin's taste for moderation appears clearly in the concluding lecture of a series on the history of philosophy, in 1828. His remarks on the Napoleonic wars—which, we learn, were in reality a prolonged battle between the ideas of democracy and aristocracy—were transcribed and published as follows:

[16] Quetelet, "Recherches sur le poids de l'homme aux différens ages," *NMAB*, 7 (1832), separate pagination, p. 2.

[17] Maurice Halbwachs, *La théorie de l'homme moyen: Essai sur Quetelet et la statistique morale* (Paris, 1913), p. 6.

[18] Victor Cousin, *Cours de philosophie: Introduction à l'histoire de la philosophie* (Paris, 1828), 13th lecture, pp. 39-40.

> What did they bring, gentlemen? Neither the one nor the other [neither aristocracy nor democracy]. Which was the conqueror? which was the vanquished at Waterloo? Gentlemen, there was none vanquished. (Applause). No, I protest that those wars had no vanquished party; the only victors were European civilization and the charter. (Unanimous and prolonged applause) Yes, messieurs, it is the charter given voluntarily by Louis XVIII, the charter maintained by Charles X, the charter destined to domination in France, and destined to subdue, I do not say enemies, she has none of them, but all those who have fallen behind French civilization; (Redoubled applause) it is the charter that is the brilliant outcome of the bloody conflict of the two systems that by now have already seen their day, to wit, absolute monarchy and the extravagances of democracy.[19]

All this, Cousin pointed out, was evidence that France had entered an era of eclecticism, of "moderation in the philosophical order" that "can do nothing during the days of crisis" and that becomes "a necessity afterwards"; of the eclecticism that belongs to no party, to no coterie, that "avows itself satisfied with its century, with its country and with the actual order of things. . . . I ask whether philosophy can fail to be eclectic, when everything around it is?"[20]

Quetelet, though never willing to express such unqualified approval for the existing order, was enthralled by Cousin's ideal of moderation and compromise. He quoted at length Cousin's observation that the spirit of a people resides in every individual, that each people must have a common type, and that the great man is he who most perfectly represents this type. While Cousin, however, stressed that great men must also preserve their individuality, thus uniting uniformity with multiplicity, Quetelet proposed an identification of greatness with the mean. He wrote that "an individual who epitomized in himself, at a given time, all the qualities of the average man, would represent at once all the greatness, beauty and goodness of that being."[21] He resolved that the mean alone can represent the ideals of a society in politics, aesthetics, and morals. "Deviations more or less great from the mean have consti-

[19] *Ibid.*, pp. 36-37.

[20] *Ibid.*, pp. 42-45.

[21] Quetelet, *Sur l'homme et le développement de ses facultés, ou essai de physique sociale* (1835; 2 vols., Brussels, 1836), vol. 2, p. 289; from Cousin, *Cours* (n. 18), 10th lecture, pp. 6-7.

tuted [for artists] ugliness in body as well as vice in morals and a state of sickness with regard to the constitution."[22] He also expressed longing for a mean philosophical and political position that could resolve social conflict and "conciliate the most advantageously the interests of the different parties."[23]

L'homme moyen, then, was not just a mathematical abstraction, but a moral ideal. Quetelet held that great inequalities of wealth and vast price fluctuations were responsible for crime and turmoil; he exalted the life of moderation, unaffected by sudden passions, and conjectured that the "higher classes" live longer than the "low people" not because of wealth or nutrition, but because of their "habits of propriety, of temperance, of passions excited less frequently and variations less sudden in their manner of existence."[24] He listed Aristotle beside Archimedes among the discoverers of the true idea of the mean for teaching that "virtue consists in a just state of equilibrium, and all our qualities, in their greatest deviations from the mean, produce only vices."[25] The average man embodied these mild virtues with mathematical precision; his faculties developed "in a just state of equilibrium, in a perfect harmony, equally distant from excesses and defects of every kind, in such a way that . . . one must consider him as the type of all that is beautiful and all that is good."[26] Quetelet even proposed a theory of the enlightened will to explain the greater regularity of moral statistics than those of births and deaths. It involved, in effect, a will to mediocrity, a tendency for the enlightened to resist the influences of external circumstances and to seek always to return to a normal and balanced state. "It is only among men entirely abandoned to the heat of their passions that one sees those sudden transitions, faithfully reflected from all the external causes that act on them."[27]

Hence the progress of civilization, the gradual triumph of mind, was equivalent to a narrowing of the limits within which the "social body" oscillated. It must inevitably be reflected in a tendency for progress to become ever more smooth and gradual, the risk of "falling into an ex-

[22] Quetelet, "Recherches sur le penchant au crime aux différens ages," *NMAB*, 7 (1832), separate pagination, p. 6.

[23] Quetelet, *Sur l'homme* (n. 21), vol. 2, pp. 296-297.

[24] *Ibid.*, vol. 1, p. 237; also Quetelet, "Penchant au crime" (n. 22), p. 44.

[25] Quetelet, *Théorie des probabilités* (Brussels, 1853), p. 49.

[26] Quetelet, *Sur l'homme* (n. 21), vol. 2, p. 287.

[27] Quetelet, *Du système social et des lois qui le régissent* (Paris, 1848), p. 97.

treme" perpetually diminishing. Quetelet epitomized the moral of his most influential book, *On Man*, as follows:

> I will finish this chapter with a final observation, which may be seen as a consequence of all that has preceded: it is, that one of the principal acts of civilization is to compress more and more the limits within which the different elements relative to man oscillate. The more that enlightenment is propagated, the more will deviations from the mean diminish; moreover, as a consequence, we tend to unite ourselves with what is beautiful and with what is good. The perfectibility of the human species is derived as a necessary consequence of all our investigations. Defects and monstrosities disappear more and more from the body; the frequency and the gravity of maladies are combatted with greater effectiveness through the progress of medical science; the moral qualities of man will meet with improvements no less tangible; and the more we advance, the less need we fear the effects and the consequences of great political upheavals and wars, the plagues of humanity.[28]

The implication of Quetelet's idealizations of the mean was that all deviation from it should be regarded as flawed, the product of error. This did not imply, however, that variation must stand outside the domain of science, for the special task of probability theory was "to establish an admirable precision where one believed there were only games of chance."[29] Science, in Victor Cousin's words, was precisely "the suppression of all anomaly, the ordered substituted for the arbitrary, reality for appearance, reason for sense and for imagination."[30] Without general laws in the physical domain, Quetelet observed, "one can imagine what dreadful chaos would be produced in the midst of these myriads of worlds circulating through space in a wholly disorganized manner, and crashing against one another." He aspired to be the Newton "of this other celestial mechanics," to find the laws that assured equilibrium in the social domain, and he, like many would-be scientific thinkers of the early nineteenth century, derived special satisfaction from contemplating the splendid noumenal order that he supposed to prevail beneath the whir of phenomena. Just as Charles Fourier busied himself calcu-

[28] Quetelet, *Sur l'homme* (n. 21), vol. 2, p. 342; also *Du système social* (n. 27), pp. 96-97.

[29] Quetelet, *Physique sociale, ou essai sur le développement des facultés de l'homme* (2 vols., Brussels, 1869), vol. 1, p. v.

[30] Cousin, *Cours* (n. 18), 8th lecture, p. 19.

lating the revenue his phalansteries would derive from increased egg production under the new system of social harmony, Quetelet announced laws of suicide by hanging in Paris and of marriages between sexagenarian women and young men in their twenties in Belgium. The most inspiring social laws were precisely those that governed the ungovernable, that subjected even lawbreaking to natural law, that exercised a perfect rule over apparent disorder. Quetelet was particularly delighted to discover that statistical laws, such as those of crime, were unaffected by the traumas of revolution, thus indicating a deep order in the social realm that was scarcely affected by the irregular march of politics.[31]

The same considerations applied to deviations from mean values. To find a law of variation was to bring it within the domain of order and rationality. Quetelet wrote:

> If one takes the trouble to examine and to bring together observations taken with care and sufficiently numerous, one will find that what was regarded as the effect of chance, is subjected to fixed principles, and that nothing escapes the laws imposed by the all-powerful onto organized beings. What we call an anomaly deviates in our eyes from the general law only because we are incapable of embracing enough things in a single glance.
>
> By placing oneself in circumstances favorable for observing, one finds that, for organized beings, all elements are subject to variation around a mean state, and that the variations to which the influence of accidental causes give birth, are ruled with such harmony and precision, that they can be classed in advance numerically and by orders of magnitude, in the limits between which they are comprised. All is foreseen, all is lawlike: only our ignorance leads us to suppose that all is subject to the whims of chance.[32]

Error was thus banished from the universe, and Quetelet accordingly resolved that the expression "law of errors" was inconsistent with the achievements of modern science. "Law of accidental causes" scarcely seemed better, and was criticized in a letter from the mathematician and statistician Jules Bienaymé, so that Quetelet eventually resolved to designate this curve simply as the binomial law.[33]

[31] Quetelet, *Du système social* (n. 27), p. 301.
[32] *Ibid.*, pp. 16-17.
[33] *Ibid.*, p. 306; Quetelet, "Sur quelques propriétés que présentent les résultats d'une série

Quetelet introduced his great innovation in the use of the probabilistic error law, its application to real variation in nature, in 1844. He did so, however, without claiming that a new understanding of this mathematical function was required. On the contrary, Quetelet, who never maintained much discipline over his own penchant for metaphor, viewed his discovery as proof that deviations from the golden mean of *l'homme moyen* were mere imperfections, even errors. Although the ultimate effect of his innovation was primarily to broaden the possibilities of mathematical statistics, and not to promote in any significant way the development of social science, Quetelet naturally conceived his work as a contribution to the latter. The introduction of the error law to the study of man revealed once again the homology between social and celestial science, confirming that the social scientist could best advance his study by looking to the methods and even the laws of astronomy.

Quetelet began his exposition of this new statistical law of man by presenting as clearly as possible the nature and uses of the error curve. First he derived the "law of possibility" from a high-order binomial, which he portrayed in terms of the expected distribution of outcomes after an enormous number of repetitions of 999 drawings from a fair urn with equal numbers of balls of two colors. Thus 500 of one and 499 of the other was most probable, 501 to 498 slightly less likely, and so on to 580-420, any result beyond which was nearly impossible. Then he observed that measurements of the height of an individual, or of the position of a star, would, if repeated in sufficient number, also distribute themselves in accordance with the error curve. The expected value of the measurements would equal the true height or position, while each measurement would be subject to numerous causes of error. Finally, he queried whether there might exist an *homme type*, a true average man, of which every real man is an imperfect replicate in the same way as a measurement of height is an effort to establish the true height. Indeed there was, he concluded. It could be seen empirically, by means of a table of the distribution of chest sizes of Scottish soldiers placed beside a table of the astronomical error function. He followed this with a similar table of the heights of young Frenchmen who had presented themselves for conscription. Again the conformity with the error curve seemed excellent, except that the numbers dropped off sharply in the neighborhood immediately above 1.57 meters, and showed a corresponding surplus be-

---

d'observations . . . ," *Bulletin de l'Académie Royale des Sciences et Belles-Lettres de Belgique*, 19, part 2 (1852), 303-317.

low that height. This, Quetelet announced, was clear evidence of fraud, and he was able to calculate the number of men whose measurements had been slightly reduced in order to gain them exemptions from their military obligations.[34]

Two years later, in 1846, Quetelet presented his great discovery in a more popular form. As in the relatively technical paper, his exposition of this law of variation appeared in a discussion of mean values; Quetelet was always more interested in mean values than in variation for its own sake. He began by distinguishing two types of mean values. An arithmetic mean, he wrote, may be calculated for the most incongruous sets of objects, but it reveals little or nothing about their collective character. When, however, the variation of a set of measurements follows the customary law of errors, then the mean value—which is also the most probable (modal) value—may be regarded as a "true" mean. Thus if one were to compute the average height of a miscellaneous assortment of houses on some street, the result would be merely an arithmetic mean, since the underlying distribution would be altogether unsystematic. If, instead, a particular house were measured repeatedly and no systematic error were made, each measurement would approximate the true value and their collective distribution would approximate the astronomical law of errors. As the measuring process was continued, the calculated mean should gradually converge with increasing precision to the true height of the house.

Quetelet then proposed that the same reasoning ought to apply to the production of copies, say of artistic works. If, for example, a thousand copies were to be made of an ancient statue, the Gladiator, their inaccuracies would doubtless be governed by the law of errors. Indeed, this could be shown empirically, in a certain sense:

> I will perhaps astonish you in saying that the experiment has been done. Yes, truly, more than a thousand copies of a statue have been measured, and though I will not assert it to be that of the Gladiator, it differs, in any event, only slightly from it: these copies were even living ones, so that the measurements were taken with all possible

[34] Quetelet, "Sur l'appréciation des documents statistiques, et en particulier sur l'appréciation des moyennes," *BCCS*, 2 (1844), 30ff. On Quetelet and the normal law, see my "The Mathematics of Society: Variation and Error in Quetelet's Statistics, *BJHS*, 18 (1985), 51-69; also Peter Buck, "From Celestial Mechanics to Social Physics . . . ," in H. N. Jahnke and M. Otte, eds., *Epistemological and Social Problems of the Sciences in the Early Nineteenth Century* (Dondrecht, 1981), pp. 19-33.

> chances of error, I will add, moreover, that the copies were subject to deformity by a host of accidental causes. One may thus expect to find here a considerable probable error.[35]

Quetelet's identification of live individuals with copies of statues conveys with unmatched clarity an impression of the way he viewed human diversity. His supposed experiment was in fact the same as that presented in 1844, the measurements of 5,738 Scottish soldiers. Their conformity to the error curve was interpreted as implying that the distribution was a genuine product of error. The soldiers had been designed according to a uniform pattern, that of the average man. They had failed to realize perfectly this imagined archetype of Scottish soldierdom on account of a host of irregularities in their development. These were so severe that the measurements exhibited a probable error exceeding an inch.

More particularly, the conformity of this distribution to the law of errors could be explained in terms of Quetelet's customary distinction between constant and perturbing causes. The constant cause was the Scottish type—Scotland's average man. The perturbing causes included nutrition, climate, and so on, all quite variable in this imperfect world, so that a host of independent small errors was generated that might either increase or decrease chest size. The development of soldiers was thus mathematically analogous to the mean of many drawings from an urn—which, after all, is simply the sum of a constant cause, the actual ratio of black to white balls, and the inescapable but unsystematic small errors generated by each individual drawing. Hence the binomial curve that regulated most games of chance should apply also to the soldiers, and since the perturbing causes were very numerous, their effects could be approximated with considerable accuracy by the error curve of De Moivre and Laplace.

Quetelet was enormously proud of his discovery that the error law governed human variability, and he applied it widely—some would say indiscriminately—in succeeding decades. He upheld it as the definitive criterion of unity of type, and he was able to refute polygenism during the sensitive period of the American Civil War by declaring authoritatively, on the basis of a few measurements of American Indians and Negroes who happened to pass through Belgium, that a single law of errors applied to the whole of humanity.[36] With perhaps less evidence, he

[35] Quetelet, *Lettres à S.A.R. le duc régnant de Saxe Coburg et Gotha sur la théorie des probabilités, appliquées aux sciences morales et politiques* (Brussels, 1846), p. 136.

[36] See Quetelet, "Sur les indiens O-Jib-Be-Wa's et les proportions de leurs corps," *Bulletin de l'Académie Royale des Sciences et Belles-Lettres de Belgique*, 13, part 1 (1846), 70-76, and

cheerfully drew bell-shaped curves to model the distribution of propensities to crime, marriage, and suicide—confirming still more impressively that crime was an attribute of the social type and not of individual deviants. In his last book, *Anthropométrie*, he drew curves giving the development of various traits over the human life cycle, and implied that these too would reflect the error curve provided the age axis was suitably transformed to reflect real and not mere chronological age. He even proposed that his great law applied to entities at a higher level of organization, and that, for example, the area of states would be distributed according to the error law if there were sufficiently many to reflect accurately the true state of things.[37]

More broadly, "the same law, so simple and so elegant, applies . . . generally to all the physical laws, we add even the moral and intellectual laws of man."[38] This was Quetelet's principal legacy to mathematical statistics. His great program of social physics, designed to form the foundation for an exact science of human societies, was only occasionally seen as persuasive, and those who did had little more idea how to implement it than their master had. His recognition that the error distribution applies not only to errors, but also to what few but Quetelet would regard as anything other than real variation was likewise taken up at first by only a few scientists; but this idea was, if not strictly true, at least accurate enough and concrete enough for his successors to do something with it. In social statistics, it was endorsed by a number of writers including the American Benjamin A. Gould, the Italians M. L. Bodio and Luigi Perozzo, Quetelet's English admirer Samuel Brown, and even, in a more qualified way, by his German critic Wilhelm Lexis.[39] For the most important and influential developments of Quetelet's new application of the error law, however, it is necessary to look to fields other than social science.

"Sur les proportions de la race noire," *ibid.*, 21, part 1 (1854), 96-100; also *Théorie* (n. 25), pp. 72-78, and *Anthropométrie, ou mesure des différentes facultés de l'homme* (1870; Brussels, 1871), p. 16.

[37] Quetelet, *Du système social* (n. 27), p. 156.

[38] Quetelet, *Anthropométrie* (n. 36), pp. 253-254.

[39] See, *inter alia*, B. A. Gould, *Investigations in the Military and Anthropological Statistics of American Soldiers* (New York, 1869); Luigi Perozzo, "Nuove applicazioni del calcolo delle probabilità allo studio dei fenomeni statistici," *Atti della R. Accademia dei Lincei, Memorie della classe di scienze morali, storiche, e filologiche*, 10 (1882), 473-503; Samuel Brown, letter of 29 Dec. 1871 in *AQP*, cahier 526; Wilhelm Lexis, "Anthropologie und Anthropometrie," in J. Conrad et al., eds., *Handwörterbuch der Staatswissenschaften* (7 vols., Jena, 2d ed., 1898-1901), vol. 1, pp. 388-409. Quetelet frequently cited authors who received his ideas favorably; see *Physique sociale* (n. 29), *passim*.

Chapter Five

SOCIAL LAW AND NATURAL SCIENCE

It was no mere accident that Quetelet's work on the error law, rather than the much more copious and mathematically sophisticated treatments of it in the literature of observational astronomy and geodesy, provided the inspiration for the most important writers on statistical mathematics of the late nineteenth century. Quetelet opened up a whole new perspective on the mathematical treatment of variation. Ironically, his perception of the wider applicability of the error law was made possible precisely by the depth of his commitment to the traditional metaphysics of error, but his readers were more interested in his results than in his opinions, and these results implied that much variation in nature was governed by this simple and elegant law. Thus, while Quetelet interpreted his discovery as confirmation that variation could be neglected in favor of the study of mean values, Maxwell and Galton, among others, saw in it a convenient and valuable tool for analyzing with mathematical precision the nature and effects of natural variation.

The mathematics of variation was instrumental for the impressive achievements of the nineteenth-century kinetic theory, including Boltzmann's reduction of the second law of thermodynamics to mechanics and probability theory. It also provided the key in biology to the quantitative study of heredity, leading eventually to what is now the most purely statistical of the natural sciences, quantitative genetics. Beyond its importance for particular natural and social sciences, however, the new understanding of the error law that derived from Quetelet's work proved essential for mathematical statistics itself. As Galton was wont to say, not in reference to Quetelet but to English statisticians and all astronomers, most scientists used the error curve in order to get rid of variation, whereas his own aim was to preserve and understand it. New techniques for handling variation were closely associated with new ideas about its nature. I show in Part Four that the mathematics of correlation, the technique that, more than any other, inspired the creation of mathematical statistics, appeared first as a principle of biological heredity, and was not recognized in mathematical abstraction as a widely applicable technique until more than a decade later.

MOLECULES AND SOCIAL PHYSICS

It is scarcely novel to associate the statistical method in physics with the social science from which that phrase was derived. James Clerk Maxwell observed in 1873 that laws of gases could never be found by following the motions and collisions of millions of independent particles, for information about individual molecules was not available and the calculations would, in any event, be impossibly complex. As an alternative, Maxwell proposed to the physicists of the British Association a new variety of social physics:

> The modern atomists have therefore adopted a method which is, I believe, new in the department of mathematical physics, though it has long been in use in the section of Statistics. When the working members of Section F get hold of a report of the Census, . . . they begin by distributing the whole population into groups, according to age, income-tax, education, religious belief, or criminal convictions. The number of individuals is far too great to allow of their tracing the history of each separately, so that, in order to reduce their labour within human limits, they concentrate their attention on a small number of artificial groups. The varying number of individuals in each group, and not the varying state of each individual, is the primary datum from which they work. . . .
>
> The equations of dynamics completely express the laws of the historical method as applied to matter, but the application of these equations implies a perfect knowledge of all the data. But the smallest portion of matter which we can subject to experiment consists of millions of molecules, not one of which ever becomes individually sensible to us. We cannot, therefore, ascertain the actual motion of any one of these molecules; so that we are obliged to abandon the strict historical method, and to adopt the statistical method of dealing with large groups of molecules. The data of the statistical method as applied to molecular science are the sums of large numbers of molecular quantities. In studying the relations between quantities of this kind, we meet with a new kind of regularity, the

> regularity of averages, which we can depend upon quite sufficiently for all practical purposes, but which can make no claim to that character of absolute precision which belongs to the laws of abstract dynamics.[1]

By the time Maxwell began discussing the analogy between statistical and thermodynamical regularities, he had become persuaded that this new approach to questions of physics yielded knowledge that, though completely reliable in practice, was necessarily imperfect and uncertain in principle. Such a belief was not built into his approach from the beginning, however, and the analogy with social science was not always raised in order to imply such uncertainty. Maxwell's friend and correspondent Peter Guthrie Tait, who shared his views on the imperfection of statistical knowledge, observed in a technical memoir at 1886 that because collisions between gas molecules break up any initial distribution of velocities and directions,

> we have at once, in place of the hopelessly complex question of the behaviour of innumerable absolutely isolated individuals, the comparatively simple statistical question of the average behaviour of the various groups of a community. This distinction is forcibly impressed even on the non-mathematical, by the extraordinary steadiness with which the numbers of such totally unpredictable, though not uncommon phenomena as suicides, twin or triple births, dead letters, &c., in any populous country, are maintained year after year.[2]

More impressive, perhaps, is the fact that Ludwig Boltzmann introduced the same analogy in 1872—independently of Maxwell, so far as we can tell. The context of its presentation was an enormously important paper on the kinetic theory delivered to the Vienna Academy of Science. In the introduction, Boltzmann sought to demonstrate the need for recourse to ratios and averages in the kinetic theory, and to establish that propositions of this sort were no less certain than those in any other area of physics. To this end, he made use of an analogy with social statistics:

> It is to be ascribed exclusively to the circumstance that even the most irregular processes yield the same average values when placed

[1] Maxwell, "Molecules," in *SP*, vol. 2, pp. 373-374.

[2] P. G. Tait, "On the Foundations of the Kinetic Theory of Gases," in *Scientific Papers* (2 vols., Cambridge, Eng., 1898-1900), vol. 2, p. 126.

under the same conditions, that we perceive wholly determinate laws in the behavior of warm objects. For the molecules of a body are indeed so numerous, and their movements are so rapid, that nothing ever becomes perceptible to us except average values. The regularity of these mean values may be compared to the astonishing constancy of the average numbers furnished by statistics, which are also derived from processes in which each individual occurrence is conditioned by a wholly incalculable collaboration of the most diverse external circumstances. The molecules are like so many individuals, having the most various states of notion, and the properties of gases only remain unaltered because the number of these molecules which on the average have a given state of motion is constant. The determination of averages is the task of the calculus of probability. It would be a mistake, however, to believe that the theory of heat involves uncertainty because the principles of probability come into application there. An incompletely proven proposition, whose correctness is for that reason problematical, is not to be confused with completely proven propositions of probability; the latter represent, like the results of every other calculus, a necessary consequence of certain premises, and are likewise confirmed by experience, whenever they are correct, if only a sufficient number of cases come under observation—which, because of the enormous number of molecular particles, is always the case in heat theory. Indeed it seems doubly necessary that they conform to conclusions with the greatest strictness.[3]

The work to which both Maxwell and Boltzmann typically referred when they discussed the regularities of social statistics was Buckle's *History of Civilization*. Maxwell had read the relevant portions about a year before he began work on the kinetic gas theory, and expressed qualified admiration;[4] subsequently, he referred routinely to Buckle when criticizing the idea that statistical propositions are wholly determinate. Boltzmann, who enjoyed comparing the uniform and lawlike behavior of gases with the regular profits gained by insurance companies, was one of Buckle's most enthusiastic admirers. He once remarked, in a lecture in 1886: "As is well-known, Buckle demonstrated statistically that if only a sufficient number of people is taken into account, then not only is the

[3] Ludwig Boltzmann, "Weitere Studien über das Wärmegleichgewicht unter Gasmolekülen," in *WA*, vol. 1, pp. 316-317.

[4] See his letter to Lewis Campbell, Dec. 1857, in *Maxwell*, p. 294.

number of natural events like death, illness, etc., perfectly constant, but also the number of so-called voluntary actions—marriages at a given age, crimes, and suicides. It occurs no differently among molecules.[5]

Doubtless it would be too brave to argue that statistical gas theory only became possible after social statistics had accustomed scientific thinkers to the possibility of stable laws of mass phenomena with no dependence on predictability of individual events. Still, the actual history of the kinetic gas theory is fully consistent with such a claim. To be sure, the idea that heat is motion has appeared intermittently throughout the period of modern science, while the drive to explain the properties of gases in terms of the movements of distinct molecules or atoms can, in a general way, be traced back to the ancient atomists. The nature of this motion was never settled, however. Heat might be an expansive, upward-tending motion, as Francis Bacon suggested, or the vibrations of an intermolecular ether, as was generally believed after the caloric theory was discredited around 1830 until the early 1850s, or the rotation of gas molecules arranged in a lattice. The kinetic theory, which derived the properties of gases from the motions and collisions of free molecules, has been restrospectively honored with a rich tradition of precursors, but none before the mid-nineteenth century had any real understanding of the dynamics of their model.

The most impressive early kinetic account of thermodynamic phenomena in gases was that in Daniel Bernoulli's *Hydrodynamica* (1738), but even he did not indicate with any clarity what kind of motion his gas molecules were supposed to undergo. John Herapath, the British scientific writer and railway engineer whose early nineteenth-century writings on gas theory seem to embrace certain features of what became the kinetic model, was thoroughly confused and inconsistent. So nearly as one can tell, he took heat to be the vibration of molecules in an otherwise stationary lattice. There is no reason to suppose that any author before Waterston and Joule began writing in the 1840s believed the rigorous laws of gas dynamics could be derived from a model of random molecular motions.[6]

[5] Boltzmann, "Der zweite Hauptsatz der mechanischen Wärmetheorie," in *PS*, p. 34; also *Lectures on Gas Theory*, Stephen Brush, trans. (1896-1898; Berkeley, 1964), p. 444.

[6] See Daniel Bernoulli, *Hydrodynamics*, trans. T. Carmody and H. Kobus (New York, 1968), pp. 226-229; also Eric Mendoza, "The Kinetic Theory of Matter, 1845-1855," *Archives internationales d'histoire des sciences*, 32 (1982), 184-220; Mendoza, "A Critical Examination of Herapath's Dynamical Theory of Gases," *BJHS*, 8 (1975), 155-165; Stephen G. Brush, *The Kind of Motion We Call Heat* (2 vols., Amsterdam, 1977); Robert Fox, *The Caloric Theory of Gases from Lavoisier to Regnault* (Oxford, 1971), *passim*.

The modern kinetic theory arose soon after the establishment of energy conservation around 1850, when it at last became clear that heat was some sort of dynamic energy, and after social statistics was well established throughout western Europe. The theory of probability was first invoked in support of the kinetic model by August Krönig in 1857. He used it to justify a calculation of the relation between molecular velocities and gas pressure, based on a simplified model in which molecules were supposed to move in parallel lines back and forth between the walls of a perfectly smooth container. Krönig said nothing of statistics, but his justification of this model was consonant with the kind of rhetoric that statistical authors had made commonplace. Since the walls of the container are uneven in comparison with the magnitude of an atom, he explained, "the path of each gas atom must be so irregular that it defies calculation. In accordance with the laws of probability, however, one can suppose, in place of this absolute irregularity, complete regularity."[7]

Rudolf Clausius, who published his ideas on the kinetic theory soon after the appearance of Krönig's paper, made virtually the same point about collisions of gas molecules with the walls of their container. Even though the angle and velocity of incidence will not generally equal those of reflection, he observed,

> yet, according to the laws of probability, we may assume that there are as many molecules whose angles fall within a certain interval, e.g. between 60° and 61°, as there are molecules whose angles of incidence have the same limits, and that, on the whole, the velocities of the molecules are not changed by the side. No difference will be produced in the final result, therefore, if we assume that for each molecule the angle and velocity of reflexion are equal to those of incidence.[8]

More compelling than these remarks on the stability of averages as testimony to the importance of social statistics for the kinetic gas theory is the story of the introduction of the error curve to gas physics. That was in fact an achievement of the first magnitude, leading to work of exceptional importance for physics but also involving an important contribution to the statistical approach. Maxwell, its author, derived his un-

[7] Quoted in Ivo Schneider, "Rudolph Clausius' Beitrag zur Einführung wahrscheinlichkeitstheoretischer Methoden in die Physik der Gase nach 1856," *Archive*, 14 (1974-75), 243.

[8] Rudolf Clausius, "The Nature of the Motion which we call Heat" (1857), reprinted in Stephen Brush, ed., *Kinetic Theory* (3 vols., New York, 1965), vol. 1, 111-134, p. 126.

derstanding of the error curve indirectly from Quetelet, and his use of it provides yet another testimony to the centrality of social science in the development of statistical thinking. He also went beyond Quetelet, however, for the importance of the error distribution formula in his theory was not exhausted by its utility for justifying or calculating a mean value.

Rudolf Clausius seems never to have fully appreciated the need to take account of deviations from mean values in studies of a statistical character. As Ivo Schneider has shown, Clausius first used assumptions of randomness of the sort presented in his earliest work on the kinetic theory while working on a meteorological problem. Meteorology was a field of frequent importance for the transmission of probability methods to the natural sciences, and also one, as is testified by an 1850 paper of the Dutch physicist Buys-Ballot, in which consideration of variation as well as averages could be seen as genuinely important.[9] Clausius, however, worked on the passage of light through the atmosphere, and invoked probability arguments to show that general principles of light transmission could be attained without solving the problem of the reflection of light by atoms.[10] Some years later, in 1862, he announced that, for the kinetic theory, "[i]f we wish to arrive at really reliable conclusions concerning this and other allied subjects, we must not be afraid of the somewhat troublesome consideration of the irregular motions." He proceeded, however, to divide the irregularities into two classes. One class, the "accidental inequalities accompanying the individual impacts," would certainly average out, and could thus be neglected. Only the "normal variations" associated with macroscopic temperature gradients needed to be considered.[11]

Maxwell began working on the dynamical theory of gases, as he then called it, early in 1859, after reading an abridgment of Clausius's second paper on gas theory, which had recently appeared in translation in the *Philosophical Magazine*. Clausius was there responding, in a significant and creative way, to an objection recently raised by Buys-Ballot, who

[9] Buys-Ballot, "On the Great Importance of Deviations from the Mean State of the Atmosphere for the Science of Meteorology," *Phil Mag* [3], 37 (1859), 42-49. See also O. B. Sheynin, "On the History of the Statistical Method in Meteorology," *Archive*, 31 (1984-85), 53-95.

[10] Clausius published this work during the late 1840s. See Ivo Schneider, "Clausius' erste Anwendung der Wahrscheinlichkeitsrechnung im Rahmen der atmosphärischen Lichtstreuung," *Archive*, 14 (1974), 143-158.

[11] Clausius, "On the Conduction of Heat by Gases," *Phil Mag* [3], 23 (1862), 417-435, 512-534, pp. 419, 422.

maintained that if gas pressure really arose from the rapid, unconstrained motion of free molecules, gases ought to interdiffuse almost instantaneously. Experience showed clearly that they did not. In defense of his theory, Clausius proposed that the molecules would collide frequently with one another, and that in this way diffusion would be greatly retarded. He went on to give a calculation of the mean free path of gas molecules, based on the simplifying assumptions of a single molecule moving through a field of stationary particles. He then asserted without proof that multiplication by a factor of 3/4 would generalize the expression to cover the case in which all particles moved with equal velocities. The entire exercise was purely theoretical, for the actual distance could not be computed so long as the volume and number density of individual molecules was entirely unknown. The paper was important for several reasons, however, not least among which is that it introduced the first probability distribution into physics. This was his formula for the distribution of path lengths, one which later became known as the Poisson distribution. Clausius, characteristically, did no more with this curve than to calculate its mean. Once again, the "laws of probability" served only to justify the supposition that the totality of motions in a confined system could be adequately represented by a single mean value.[12]

"As soon as I became acquainted with the investigations of Clausius," Maxwell later wrote, "I endeavoured to ascertain" the law of equilibrium velocity distribution.[13] Within a few months of the publication of this paper in England, Maxwell had completed most of the work for his landmark paper introducing that distribution, which he read to the British Association late in 1859. On May 30 of that year he had written a letter to his friend George Gabriel Stokes to inquire whether Stokes knew of any experiments that could be compared with some predictions made from the model. In particular, Maxwell had derived from the kinetic theory the principle that gaseous friction should be independent of density. This proposition violated his intuition and he was accordingly doubtful that the model could be upheld. Even the published paper, which appeared early in 1860, intimated that its author expected the theory eventually to be disproved.

Evidently Maxwell took up work on the kinetic theory not because he

[12] Clausius, "On the Mean Lengths of the Paths Described by the Separate Molecules of Gaseous Bodies" (1858), in Brush, *Kinetic Theory* (n. 8), vol. 1, pp. 135-147.

[13] Maxwell, "On the Dynamical Evidence of the Molecular Constitution of Bodies," in *SP*, vol. 2, p. 427.

thought it was true, but because he had taken an interest in working out its mathematical implications. He told Stokes that although he had been unable to determine whether the theory was valid, yet "as I found myself able and willing to deduce the laws of motion of systems of particles acting on each other by impact, I have done so as an exercise in mechanics." He also remarked: "I have been rather diffuse on gases but I have taken to the subject for mathematical work lately and I am getting fond of it and require to be snubbed a little by experiments."[14] In short, Maxwell took up the kinetic theory as an exercise in rational mechanics. He did not present the mathematics to Stokes, but he did offer a new formula for mean free path that clearly was derived from the supposition that molecular velocities were distributed according to the error law. Moreover, the general mechanical model that made up the first half of his published paper consisted almost entirely of propositions that were dependent on this distribution law. Evidently Maxwell recognized the applicability of the law of errors to molecular velocities at the very beginning of his work on the kinetic theory. Probably the opportunity for mathematical work presented by the error law had inspired him to begin work on this model in the first place.

How, then, did Maxwell become convinced that the molecular velocity distribution would conform to the astronomer's error law? It is now clear that the inspiration came from the ideas of Quetelet, and more particularly from an essay review of Quetelet's 1846 work on probability and its applications written by John Herschel. More than two decades ago Charles Gillispie pointed out the similitude of the approaches to probability to be found in this review and in Maxwell's paper, and Stephen Brush subsequently recognized that the formal derivation of the error law given by Maxwell was in every important respect identical to the one introduced by Herschel in this essay.[15] Since Maxwell did not discuss or even acknowledge Herschel and may not have remembered where he had encountered this derivation, it is difficult to know just how he decided that its argument could be extended to mol-

[14] See Joseph Larmor, ed., *Memoir and Scientific Correspondence of the Late Sir George Gabriel Stokes* (2 vols., Cambridge, 1907), vol. 2, p. 10.

[15] C. C. Gillispie, "Intellectual Factors in the Background of Analysis by Probabilities," in A. C. Crombie, ed., *Scientific Change* (New York, 1963), pp. 431-453; Brush, *Heat* (n. 6), esp. pp. 184-187. See also C.W.F. Everitt, "Maxwell," in *DSB*, vol. 9, pp. 198-230; Elizabeth Wolfe Garber, "Aspects of the Introduction of Probability into Physics," *Centaurus*, 17 (1972), 11-39; Theodore M. Porter, "A Statistical Survey of Gases: Maxwell's Social Physics," *HSPS*, 12 (1981), 77-116.

ecules. Still, Herschel's essay points unambiguously to a crucial connection between the ideas of social statistics and the most successful application of statistical ideas to physics during the nineteenth century, the kinetic gas theory.

Herschel had not been actively involved in social statistics or insurance, but in other respects his research aims were highly similar to Quetelet's. He was the leading practitioner of the quantitative natural-historical sciences in Britain and, in the eyes of his countrymen, probably the most eminent man of science of his time. What Herschel wrote of Quetelet might equally have been applied to himself:

> No one has exerted himself to better effect in the collection and scientific combination of physical data in those departments which depend for their progress on the accumulation of such data in vast and voluminous masses, spreading over many succeeding years, and gathered from extensive geographical districts,—such as Terrestrial Magnetism, Meteorology, the influence of climates on the periodical phenomena of animal and vegetable life, and statistics in all the branches of that multifarious science, political, moral, and social.[16]

These were the problems that concerned Herschel, and his essay made it clear that the methods of statistics enjoyed a range of applicability extending far beyond the social domain.

Herschel followed Laplace and De Morgan in interpreting probability subjectively as the logic of belief about individual events which are imperfectly understood. The idea of chance, he wrote, enters the reasonings of probability only "as the expression of our ignorance of agents, arrangements and motives, but with the express view to its exclusion from their results." He adopted, however, a strictly empirical line in applying his subject to real phenomena. Commending John Stuart Mill for his recognition that true science involved successful prediction rather than the intuitive comprehension of things in themselves, he argued that probability was the clearest example of a science upon which the "metaphysical idea of Causation" had no bearing. The term "cause," after all, "simply expresses the occasion for the more or less frequent occurrence" of a result; one cannot say that events conform them-

[16] [John Herschel], "Quetelet on Probabilities," *Edinburgh Review*, 92 (1850), 1-57, p. 14.

selves to the laws of probability, for "the laws of probability, as acknowledged by us, are framed in hypothetical accordance with events."[17]

Herschel's adherence to this positivist line did not diminish his expectations of the results that could be attained through empirical research. He opened his essay with the profound observation: "experience has been declared, with equal truth and poetry, to adopt occasionally the tone, and attain to something like the certainty, of prophecy."[18] Herschel identified statistics as a compelling instance of successful prophecy, and praised the flourishing insurance industry for its success in diminishing the impact of chance on human affairs. Casting Bernoulli's law of large numbers as the central truth of probability theory, he suggested that probability in turn provided a model for empirical scientific reasoning. Probability, he wrote, may be contemplated "as a practical auxiliary of the inductive philosophy. . . . Its use as such depends on that mutual destruction of accidental deviations from the regular results of permanent causes which always takes place when very numerous instances are brought into comparison."[19] The destruction of accidental irregularities was most conspicuous in the field of social and moral statistics, whose importance for the improvement of administration Herschel did not neglect to mention. He was deeply impressed by statistical regularities, which indicated the possibility of a science of man governing aggregates without constraining individuals. Births and marriages, he pointed out, are "free as air in individual cases," yet they

> seem to be regulated with a precision, where masses are concerned, clearly proving the existence of relations among the acting causes so determinate, that there is evidently nothing but the intricacy of their mode of action to prevent their being subjected to exact calculation, and tested by appeal to fact. Taken in the mass, and in reference both to the physical and moral laws of his existence, the boasted freedom of man disappears; and hardly an action of his life can be named which usages, conventions, and the stern necessities of his being, do not appear to enjoin on him as inevitable, rather than to leave to the free determination of his choice.[20]

Most important, Herschel was fully convinced by Quetelet's application of the error law to real variation in nature, and he took some care

[17] *Ibid.*, pp. 4-5, 29-30.
[18] *Ibid.*, p. 1.
[19] *Ibid.*, p. 29.
[20] *Ibid.*, p. 42.

to explain it to his readers. Like Quetelet, he introduced the idea in the context of a distinction between two kinds of mean values, arguing that a "true mean" can exist only where deviations are subject to the law of errors. Quetelet's distinction between true means and arithmetic means became for him one between means and averages, but his main point was the same:

> An average may exist of the most different objects, as of the heights of houses in a town, or the sizes of books in a library. It may be convenient, to convey a general notion of the things averaged; but involves no conception of a natural and recognisable central magnitude, all differences from which ought to be regarded as deviations from a standard. The notion of a mean, on the other hand, does imply such a conception, standing distinguished from an average by this very feature, viz. the regular march of the groups, increasing to a maximum and thence again diminishing. An average gives us no assurance that the future will be like the past. A mean may be reckoned on with the most implicit confidence. All the philosophical value of statistical results depends on a due appreciation of this distinction, and acceptance of its consequences.[21]

Mean values, then, only took on genuine scientific worth when they represented a type, the deviations from which were distributed in the characteristic form assumed by error. Herschel argued that statistical regularity would only prevail for these true means, and implied that this, happily, was the case for most natural and social phenomena.

In this way Herschel communicated to Maxwell and other British readers Quetelet's belief in the universality of the error law, which he reaffirmed with his own considerable authority. Herschel did more than this, however, for his derivation of the error law—by which we are able to recognize his influence on Maxwell—was presented as an original one, and was quite different in character from the standard argument about the limit of the binomial developed by De Moivre and repeated by Quetelet. Herschel posited nothing about the nature of the constituent deviations, but instead made assumptions about the properties of the error function as a whole. He supposed first that an error of a given magnitude in one direction is precisely as likely as an error of the same magnitude in another direction, and hence that the error function is spherically symmetric. Next he hypothesized that the error components

[21] *Ibid.*, p. 23.

along perpendicular axes are strictly independent of one another—that is, the probability of an error (x_1, y_1) is equal to the product of the probabilities of the separate components. With these two premises it followed directly that the error function must have the form Ae^{-cx^2}, where A and *c* are arbitrary, positive constants. This was the formula for the error component along a single axis which, as Herschel recognized later, could easily be compounded to yield the solution for two or more dimensions.[22]

Herschel's derivation was described in terms of the errors committed when balls are dropped at a target, but there was nothing in it restricting it to this problem. Indeed, Herschel accepted Quetelet's view that the error law was to be found everywhere, applying equally to the distribution of human heights and to errors of observation in astronomy, and his derivation was intended to offer at once the universality and simplicity appropriate for a function found so frequently in such a wide variety of contexts. To be sure, the derivation could not be applied directly to scalar quantities like height, for which the assumption of independence of perpendicular components would have no meaning. For that matter, the persuasiveness of Herschel's reasoning even with respect to the narrow domain of error did not go unchallenged. While the subjectivist W. F. Donkin repeated Herschel's argument,[23] frequentists like R. L. Ellis found in it the same lack of good sense that bolstered the mistaken faith in *a posteriori* probabilities. A matter like this is not subject to *a priori* proof, Ellis maintained, and though the error law may somehow represent our ignorance, there is no basis for believing that abstract reasoning "can lead us to assumptions which correspond to and represent outward realities." In particular, "there is no shadow of reason for supposing that the occurrence of a deviation in one direction is independent of that of a deviation in another, whether the two directions are at right angles or not." Herschel had been misled by a mistaken analogy with the composition of forces in mechanics to assume *a priori* what could only be established from experience of actual frequencies.[24]

[22] *Ibid.*, pp. 19-20. See also Herschel, "On the Estimation of Skill in Target-Shooting," in *Familiar Lectures on Scientific Subjects* (London, 1866).

[23] W. F. Donkin, "On an Analogy Relating to the Theory of Probabilities and on the Principle of the Method of Least Squares," *Quarterly Journal of Pure and Applied Mathematics*, 1 (1857), 152-162.

[24] R. L. Ellis, "Remarks on an alleged proof of the 'Method of Least Squares' contained in a late Number of the *Edinburgh Review*," *Phil Mag* [3], 37 (1850), 321-328, pp. 325-326. See also Ellis, "On the Method of Least Squares," *TPSC*, 8 (1849), 204-219.

Still, Maxwell found this exceptionally abstract derivation convincing, and reproduced it almost exactly in his first paper on the kinetic theory. By the time he saw the paper by Clausius that inspired his interest in the kinetic model, he had encountered Hershel's essay review of Quetelet at least twice. Maxwell read it when it first appeared anonymously in the *Edinburgh Review*, and forthwith dispatched a letter to Lewis Campbell explaining that probability was "the true Logic for this world." Maxwell remarked again on the merits of Herschel's work when he read the volume of his collected essays which appeared in 1857.[25] His knowledge of probability and his acquaintance with statistical reasoning, was not, of course, limited to this essay. He had read Laplace's *Théorie analytique* and had been a student of J. D. Forbes at Edinburgh. Forbes had been active during the late 1830s in anthropological statistics and was a close acquaintance of Quetelet, from whom he solicited a testimonial for the chair in natural philosophy that he subsequently occupied.[26] He prepared his critique of the probability argument about double stars while Maxwell was at Edinburgh, and read it at a meeting of the British Association that Maxwell attended. Maxwell was an admirer of George Boole's writings, and had read Mill's *Logic* as well as the general introduction to Buckle's *History of Civilization*. It is the Herschel essay, however, that is imprinted most distinctly on his version of the kinetic gas theory.

Maxwell's introduction of the molecular velocity distribution in his 1859 paper was prefaced by three propositions giving the general laws for collisions between elastic particles of different mass, and showing that such collisions would quickly obliterate any trace of the initial arrangement of the system. Arguing in characteristic statistical fashion, he then proposed that this very confusion implies the existence of stable regularities that must prevail in the mass.

> If a great many equal spherical particles were in motion in a perfectly elastic vessel, collisions would take place among the particles, and their velocities would be altered at every collision; so that after a certain time the *vis visa* will be divided among the particles according to some regular law, the average number of particles

[25] See *Maxwell*, pp. 138-144, 294.

[26] See cahier 1028, *AQP*; also J. C. Shairp, P. G. Tait, A. Adams-Reilly, *Life and Letters of James David Forbes, F.R.S.* (London, 1873), p. 123 and *passim*.

> whose velocity lies between certain limits being ascertainable, though the velocity of each particle changes at every collision.[27]

He then gave, as Proposition 4, a derivation identical to that invented by Herschel. Maxwell premised first that the velocity along any one coordinate is independent of that among each of the other two perpendicular coordinates and, second, that the density of the distribution is spherically symmetric. The independence assumption implied that the total distribution was the product of three independent terms, $f(x)f(y)f(z)$, and the symmetry assumption implied that this product must be a function of the total magnitude of the velocity only—or, equivalently since it is a scalar, to its square. That is, in Maxwell's notation,

$$f(x)f(y)f(z) = \phi(x^2 + y^2 + z^2)$$

The solution to this was the exponential error curve[28]

$$F = Ae^{-(x^2 + y^2 + z^2)}.$$

The error curve was introduced to the kinetic theory, then, on the basis of a derivation so abstract as to be applicable without modification to telescopic observations, balls dropped at a target, and the distribution of velocities in a system of rigid, elastic molecules. This is not the kind of argument that would emerge from a detailed study of the collisions that a molecule in a gaseous system might undergo. It was instead the result of an intuition. Possibly Maxwell had been prepared to recognize the significance of a distribution formula for the kinetic model by his previous work on the stability of Saturn's rings, which in certain respects had ended unsatisfactorily.[29] In any event, it seems probable that the velocity distribution formula was his very first result in the kinetic theory, and the cause of the sudden interest he took in it.

Maxwell's work shows, rather more compellingly than Quetelet's, that the same mathematics can lead to radically different results when applied to a new object. Quetelet had been able to do no more with the error curve as a law of natural variation than to assert its applicability to heights, penchants for crime and suicide, and so on. When incorporated into the kinetic theory, however, the error curve could be put to work to yield powerful new conclusions and methods. Assuming always

[27] James Clerk Maxwell, "Illustrations of the Dynamical Theory of Gases," in *SP*, vol. 1, p. 380.

[28] *Ibid*.

[29] See Stephen Brush, C.W.F. Everitt, and Elizabeth Garber, *Maxwell on Saturn's Rings* (Cambridge, Mass., 1983).

that this velocity distribution was stable, the distribution of relative velocities could be found through a simple integration, which Maxwell carried out in his Proposition 5. He also calculated the joint distribution of velocities when particles of different masses were mixed together, based on the result of his Proposition 6 that the mean energy of particles must be the same, irrespective of their relative masses. In the same way, Maxwell calculated rates of collision and mean free path distances for systems consisting either of one or two kinds of particles. All this was accomplished with elegant combinatorial mathematics of a sort that few if any previous writers had found reason to use. A few years later, Maxwell employed his distribution formula to investigate transport phenomena in gases.

To Maxwell belongs the credit for first introducing the explicit consideration of probability distributions into physics. Indeed, he was arguably the first to use distribution formulas in a significant and productive way in any science. The most impressive results of the kinetic theory during the nineteenth century were closely associated with Maxwell's distribution. Most notable among these was Ludwig Boltzmann's demonstration that the second law of thermodynamics could be understood in terms of mechanics and probability theory. At the time Maxwell began work on the kinetic theory, the second law was a purely thermodynamic principle, expressed entirely in thermodynamic terms—specifically heat, temperature, and entropy. The second law expressed the observed tendency for heat to flow from warm to cold bodies, or, more precisely, the amount of mechanical work that could be performed when this passage of heat was harnessed to an idealized, reversible engine. Boltzmann felt that this formulation by Carnot and Clausius lacked conceptual clarity, which could only be achieved through consideration of the mechanics of heat. Interestingly, his reformulation of the second law was closely tied to his search for a satisfying proof that the velocity distribution of molecules must be governed by Maxwell's formula.

Already in 1866 Maxwell had recognized that there might be problems with his first derivation of the velocity distribution. Since the "assumption that the probability of a molecule having a velocity resolved parallel to *x* lying within given limits is not in any way affected by the knowledge that the molecule has a given velocity resolved parallel to *y*" now appeared "precarious," he sought to "determine the form of the function in a different manner."[30] His new derivation, which formed the starting point for much of Boltzmann's work, presupposed that the tremendous frequency of collisions would produce and maintain an equilibrium state, in which both the velocity distribution and the rate of collisions of every specified type would be stable. Since, Maxwell argued, there is no reason for transformation cycles to run preferentially in one direction, the number of collisions involving two molecules with initial velocities v_1 and v_2 and final velocities v_1' and v_2' must equal the number of collisions of the opposite sort. Maxwell showed that this equality would prevail only if the molecular velocities were distributed in accordance with the error law.

Around the time Maxwell published this second paper on the kinetic theory, the young Austrian Ludwig Boltzmann became familiar with Maxwell's striking new approach to thermodynamics. In 1866 he had written a paper on the mechanical meaning of the second law in which he modestly claimed "to give a purely analytical, completely general proof of the second law of thermodynamics, as well as to discover the theorem in mechanics that corresponds to it";[31] but he quickly recognized both the power and the necessity of Maxwell's new statistical approach for studies of thermal phenomena in mechanical terms. Soon he had extended the Maxwell distribution law to complex, polyatomic molecules subject to external forces, and then in 1871 he succeeded in formulating Clausius's thermodynamic concept of entropy in terms of distributions of molecular velocities.[32]

Armed with the generalized distribution law, Boltzmann returned to the problem of reducing the second law to mechanics in the remarkable paper of 1872 whose preface, on the relations of probability mathematics, statistics, and scientific certainty was quoted above. In this paper, Boltzmann defined a quantity E (subsequently immortalized as H), a function of the distribution of molecular velocities, which he showed to be minimized when the Maxwell-Boltzmann distribution prevailed. Assuming, as Maxwell had, that the collision rate between molecules within any specified energy range varies in proportion to the product of their frequencies in the entire population of molecules, Boltzmann ar-

[30] Maxwell, "The Dynamical Theory of Gases" (1867), in *Papers*, vol. 2, p. 43.

[31] Quoted in Martin Klein, "The Development of Boltzmann's Statistical Ideas," *Acta Physica Austriaca*, Suppl. X (1973), 53-106, p. 57.

[32] Edward E. Daub, "Probability and Thermodynamics: The Reduction of the Second Law," *Isis*, 60 (1969), 318-330.

gued from symmetry to show that the derivative of E must always be negative unless the error-curve distribution prevails, when it will be zero. That is, he demonstrated not only that Maxwell's distribution was stable, but also that any other distribution must converge to it. Even more, since the quantity E was proportional to the negative of his expression for entropy, the tendency for entropy to reach a maximum was equivalent to the tendency for a system to approach the Maxwell distribution.[33]

The relation between probability theory and the second law was brought out more clearly in 1877, when Boltzmann applied a combinatorial model to the distribution of molecular energies. He had first introduced combinatorics to the kinetic theory in 1868, because he was concerned that for a system with a finite number of molecules and a fixed total energy, the energy of the last molecule must be completely determined by that of the other n-l molecules. Hence the energies of colliding molecules would not be fully independent, and Maxwell's 1867 derivation could not be fully rigorous.[34] The problem of distributing a fixed amount of energy among a finite number of molecules was similar to a problem solved by Laplace to support his calculation of the probability that the sums of the angles made by the planes of the orbits of the planets with the ecliptic would be as small as it is, had the orbits been produced by chance. Posed abstractly, in terms of the urn model, the query was: "An urn being supposed to contain n + l balls, marked with the numbers 0, 1, 2, 3, . . . n, one draws a ball from it and places it back in the urn after drawing. One asks the probability that after i drawings the sum of the numbers drawn will be equal to s."[35] Boltzmann constrained the system to a fixed total value and sought the probability that any given molecule would have a certain energy.

To do this, he was obliged to divide the total energy into a finite number of intervals of width ε, and to suppose every energy interval to be equipossible, *a priori*, for a given molecule, up to the total energy available. The actual probability that an arbitrary molecule has energy $n\varepsilon$ is then given by a fraction whose denominator is the total number of combinations satisfying the energy constraint and whose numerator is the

[33] Boltzmann, "Weitere Studien" (n. 3).

[34] Boltzmann, "Studien über das Gleichgewicht der lebendigen Kraft zwischen bewegten materiellen Punkten" (1868), in WA, vol. 1, pp. 80-81.

[35] Laplace, *Théorie analytique des probabilités* (Paris, 2d ed., 1820), in *Oeuvres*, vol. 10, p. 257. See also Oettinger, "Untersuchungen über die Wahrscheinlichkeitsrechnung," in Crelles, *Journal für die reine und angewandte Mathematik*, 26 (1843), 34 (1847), and 36 (1848), esp. vol. 26, 311-332.

number of these for which the given molecule occupies the prescribed energy level. Already in 1868 Boltzmann was able to derive the Maxwell velocity distribution following this procedure. In 1877 he approached the problem with a broader perspective, and sought not just the probability that any given molecule will have some particular energy, but that the entire system will be characterized by a given energy distribution.

Once again, Boltzmann assumed that every permissible "complexion"—that is, every assignment of energy levels to the individual molecules consistent with a fixed total energy—was equipossible. The problem was to calculate the probabilities of the various possible "state distributions" $(w_o, w_1, w_2, \ldots w_p)$, where each w_i designates the number of molecules with energy level *i*. Using some very sophisticated combinatorial mathematics along with the familiar Stirling approximation, Boltzmann found that this probability could be given in terms of a "permutability measure," which was a special form of the H-function he had discussed five years earlier. Since he had already shown that the H-function decreases continuously until the system reaches equilibrium at Maxwell's distribution, he announced the problem solved. "Thus we can see that it can justifiably be said: that distribution which is the most probable of all, corresponds also to the state of thermal equilibrium. For if an urn is filled with slips in the manner indicated above, then it will be most probable that the distribution corresponding to thermal equilibrium will be written on the slips that are drawn."[36] The second law of thermodynamics was equivalent to a tendency for a system of molecular velocities to approach its most probable state, the Maxwell-Boltzmann distribution, and entropy was proportional to the logarithm of probability.[37]

GALTON AND THE REALITY OF VARIATION

Like Maxwell and Boltzmann, Francis Galton was not content simply to assert, with Quetelet, the universal applicability of the error law, and he forthrightly opposed the social physicist's interpretation of this curve as proof that everything exceptional could best be interpreted as a flaw.

[36] Boltzmann, "Über die Beziehung zwischen dem zweiten Hauptsatze der mechanischen Wärmegleichgewicht und der Wahrscheinlichkeitsrechnung respektive den Sätzen über das Wärmegleichgewicht," in WA, vol. 2, p. 193.

[37] See Klein, "Boltzmann's Statistical Ideas" (n. 31), and Thomas S. Kuhn, *Black-Body Theory and the Quantum Discontinuity*, 1894-1912 (Oxford, 1978), chap. 2, esp. pp. 46-54.

Galton was the first to use the statistical methods of error analysis in order to analyze real variation, and his achievements as a biological and statistical theorist were due in no small measure to the interest he took in variation for its own sake. As Victor Hilts and Ruth Cowan have shown, Galton seems always to have been more interested in the exceptional than in the average: "Some thorough-going democrats may look with complacency on a mob of mediocrities, but to most other persons they are the reverse of attractive."[38] The statistical techniques of astronomers, which Quetelet had sought to imitate, seemed to Galton wholly unsuitable, since they were designed to eliminate errors and deviations, the very objects that he most wished to study and preserve. Galton admired Quetelet's use of the error law, perhaps precisely because he paid no attention to Quetelet's interpretation of it. Galton understood the error law as an invaluable means for taking account of natural variation, and he was accordingly critical of contemporary statists who, because of their infatuation with averages, failed to make use of it. He castigated them for their neglect of the natural diversity that makes society interesting in a noteworthy passage in *Natural Inheritance* entitled "The Charms of Statistics": "It is difficult to understand why statisticians commonly limit their inquiries to Averages, and do not revel in more comprehensive views. Their souls seem as dull to the charm of variety as that of the native of one of our flat English counties, whose retrospect of Switzerland was that, if its mountains could be thrown into its lakes, two nuisances would be got rid of at once."[39]

Galton's most important work, as well as his sustained commitment to the study of human variety, derived from an ideal that differed in one other important respect from the aims and outlook of nineteenth-century statisticians. Galton was the founder of eugenics, the evolutionary doctrine that the condition of mankind can most effectively be improved through a scientifically directed process of controlled breeding. Although public health reform movements had led men like William Farr to give attention to the biological determinants of health and even of achievement,[40] the usual aim of statistics as social science had been to

[38] Galton, "President's Address," *JAI*, 18 (1889), 401-419, p. 407. See also Victor L. Hilts, "Statistics and Social Science," in Ronald N. Giere, Richard S. Westfall, eds., *Foundations of Scientific Method: The Nineteenth Century* (Bloomington, 1973), pp. 206-233; Ruth S. Cowan, "Francis Galton's Statistical Ideas: The Influence of Eugenics," *Isis*, 63 (1972), 509-528.

[39] Galton, *Natural Inheritance* (New York, 1889), p. 62.

[40] See Victor L. Hilts, "William Farr (1807-1883) and the 'Human Unit,' " *Victorian Studies*, 14 (1970), 143-150.

improve men by providing them with education and with homes and workplaces free of physical and moral contamination. Galton rejected the principle that human character was determined by environment.

In this sense, at least, Galton can accurately be called a conservative. Although he felt nothing but disdain for those members of the aristocracy who had neither been "nobly educated" nor had "any eminent kinsmen, within three degrees," he firmly believed that men are not "of equal value, as social units, equally capable of voting, and the rest." Galton glibly assumed that "social worth" among all but the intellectual and scientific classes was distributed roughly in the same proportions as income, and he rated professionals above tradesmen, shopkeepers above artisans, and the latter above unskilled workers, in innate merit. These gradations, he insisted, reflected differences of biology, not merely of upbringing—at least in liberal England. "There can be no doubt but that the upper classes of a nation like our own, which are largely and continually recruited by selections from below, are by far the most productive of natural ability. The lower classes are, in truth, the 'residuum'."[41] It was Galton who coined the terminology of the nature-nurture debate, and he held steadfastly to the position that inherited qualities provided by nature were by much the most important in determining the achievement of individuals. He sought to support this opinion with a variety of studies, and felt it to be confirmed most decisively by a comparison of identical and fraternal twins. The uncanny resemblance between identical twins in matters of behavior as well as appearance, as revealed by a litany of anecdotes, seemed to imply that even complex abilities were largely determined by heredity.[42]

Galton cannot, however, be construed as simply a conservative, or his eugenic creed as a response to any resurgence of conservatism emerging in opposition to the reform movements leading up to the franchise extension of 1867. To begin, there is little evidence that he ever took an active interest in ordinary political affairs,[43] and still less that his political

[41] Galton, *English Men of Science: Their Nature and Nurture* (1874; New York, 1875), p. 17. See also Galton, "Hereditary Improvement," *Fraser's Magazine*, n.s. 7 (1873), 116-130, p. 127; Galton, *Hereditary Genius: An Inquiry into Its Laws and Consequences* (London, 1869), pp. 86-87.

[42] Galton, "The History of Twins as a Criterion of the Relative Powers of Nature and Nurture," *JAI*, 5 (1876), 391-406; see also Stephen J. Gould, *The Mismeasure of Man* (New York, 1981); Daniel J. Kevles, *In the Name of Eugenics: Genetics and the Uses of Human Heredity* (New York, 1985).

[43] See, however, Ruth Schwartz Cowan, "Sir Francis Galton and the Study of Heredity in the Nineteenth Century" (Ph.D. Dissertation, Johns Hopkins University, 1969), p. 56, who observes that Galton already was voting Conservative as a student at Cambridge.

views underwent any significant shift during the years preceding his first publication on "hereditary genius" in 1865. That his views were untimely is strongly suggested by his failure to win any real support for eugenics until the 1890s. Galton's ideology was too unorthodox, its implications too revolutionary, for it to be seen as an ordinary contribution to political discourse. To be sure, his object was not to overturn social institutions, but to reinvigorate them, by replacing blind faith in a moribund tradition with reliance on the growing powers of science. But to make talent rather than property the basis of nobility, or to replace the existing House of Lords with an upper legislative chamber of hereditary ability, would be to transform the entire basis of aristocracy, and not merely to change its personnel. Galton was not really a conservative, but a deeply committed reformer whose tastes ran to conservatism only in certain respects.

Galton's biography suggests that his dedication to eugenics derived at least as much from Quaker idealism and the Victorian intellectual aristocracy's reverence for science as from social or political conservatism.[44] He was descended from a line of Birmingham Quakers whose sizable fortune was derived from banking and the manufacture of arms, among other things. The family was becoming increasingly genteel during the early nineteenth century; his father had joined the Anglican church, and his two older brothers became landed gentlemen, devoting their lives to hunting and other forms of refined idleness. Francis Galton seems to have feared a similar destiny, which he was not content to follow. He was the product of some remarkable educational efforts on the part of his elder sisters, leading to the acquisition of certain skills at a very tender age, which inspired Lewis Terman to assign him an unparalleled intelligence quotient. His mother aspired for him to become, like her father Erasmus Darwin, an eminent physician, and although he found the medical profession no more appealing than had his older cousin, Charles Darwin, he gave evidence already during his youth of great personal ambition. In particular, Galton was driven during adolescence and early adulthood by an overpowering desire "to do good."

At Cambridge, Galton suffered two debilitating mental breakdowns, which prevented him from taking honors, and he seems then to have begun to feel doubts about his own abilities. What was worse, the idealistic and ambitious young man was unable to find a purpose worthy of

[44] See MacKenzie; also Noel Annan, "The Intellectual Aristocracy," in J. H. Plumb, ed., *Studies in Social History* (London, 1955), pp. 241-267.

his energies, and when his father died, leaving him a considerable inheritance that freed him from the obligation to study medicine, his own aimlessness became an intolerable burden. The period that followed was one of "wanderings," in the fullest sense. He traveled through the Middle East, without any particular goal or destination, and discovered, as he later observed, that experience of new cultures tends to upset the assumptions of the traveler. In particular, he began to question his religious faith. While at Cambridge, he had written poetry lamenting the failure of his materialistic society to live up to Christian ideals, and expressing a desire to be a citizen "of some state, modelled after Plato's scheme, and overruled by Christianity."[45] By this time he had become persuaded that the era of Christianity was past, that the Christian faith lacked the power to guide a life. In contrast, he found in Egypt reason to be "much impressed by the nobler aspects of Mussulman civilisation, especially, I may say, with the manly conformity of their every-day practice to their creed, which contrasts sharply with what we see among most Europeans, who profess extreme unworldliness and humiliation on one day of the week, and act in a worldly and masterful manner during the remaining six."[46]

Science did not cause Galton's crisis of faith, but it did make itself available as an alternative to Christianity. In Egypt, Galton met a French St. Simonian, Arnaud Bey, who proposed a scientific object for his travels, and from 1845 to 1860 his life was increasingly shaped by his activities in the Royal Geographical Society—Galton was among the most intrepid African explorers[47]—and by his meteorological researches. Science, having blessed Galton's wanderlust with the dignity of purpose, at last gave him an object fully worthy of his talents and philanthropic longings in the form of the biological principle of evolution. Galton's reaction to the *Origin of Species* was correspondingly inflated. He wrote to his cousin a decade later:

[45] Quoted in *Galton*, vol. 1, p. 177. See also Galton, *Men of Science* (n. 41), p. 218; Ruth Cowan, "Nature and Nurture: The Interplay of Biology and Politics in the Work of Francis Galton," *Studies in History of Biology*, 1 (1974), 133-208; Kevles, *Eugenics* (n. 42); Raymond E. Fancher, "Biographical Origins of Francis Galton's Psychology," *Isis*, 74 (1983), 227-233.

[46] Galton, *Inquiries into Human Faculty and its Development* (1883; London, 2d ed., 1907), p. 154. See also Galton, *Memories of My Life* (London, 3d ed., 1909), p. 88, where Galton noted a sharp contrast between his own carousing and Moslem piety. Galton's intellectual development during his years of travel is insightfully presented by Derek W. Forrest, *Francis Galton: The Life and Work of a Victorian Genius* (London, 1974).

[47] On the relation of Galton's African travels to his later work, see Raymond E. Fancher, "Francis Galton's African Ethnography and Its Role in the Development of his Psychology," *BJHS*, 16 (1983), 67-79.

> I always think of you in the same way as converts from barbarism think of the teacher who first relieved them from the intolerable burden of their superstition. I used to be wretched under the weight of the old-fashioned "argument from design" of which I felt though I was unable to prove to myself the worthlessness. Consequently the appearance of your "Origin of the Species" formed a real crisis in my life; your book drove away the constraint of my old superstition as if it had been a nightmare and was the first to give me Freedom of thought.[48]

This, clearly, was not the reluctant concession of an admirer of Christianity to the weight of scientific counterevidence to his faith. Galton's disenchantment with the Christian religion derived not from doubts concerning the truth of its dogmas, or from scruples about the humaneness of its ethical code, but from a conviction of its ineffectuality in the industrial society that was Victorian England. The argument from design, this intellectual justification for Christianity, was for Galton the last support of the old faith to crumble. By the 1860s Galton had come to regard science as a robust and manly replacement for an enfeebled Anglicanism, and his eagerness to interpret the theory of evolution as a refutation of Christianity was born of his belief that this ancient religion must be demolished to make way for the future. Darwinism was a decisive defeat of Christianity by science, a conquest of the old order by the new.

Thus Galton, like many of the intellectuals who came forward to attack the old religious order during the decades after 1859, saw in evolution a rallying cry for the reconstruction of society in the image of science. He stated plainly that his own distinctive contribution to this movement, eugenics, was a substitute religion—or rather, as he would have it, the true religion. Galton rhapsodized about the joy and serenity, comparable to that of religious contemplation, to be derived from observing the solidarity of the fabric of nature, and from furthering the processes of race improvement revealed by the theory of evolution to be the goal of nature itself. He also called for "the establishment of a sort of scientific priesthood throughout the kingdom, whose high duties would have reference to the health and well-being of the nation in its broadest sense, and whose emoluments and social position would be made commensurate with the importance and variety of their func-

[48] See facsimile of this letter in *Galton*, vol. 1, plate 2.

tion."[49] The need for this priesthood he justified with an argument about the increasing complexity of society that had already endeared itself to several generations of social scientists and would-be technocrats.

> The average culture of mankind is become so much higher than it was, and the branches of knowledge and history so various and extended, that few are capable even of comprehending the exigencies of our modern civilisation; much less of fulfilling them. We are living in a sort of intellectual anarchy, for the want of master minds. The general intellectual capacity of our leaders requires to be raised, and also to be differentiated. We want abler commanders, statesmen, thinkers, inventors, and artists. The natural qualifications of our race are no greater than they used to be in semi-barbarous times, though the conditions amid which we are born are vastly more complex than of old. The foremost minds of the present day seem to stagger and halt under an intellectual load too heavy for their powers.[50]

The theory of evolution by natural selection provided the context in which statistical biology was introduced, and within which it has since been most fruitfully developed. Darwin's own ideas, though rightly regarded as crucial for the development of what Ernst Mayr calls "population thinking" in the biological sciences, can only in retrospect be construed as statistical. Perhaps it only begs the question to point out that Darwin claimed to be unstatistical by disposition, and that there is little evidence of any influence on his work from social statistics beyond the well-known inspiration he received from a completely different aspect of T. R. Malthus's theory of population. But even though Darwin's theory involved the production of variation by causes that were poorly understood in general and wholly ignored in detail, and the differential survival over the long run of certain variants, Darwin never developed anything like a quantitative model of evolutionary change.[51] Neither his

[49] Galton, *English Men of Science* (n. 41), p. 195. Galton often raised the possibility of a scientific priesthood, in whose work statistics would have a prominent place. On the religious character of eugenics, see Galton, "The Part of Religion in Human Evolution," *National Review*, 23 (1894), 755-763.

[50] Galton, "Hereditary Talent and Character," *Macmillan's Magazine*, 12 (1865), 157-166, 318-327, p. 166.

[51] On Darwin and probabilistic thinking, see M.J.S. Hodge, "Law, Cause, Chance, Adaptation and Species in Darwinian Theory in the 1830s, with a Postscript on the 1930s," in Michael Heidelberger et al., eds., *Probability since 1800: Interdisciplinary Studies of Scientific Development*, *Report Wissenschaftsforschung*, 25 (Bielefeld, 1983), 287-329. In 1901, when Karl Pearson and W.F.R. Weldon were setting up, in collaboration with Galton, the journal

theory nor Gregor Mendel's combinational model of hereditary transmission can be regarded as statistical in the sense used here—that is, employing a mode of reasoning based on stable numerical frequencies.[52]

The statistical approach was introduced to biology by Galton, who was Darwin's cousin. When, as an old man, Galton looked back on his career, his tendency to approach problems statistically seemed so thoroughly ingrained in his character that he attributed it to a genetic inheritance—from the Galton side of the family, not the Darwins. A variety of anecdotes survive telling of how young Galton employed a sextant to ascertain the dimensions of some attractive African women, and of how he invented a series of concealed devices for registering fidgets and other unconscious acts as an index of the dullness of meetings. Perhaps, however, it is unnecessary to assume an unusually vigorous statistical gemmule in the Galton ancestry to account for this aspect of his character. Galton reached maturity at the peak of the European statistical movement, and his early numerical studies of meteorological and geographical statistics were, in an extended sense, part of it. It is perhaps also relevant that his father, a successful businessman and banker, inculcated in Galton and his brothers the most fastidious accounting habits—as the regular financial reports in the son's letters from Cambridge plainly reveal.[53]

Although Galton wrote a paper in 1873 for the Royal Statistical Society (successor to the Statistical Society of London), he was skeptical of the scientific worth of the society's activity. In 1877 he proposed that Section F, statistics and economic science, be expelled from the British Association. "Usage has drawn a strong distinction," he wrote, "between knowledge in its generality and science, confining the latter in its strictest sense to precise measurements and definite laws, which lead by

Biometrika, an effort was made by Pearson to establish that statistics was the natural fulfillment of Darwin's evolutionary approach, based on the supposition that since the study of evolution was properly a statistical affair, Darwin's thought must have been deeply statistical. Galton reported to Pearson a negative conclusion, in which both Frank and Leonard Darwin had concurred, that their father had a "non-statistical" mind. "I fear you must take it as a fact that Darwin had no liking for statistics." See letters from Galton to Pearson, 4 Feb. 1901 and 8 Feb. 1901, in *FGP*, file 245/18 E; also Pearson to Galton, 3 July 1901 and 10 July 1901 in file 293 E. On the other hand, George Darwin had written to Galton in a letter some 25 years earlier that he and Galton shared "a common family weakness for statistics" (4 Jan. 1875, file 190 A).

[52] It is clear, however, that chance was an important issue in Darwin's theory of evolution from the beginning; see Silvan S. Schweber, "The Origin of the Origin Revisited," *Journal of the History of Biology*, 10 (1977), 229-316; Schweber, "Aspects of Probabilistic Thought in Great Britain During the 19th Century: Darwin and Maxwell," in Heidelberger, *Probability* (n. 51), pp. 41-96.

[53] See Cowan, "Nature and Nurture" (n. 45); also *Galton*, vol. 1, p. 107.

such exact processes of reasoning to their results, that all minds are obliged to accept the latter as true."[54] Galton believed in the need for higher mathematics in statistics, and saw little evidence that members of the statistical societies were interested in the mathematical aspects of their discipline. For that reason, he thought the old Section F could most appropriately be transferred to the Social Science Congress.

Nevertheless, Galton was not untouched by the preconceptions and beliefs of the social fact gatherers. It is significant that his program for the quantitative study of heredity was originally founded on a "simile" which explicated the process of hereditary transmission in terms of social processes.[55] Galton took as his starting point the principle of statistical regularity developed by Quetelet and the British statists. In his work it appeared in a slightly more abstract form as the "axiom of statistics that large samples taken out of the same population at random are statistically similar."[56] Galton did not characterize this similarity mathematically, or derive it from probability theory. Instead, he explained it in terms of an easily understood analogy. He proposed, for example, that the selection of genetic material to be passed on from parents to offspring was similar to a process of "indiscriminate conscription: thus, if a large army be drawn from the provinces of a country by a general conscription, its constitution, according to the laws of chance, will reflect with surprising precision, the qualities of the population whence it was taken; each village will be found to furnish a contingent, and the composition of the army will be sensibly the same as if it had been due to a system of immediate representation from the several villages."[57]

Galton introduced this important simile between the phenomena of biological inheritance and the statistical behavior of a free society in 1869, only a year or two before Maxwell and Boltzmann proposed the same analogy for physics. The nearly perfect coincidence is perhaps just luck, but their independence in proposing this analogy seems clear, and is at least highly suggestive. Seemingly without exception, those who applied statistical thinking to any of the sciences during the second half of the nineteenth century thought in terms of analogies with the social science of statistics.

[54] Francis Galton, "Considerations Adverse to the Maintenance of Section F . . . ," *JRSS*, 40 (1877) 468-473, p. 471.

[55] See the discussion of Galton in chapter 9.

[56] Galton, "Family Likeness in Stature," *PRSL*, 40 (1886), 42-73, p. 43.

[57] Galton, "On Blood-relationship," *PRSL*, 20 (1871-72), 394-402, pp. 397-398.

Galton also used the methods of the social statisticians to debunk those conventional religious beliefs whose unscientific character seemed to him to render them untenable in the modern age. The most notorious contribution to this crusade was his remarkable "Statistical Inquiries into the Efficacy of Prayer," a paper in which he sought to resolve scientifically whether piety or prayer brought any objective advantages to its intended beneficiaries. Its effect was to crush mystical piety under a heap of miscellaneous statistical facts. Sovereigns, whose lives were the object of regular prayerful appeal of whole nations, proved to live no longer than other members of the prosperous classes. Galton supposed that clergymen could be expected to plead on occasion for their own health, yet found that their life span was similar to that of physicians or attorneys. The final blow was struck by the practice of insurance companies, which, it seems, did not distinguish between the lives of the pious and the worldly. They even offered slave vessels the same advantageous rates as missionary ships. "If prayerful habits had influence on temporal success, it is very probable, as we must again repeat, that insurance offices, of at least some descriptions, would long ago have discovered and made allowance for it. It would be most unwise, from a business point of view, to allow the devout, supposing their greater longevity even probable, to obtain annuities at the same low rates as the profane."[58]

The statistical studies that Galton performed to promote and facilitate eugenic intrusion into human heredity were fundamentally dependent on Quetelet's error law. There are some respects in which his dedication to this and other statistical laws revealed concerns similar to that of the social statisticians. Galton was a man of order, and he displayed a special interest in subjecting the irrational and inexplicable to scientific principle. He was himself burdened with an unstable psyche, reflected in his idea that "the tableland of sanity upon which most of us dwell, is small in area, with unfenced precipices on every side, over any one of which we may fall." He once performed some introspective experiments on paranoia and fetishism whose success was so great that he had difficulty regaining a normal perspective.[59] He was fascinated by the phenomena of mental imagery, to which he believed a certain portion of the popu-

[58] Galton, "Statistical Inquiries into the Efficacy of Prayer," *Fortnightly Review*, n.s. 12 (1872), 125-135, p. 134. Galton's statistical argument on prayer was criticized by George John Romanes, *Christian Prayer and General Laws* (London, 1874), pp. 253-268.

[59] Galton, *Memories* (n. 46), pp. 38, 276-277.

lation was congenitally susceptible. He reported that whenever the climate of opinion becomes favorable to supernaturalism, these normal mental processes are misinterpreted, and seers of vision appear in every community. Galton took great pains to establish a quantitative study of phenomena like these, and thereby to show how a multitude of subconscious mental processes "admit of being dragged into light, recorded and treated statistically, and how the obscurity that attends the initial steps of our thoughts can thus be pierced and dissipated."[60]

Galton was rarely satisfied with counting alone, and he brought even to these psychological studies the aims of the hereditary quantifier, as Galton would have it, as well as the quantifier of heredity. Though minimally proficient at algebraic operations, Galton had received training in mathematics at Cambridge and had acquired, like so many Cambridge graduates, a sound geometrical intuition as well as an ability to set up problems correctly even when he was unable to solve them. Since the statistical study of heredity required techniques for taking account of variation, Galton was pleased to have available the astronomical error law. He insisted, however, that in biological studies it was absurd to use such expressions as "probable error," since variation in this domain was genuine, and not the mere product of error.[61] For eugenic purposes, it was natural to look upon the exceptional as more interesting and important than the average.

Galton may have encountered the error function in his meteorological researches at the Kew Observatory during the 1850s, but he became interested in learning about its use only after he became aware of its applicability as a law of distribution.[62] In 1861 his friend and fellow geographer William Spottiswoode published a paper that made use of the error law as a definition type. His object was to ascertain whether a certain group of Asian mountain ranges had been produced by a single cause—subject, as ever, to numerous incidental sources of variation. Invoking Quetelet on the principles of type and of natural variation, he proposed that the discrepancies in the orientation of the principal axes of these ranges from their mean ought to be governed by the astronomical law of

[60] Galton, *Human Faculty* (n. 46), p. 145; also p. 121 and "The Visions of Sane Persons," *Fortnightly Review*, n.s. 29 (1881), 729-740.

[61] Galton, *Natural Inheritance* (n. 39), p. 58.

[62] After learning of Spottiswoode's use of the error law, discussed below, he promptly looked up George Biddell Airy's *On the Algebraical and Numerical Theory of Errors of Observation* (London, 1861). There is no evidence that he consulted such works to support his meteorological researches.

errors. Spottiswoode was genuinely interested in his Asian mountains, but he regarded his brief memoir primarily as a methodological contribution to natural science. He saw Quetelet's use of the error law as the basis for a general test which could be applied to identify those phenomena for which a common cause could be sought with reasonable expectation of success. Spottiswoode was, consequently, pleased to explain to Galton "the far-reaching application of that extraordinarily beautiful law."[63]

Galton was deeply impressed by Spottiswoode's exposition of the error law, of which he first learned in 1860 or 1861 at about the time Quetelet journeyed to London for the International Statistical Congress there. He met Quetelet at that time, and consulted his book of 1846 on the methods of statistics before he began using the error curve, in 1869. Galton remained one of the most loyal partisans of the error law throughout his life. Even though he was among the first to propose an alternative distribution function, the so-called log-normal, in connection with a certain class of data, that formula involved no rejection of the conventional error law. Galton simply believed that sensations governed by Fechner's law—according to which the perceived force of a given alteration in a stimulus is inversely proportional to the total level of that stimulus—could be expected to reflect the same structure of deviations in the error distribution function. Accordingly, a geometric rather than an arithmetic mean should be used, and the treatment of variation similarly modified.[64]

Despite his enthusiasm for the exceptional, Galton often used the error curve precisely in the fashion developed by Quetelet and Spottiswoode, as a definition of type. He sought, for example, to ascertain whether the basic forms of fingerprints represented genuine differences of type by determining if the variation within each were governed by Quetelet's law.[65] He also based his use of the remarkable technique of composite photography on the principle that deviations from any given type must be distributed in the same way as errors of observation. So long as deviations from type were accidental, flaws and eccentricities

[63] Galton, *Memories* (n. 46), p. 304. See William Spottiswoode, "On Typical Mountain Ranges: An Application of the Calculus of Probabilities to Physical Geography," *Journal of the Royal Geographical Society*," 31 (1861), 149-154. Spottiswoode's source for a geometrical approach to mountain formation was the Petersburg academician O.H.W. Abich.

[64] See Francis Galton and Donald MacAlister, "The Geometric Mean, in Vital and Social Statistics," *PRSL*, 29 (1879), 365-376.

[65] Galton, *Finger Prints* (London, 1892), p. 19.

would average out, and the result of superimposing photographic images would be a picture very similar to the type itself. That is, the technique could be used to discover Quetelet's average man, although Galton did not want to take the average over a whole population, but rather over the individuals belonging to a distinctive human type. By taking numerous, rapid photographs of perfectly aligned individuals on a single plate of film, Galton believed he had uncovered the pure type of the criminal, the consumptive, and so on. Notwithstanding the beauty of the exceptional, Galton's composites were always "better looking than their components, because the averaged portrait of many persons is free from the irregularities that variously blemish the looks of each of them."[66]

Interestingly, Galton held to this definition of type even when it presented severe—and, as Pearson believed, illusory—obstacles to the attainment of eugenic objectives. Thus Galton began in the 1880s to identify the concept of race with a distribution governed by the error law and, for that reason, tending always to remain stable. "It is the essential notion of a race that there should be some ideal typical form from which the individuals may deviate in all directions, and towards which their descendants will continue to cluster."[67] Since the type was defined by a stable point, and not merely as the mean of an arbitrary distribution, deviations were only imperfectly perpetuated. Galton became persuaded that evolution was based on discontinuous variation, or "sports," and that no significant progress towards a eugenic society could be achieved using natural or continuous variation. Sports, however, represented new types, or new centers of stability, towards which the offspring of their own exceptional progeny would in turn tend to regress.[68]

Galton began using the error law in his first book on biological heredity, the notorious *Hereditary Genius* of 1869. Like Quetelet, he stressed its universality. Since "so much has been published in recent years about statistical deductions," Galton began, every reader should

[66] Galton, "Composite Portraits," *Nature*, 18 (1878), 97-100, p. 98; see also *Human Faculty* (n. 46), p. 230.

[67] Galton, *Human Faculty* (n. 46), p. 10. See also "Regression towards Mediocrity in Hereditary Stature," *JAI*, 15 (1886), 246-263; "Pedigree Moth-breeding, as a means of verifying certain important constants in the General Theory of Heredity," in *Transactions of the Entomological Society of London* (1887), pp. 19-28.

[68] Galton believed that the different forms of fingerprints were instances of variant types; see "The Patterns in Thumb and Finger Marks," *Phil Trans*, B, 182 (1892), 1-23. On the wider issue, see William B. Provine, *The Origins of Theoretical Population Genetics* (Chicago, 1971).

be prepared to concede that the average height of a large, isolated population will be constant from year to year. Indeed, "the experience of modern statistics" shows that not only the mean height, but also the fraction of the population in any given range of heights, must likewise be very nearly uniform over time. The variation of stature, he went on, will be governed by a particular law, the "law of deviation from an average." This law is "perfectly general in its application," as all of statistics demonstrates. "Wherever there is a large number of similar events, each due to the resultant influences of the same variable conditions, two effects will follow. First, the average value of those events will be constant, and, secondly, the deviations of the several events from the average, will be governed by this law."[69] The error law was not confined to the physical domain, for if the error curve expresses the distribution of stature, "then it will be true as regards every other physical feature—as circumference of head, size of brain, weight of grey matter, number of brain fibres, &c.; and thence, by a step on which no physiologist will hesitate, as regards mental capacity."[70]

Hereditary Genius was intended to establish that exceptional ability in a variety of pursuits from music, justice, and statesmanship to wrestling and rowing was inherited. Galton argued that whereas a moderate talent might on occasion be held back by deprived circumstances, and a mediocre one elevated to high office by education or background, exceptional hereditary ability was both necessary and sufficient for the attainment of real eminence in a given field. Hence he decided that his case would be adequately established if he could show, which he did, that the most distinguished judges, statesmen, and wrestlers were far more likely to be near relatives of other eminences in these activities than was an ordinary member of the population. His argument required no use of mathematical formalism more elaborate than the most straightforward calculation of numerical probabilities.

Galton, however, saw in the error curve both the possibility of increasing the impressiveness of his study and a technique for quantifying a range of attributes that previously had resisted exact investigation. "Whenever we have grounds for believing the law of frequency of error to apply, we may work backwards, and, from the relative frequency of occurrence of various magnitudes, derive a knowledge of the true rela-

[69] Galton, *Hereditary Genius* (n. 41), pp. 26-29.
[70] *Ibid.*, pp. 31-32.

tive values of those magnitudes, expressed in units of probable error,"[71] he wrote in 1875. In *Hereditary Genius*, Galton assumed that 250 of every million men are properly eminent. This group corresponded to a tiny area at the right end of the error curve. Galton then divided the axis between this point and the mean into five equal distances and, using this small interval as his unit, marked one more class above minimal eminence and seven below mediocrity. These fourteen intervals represented the grades of intelligence or statesmanship, as defined by mathematics.

According to this scheme, the classes immediately above and below the mean—which Galton labeled A and *a*, respectively—each contained about 257,000 men per million. The next higher grade of intelligence, *B*, and the next lower grade of stupidity, *b*, numbered 161,000 each. The density declined at an accelerated rate so that the first eminent category, *F* (and the corresponding state of abject idiocy, *f*), contained only 233 men per million. Beyond the seventh division at each end of the curve, and extending to infinity, were X and *x*, each with a mere one man per million.

> It will, I trust, be clearly understood that the numbers of men in the several classes in my table depend on no uncertain hypothesis. They are determined by the assured law of deviations from an average. It is an absolute fact that if we pick out of each million the one man who is naturally the ablest, and also the one man who is the most stupid, and divide the remaining 999,998 men into fourteen classes, the average ability in each being separated from that of its neighbours by *equal grades*, then the numbers in each of these classes will, on the average of many millions, be as stated in the table. . . . Thus the rarity of commanding ability, and the vast abundance of mediocrity, is no accident, but follows of necessity, from the very nature of these things.[72]

Although this introduction of the error law yielded no concrete results in *Hereditary Genius*, and added little if anything to its persuasiveness, the mathematics of error analysis was itself enriched by its application to elusive objects such as intelligence. Because he was unable to measure these traits directly, he was obliged to have recourse to statistical units, the great ancestors of which were these fourteen classes of intelligence.

[71] Galton, "Statistics by Intercomparison, with Remarks on the Law of Frequency of Error," *Phil Mag* [4], 49 (1875), 33-46, p. 37.

[72] Galton, *Hereditary Genius* (n. 41), pp. 34-35.

Statistical units proved indispensable for Galton's subsequent quantitative study of inheritance and his method of intercomparison.

Galton's use of statistical units in connection with the analysis of objects that could be ordered by rank but not directly measured became more sophisticated during the early 1870s. In 1874 he published his study of *English Men of Science*, an effort to identify those traits which contribute to great achievement in various scientific fields. His material consisted, for the most part, not of direct measurements, but of responses to a survey that he distributed among his scientific associates. Galton had intended to arrange the replies to his survey of these men in order according to degree of religious piety, energy, business sense, and so on, much as he later did with reports of ability to form mental images,[73] and from this ordering by rank to place each man in the appropriate class of ability as defined by relative rarity and the error law. He hoped to supplement this information with the judgments of biographers of scientists and other great men, to whom he appealed to rank their subjects on his statistical scale, "in respect to every quality that is discussed."[74] From this information, Galton planned to find relations between each of these qualities and scientific achievement by field. It is significant that Galton's first "correlation diagram," a chart plotting head size against weight, which Galton never published, was prepared in conjunction with this study of English scientists.[75]

In the event, the data proved too unreliable to be handled by rigorous techniques. "It had been my wish to work up the materials I possess with much minuteness; but some months of careful labor made it clear to me that they were not sufficient to bear a more strict or elaborate treatment than I have now given to them."[76] Galton found it impossible to attach reliable ratings to the replies, or to distinguish between the *F*'s and G's of scientific ability. As a consequence, the most important statistical result of this inquiry was not the analysis of scientific character, but the statistical technique that Galton had devised to enable him to draw exact conclusions from this data as reliably and efficiently as possible. This was his method of "statistics by intercomparison," which he was able to

[73] Galton, "Mental Imagery," *Fortnightly Review*, n.s. 18 (1880), pp. 312-324.

[74] Galton, "On Men of Science, Their Nature and Nurture," *PRI*, 7 (1874), 227-236, p. 234.

[75] See Victor L. Hilts, "A Guide to Francis Galton's *English Men of Science*," *Transactions of the American Philosophical Society*, 65, part 5 (1975), 25-26.

[76] Galton, *English Men of Science* (n. 41), p. vii.

apply to his data on men of science and which he published separately the next year.

The method of intercomparison was expressly designed to bring attributes that could be ranked but not measured within the purview of statistical analysis. It was "a method for obtaining simple statistical results which has the merit of being applicable to a multitude of objects lying outside the present limits of statistical inquiry."[77] Having recently read Fechner's *Elemente der Psychophysik*, Galton recognized that only a slight increase in inaccuracy would be introduced if the median were substituted for the mean—and it was, for his purposes, far more convenient.[78] He also abandoned the whimsical notion of defining a statistical unit as one-fifth the distance between the mean and the threshold of genius. Instead he adopted the customary measure of the width of distributions used by astronomers, the "probable error" (or median error). This was, in any study based on ordering by rank, much easier to learn than the "dispersion" ($\sqrt{2}$ times the modern standard deviation), which also was occasionally used by contemporaries to measure the width of a distribution.

Using the method of intercomparison, it no longer was necessary to measure directly every individual in a group, and record all pieces of information separately. The presumptive applicability of the error law implied that only two pieces of information need be known in order to characterize the entire distribution. Hence everything necessary could be learned simply by arranging the group in order, beginning with those possessing the lowest degree of the attribute in question and proceeding to the highest. The middlemost individual in the series will then represent its mean—"in at least one of the many senses in which that term may be used."[79] Galton later adopted Cournot's term "median" to describe this middle term, but he was aware from the beginning that expected mean and median were identical so long as the normal law prevailed. He went on to point out that the individuals one-fourth and three-fourths of the way along the curve would represent the probable error of the series. These two values, mean and quartile, were sufficient to characterize or compare populations. Galton used the architectural

[77] Galton, "Intercomparison" (n. 71), p. 33.

[78] Gustav Theodor Fechner, *Elemente der Psychophysik* (2 vols., Leipzig, 2d unaltered ed., 1907), vol. 1, pp. 120-129 and *passim*. Galton referred very favorably to this book in a letter of 1875; see *Galton*, vol. 3B, p. 464.

[79] Galton, "Intercomparison" (n. 71), p. 34.

term "ogive" to designate the curve generated by this procedure. He frequently described its shape by imagining a series of men whose heights are normally distributed arranged in a row in order of stature. The curve that touches the top of each head will be an ogive; it will slope sharply near the ends, and very gradually at the middle.

Although the method of intercomparison proved difficult to use fruitfully in connection with self-characterizations given by prominent men of science, it still had certain noteworthy advantages over ordinary statistical procedures. It was far more convenient for rank data than the technique in *Hereditary Genius.* Since the ogive is completely determined as soon as two points on it are known, an anthropologist could infer the complete distribution of heights in a tribe using only two measurements, if only the barbarian chief could be persuaded to assemble his men in order of stature. Again, if the anthropologist wished to learn the "strength of pull" of these same aborigines, he was no longer obliged to carry around an elaborate device capable of registering continuous magnitudes. Instead, the tribesmen need only be asked to try their strength against two bows of known stiffness. When the percentage that can draw each bow is known, the corresponding ogive will give the distribution of strengths in the entire tribe.[80]

These anthropological procedures, recommended by Galton in 1874 in the British Association handbook *Notes and Queries in Anthropology*, did not exhaust the practical uses of analysis by intercomparison. Galton later conceived a yet more ingenious use of his method of fitting ogives—this one a contribution to one of the most time-honored beneficiaries of probabilistic analysis, the theory of elections. If a legislative body is seeking to reach a decision on some matter subject to a continuous range of choices, such as the allocation of money, the representatives need not haggle endlessly in order to locate the true point of consensus, which is the median. Instead they can simply take a recorded vote on any two amounts that fall within the range of variation. Thus if 60 percent think that at least A pounds should be allocated, while 25 percent favor at least *B* pounds, the ogive passing through these points need only be constructed and the median can be ascertained directly.[81]

The technique of intercomparison reinforced Galton's awareness of the utility of units of probable error, which were indispensable to him when he began the work that led him to the statistical method of cor-

[80] Galton, "President's Address" (n. 38), p. 411.
[81] Galton, "The Median Estimate," *BAAS*, 59 (1899), 638-640.

relation. Still, the ogive was an outgrowth of the error curve. The information it conveyed could equally be expressed in terms of the error distribution, and Galton's belief in its universality was in this sense derived directly from Quetelet. That faith, shared by Maxwell and Francis Edgeworth as well, was instrumental in the most important work in statistical mathematics of the nineteenth century. Quetelet's idea was thus of signal importance. His influence was not wholly due to the weight of his logic. Galton and Maxwell did not simply go beyond him; they also differed with him in some important matters of interpretation. Still, the submergence of local disorder before general principle—Quetelet's version of the invisible hand—gained wide acceptance because the idea of "Order in Apparent Chaos" was as welcome to scientific thinkers in Great Britain and Germany as in Belgium and France. The nature of this conviction was brought out most clearly in Francis Galton's book *Natural Inheritance*.

> I know of scarcely anything so apt to impress the imagination as the wonderful form of cosmic order expressed by the "Law of Frequency of Error." The law would have been personified by the Greeks and deified, if they had known of it. It reigns with serenity and in complete self-effacement amidst the wildest confusion. The huger the mob, and the greater the apparent anarchy, the more perfect is its sway. It is the supreme law of Unreason. Whenever a large sample of chaotic elements are taken in hand and marshalled in the order of their magnitude, an unsuspected and most beautiful form of regularity proves to have been latent all along. The tops of the marshalled row form a flowing curve of invariable proportions; and each element, as it is sorted into place, finds, as it were, a preordained niche, accurately adapted to fit it.[82]

[82] Galton, *Natural Inheritance* (n. 39), p. 66, adapted from "President's Address," *JAI*, 15 (1886), 489-499.

PART THREE

Through the great beneficence of Providence, what is given to be foreseen in the general sphere of masses, escapes us in the confined sphere of individuals; what statistics indicates to be a definite law for the country, or even for the town, cannot be discerned in hearth and home. —VALENTIN SMITH (1853)

What, statistics prove that there are laws of history? Laws? Yes, it proves how mean and disgustingly uniform the masses are: is one to call laws the effect of inertia, stupidity, aping, love and hunger? Well, we will admit it, but with that the following proposition is also sure: so far as there are laws in history, laws are worth nothing and history is worth nothing. —FRIEDRICH NIETZSCHE (1874)

Number is but another name for diversity.
—W. S. JEVONS (1874)

THE SCIENCE OF UNCERTAINTY

The acceptance of indeterminism constitutes one of the most striking changes of modern scientific thought. With few exceptions, scientists and philosophers previous to the late nineteenth century would have agreed with Augustus De Morgan that to say an event occurs by chance is to say that it occurs for no reason at all. It is to speak nonsense, to confuse a mere word with a cause. Mathematical probability, "far from proceeding out of any admission of events happening by chance, is a consequence of the directly opposite belief; for, if preceding occurrences had been perfectly fortuitous, the arrival of one event would furnish no probability whatever for the repetition of the same under similar circumstances."[1] The view of Quetelet and Buckle that the regularities of statistics constituted laws pointed in the same direction. The inability of the social scientist to predict in detail the behavior of individuals was of no more consequence than minute puffs of wind that can alter the outcome of individual coin tosses.

Yet Quetelet's statistical laws seemed to some not wholly satisfactory, even at the time. Statistical regularities, it was argued, prove nothing about the causes of things, or they embrace acts so diverse as to be almost uninterpretable. Most significantly, it was often pointed out, especially by physicians, that statistics is quite incapable of justifying conclusions about individuals. Quetelet sought to escape the force of these objections by treating social physics as a science of society, and not merely of individuals, but only the most dedicated partisans of statistical social science were willing to concede that a mere mass regularity could constitute a natural law. Those less inclined to grant this were roused from their lethargy by the claim that statistical laws had shown human free will to be illusory. The defense of human freedom inspired a wide-ranging revaluation of statistical thought during the late nineteenth century. Statistics came to be seen not as the method of physical science, applied to society, but as a new scientific strategy, more problematical in many respects than the old, but also one with great promise. If the analogies between statistics and physics survived, it was not because of the success

[1] [Augustus de Morgan], "Quetelet on Probabilities," *Quarterly Journal of Education*, 4 (1832), 101-110, p. 102.

of Quetelet's social physics, but because certain crucial areas of physics soon came to be seen by many as merely statistical. Maxwell's radical new argument that the second law of thermodynamics could be no more than a statement of probability adumbrated a new understanding of physics. Significantly, it was put forth as a defense of metaphysical freedom, in opposition both to the alleged statistical disproof of free will and to other contemporary arguments against the possibility of freedom based on energy conservation and brain physiology.

Maxwell's line of thinking led eventually to C. S. Peirce's depiction of a universe of chance and, less directly, to the discovery by quantum physicists of the 1920s that the most fundamental particles of nature exhibit irreducible chance in their movements and interactions. In the aftermath of this stunning new development in physics, the indeterminism of nature has become almost a part of common knowledge. Its interest and significance can hardly be gainsaid. Clearly, chance is recognized as a fundamental aspect of the world in a way that it was not before.

Still, this story is not simply one of a new appreciation of the empire of chance. Randomness first attained real standing in scientific thought not as a source of massive uncertainty, but as a small-scale component of an overarching order. The recognition of chance stemmed not from the weakness of science, but from its strength—or rather, its aggressive imperialism, the drive to extend scientific determinism into a domain that had previously been seen by most as the realm of inscrutable whimsy. Even Quetelet was initially surprised and shocked at the regularities of crime and suicide. In many ways, the drive to find reliable laws of social phenomena has not succeeded. But its failure has been most pronounced not where phenomena are truly random, but where unknown or unmeasurable causes are perpetually shifting the probabilities of events in ways that cannot be predicted. The indeterminism of probability is so reliable and highly structured that randomness seems to disappear from the end result. However great the metaphysical interest of a universe of chance, it is at least equally significant from the prosaic standpoint of the historian that chance has proved no obstacle to scientific prediction and control.[2]

[2] On these matters, see Ian Hacking, *The Taming of Change* (forthcoming, 1988?); also Lorenz Krüger, "Philosophical Arguments for and against Probabilism," in *Prob Rev*.

Chapter Six

STATISTICAL LAW AND HUMAN FREEDOM

Sharp criticism of the pretensions of statistics began with the science itself. Already in the early nineteenth century, the statistical approach was attacked on the ground that mere statistical tables cannot demonstrate causality, or that mathematical probability presupposes (what is unthinkable) the occurrence of events wholly by chance. The intent of these early critics was not to suggest the inadequacy of causal laws in social science, but to reject the scientific validity of statistics. But statistics survived, and when the same arguments were rehearsed later in the nineteenth century, they were meant to imply that these shortcomings of statistics were intrinsic to its subject matter, and not flaws in its method.

The new interpretation of statistics that emerged during the 1860s and 1870s was tied to a view of society in which variation was seen as much more vital. Statistical determinism became untenable precisely when social thinkers who used numbers became unwilling to overlook the diversity of the component individuals in society, and hence denied that regularities in the collective society could justify any particular conclusions about its members. The connection between this interest in human heterogeneity and statistical uncertainty is exemplified by the remarks of a certain French statistician, Bourdin, who stood up at the 1869 meetings of the International Statistical Congress to announce that man "does not exist as a unitary being." Statistical results might be certain, positive, and absolute in the domain of the physical sciences, he conceded, but in respect to society that science could attain to no more than averages. These results, he went on, "undeniably have a certain value," but they are, "like the science itself, essentially variable."[1] Quetelet, *eminence grise* of the Congress and chairman of this particular session, protested that statistical research could yield "admirable laws" for

[1] "Théorie de la statistique et application des donnés statistiques," in *Compte Rendu de la septième session du Congrès international de statistique* (The Hague, 1870), 33-165, pp. 52-53.

society as for nature, but Bourdin spoke for an emerging consensus against which Quetelet's protests were unavailing.

The influence of these social discussions on natural science and philosophy will be seen in chapter 7, where we find them debated again in the more structured context of gas physics. This line of thought culminated, in a sense, in the work of Charles Sanders Peirce, who made the spontaneous production of chance variation an indispensable condition of evolutionary change, whether physical, biological, or social. They also bore fruit in the growing interest in the analysis of variation, and not merely of mean values, evinced by the late-century mathematical statisticians. To be sure, Galton gave little attention to the debates on human freedom, but Francis Edgeworth was closely familiar with them, and Wilhelm Lexis's important work on dispersion can only be understood in the context of this tradition.

THE OPPONENTS OF STATISTICS

Champions of statistics of Quetelet's generation preferred not to view it as different from other sciences, but they did see in the strategy of mass observation a singular virtue that permitted the methods of natural science to be applied to social matters. The great merit of statistics was that it eliminated perturbations by ignoring individuals and letting their unpredictable activities average out. That this was genuinely novel is confirmed by the writings of the leading critics of statistics, most of them positivistically oriented Frenchmen, who rejected precisely this feature of it. The method of averaging, they argued, does not clarify, but confuses by mixing together things that are fundamentally different. Any social science that views the differences among individuals as random, they argued, is irremediably flawed. Auguste Comte, the social scientist, argued that one must analyze carefully in order to establish causes and recognize their heterogeneous effects on different parts of the population. Medical opponents held that because statistical generalizations could not be applied to individuals, they were useless and even immoral.

As early as 1803, before statistics had even become associated specifically with numbers, Jean-Baptiste Say argued vigorously that nothing like a reliable science could be formed from it. It would be absurd, he explained, to base political economy on the miscellaneous numbers and descriptions in terms of which statistics was defined. Statistics was derived from the Latin *status* ("*état, situation*"); it was a highly detailed geography that could do no more than "reveal the state of production and of consumption of one or several nations at a designated point in time (*époque*, or for several successive points in time."[2] This might have some interest, but it was subject to instant obsolescence. Sinclair's great statistical compilation for Scotland, Say remarked, probably ceased to be accurate within a year of its publication. Statistics was the most transitory kind of knowledge, and for that reason Say thought it preposterous to speak of "laws" of mortality or of population derived from statistics. He wrote: "Variable relations are not laws: they change."[3]

Long before Quetelet invented his doctrine of statistical uniformity, which may be seen in part as an answer to objections like this, the prominent Italian statistical writer Melchiorre Gioja had countered Say with the observation that most statistical objects vary slowly, if at all, and hence that its results remain approximately valid for an appreciable period.[4] Evidently the defense of statistics moved Say slightly, for Quetelet was able to procure a letter from him indicating that statistics could contribute to knowledge of society if pursued competently.[5] Fundamentally, however, Say's conception of political economy was ill-suited to offer statistics a place of any significance. Say argued that there are two kinds of facts: general or constant facts, arising from the action of laws of nature in well-defined cases, and particular, or variable, facts, also the result of inviolable laws, but modified by special circumstances that conceal the true underlying relations. Facts of the second sort, which constitute the subject matter of statistics, are like the jets of water in a fountain that can counter the force of gravity and suspend a heavy object in equilibrium. They may be curious and amusing, but they make a poor basis for science. Say viewed Adam Smith's *Wealth of Nations* as a confused mixture of political economy and statistics; his declared in-

[2] Jean-Baptiste Say, *Traité d'économie politique, ou simple exposition de la manière dont se forment, se distribuent, et se consomment les richesses* (2 vols., Paris, 1803), vol. 1, p. v.

[3] Say, "De l'objet et de l'utilité des statistiques," *Revue encyclopédique*, 35 (1827), 529-553, pp. 547-548; also *Cours complet d'économie politique pratique* (2 vols., 2d ed., 1852; repr. Osnabrück, 1966), vol. 2, p. 486; review of Joseph Lowe, "The Present State of England," in *Revue encyclopédique*, 18 (1823), 312-324.

[4] Melchiorre Gioja, *Filosofia della statistica* (1826; Mendrisio, 1839), pp. 7-16; also *Indole, estensione, vantaggi della statistica* (Milan, 1809).

[5] See Quetelet, *Physique sociale, ou essai sur le développement des facultés de l'homme* (2 vols., Brussels, 1869), vol. 1, p. 104; vol. 2, p. 447. The letter was written in 1832.

tention was to secure the foundations for a science of economy based solely on general or reliable facts.[6]

Naturally, Say denied that his was a purely deductive science. An enlightened *idéologue*, he eschewed, in the customary and acceptable manner, that spirit of system which had given rise to scholasticism and to the Cartesian vortices.[7] Political economy could not, however, be based on the grab-bag of facts presented by statistics, but only on selected experimental facts chosen especially for the pristine clarity that arises from the absence of modifying factors and for the availability of complete knowledge about them.[8] Say conceded that economic phenomena are often formidably complex, and he disavowed on behalf of political economy any pretense to explain them completely. For that reason, it seemed all the more absurd to suppose that they explain themselves. The truth must be discovered through exact analysis, and facts only take on meaning when elucidated by reason according to economic principles. Hence political economy is logically prior to statistics, and is essential to raise understanding above the "dangerous empiricism" that "applies the same methods to opposite cases which it believes similar." The truths of political economy need not be tested against undigested facts. For example, it is a law that interest is proportional to risk, and though a particular lender may not know the risk, or some other circumstance may confuse an individual case, the general law remains, "and must recover its whole empire at the moment when the causes of perturbation, themselves the effect of some other general law, cease to act."[9]

At the same time, and for identical reasons, Say denied that political economy could attain greater exactitude through mathematics. The law of supply and demand is universal and invariable, but the price of a bottle of wine cannot be predicted from it, for it will fluctuate with availability of capital, changing expectations and tastes, weather, and a host of other causes. For this reason, experience furnishes no fixed value to which calculation can be applied.[10] Even if the problem of fluctuations could be overcome, most of the numbers presented by statistics are seriously misleading, for they lump together facts of fundamentally dif-

[6] Say, *Traité* (n. 2), pp. iv-ix.
[7] Say, *Traité* (6th ed., 1841; repr. Osnabrück, 1966), vol. 1, pp. 15-16.
[8] Say, *Traité* (2d ed., Paris, 1814), vol. 1, pp. xxv-xxvii; 6th ed., vol. 1, pp. 6-7.
[9] *Ibid.*, 1st ed., vol. 1, pp. viii-ix.
[10] *Ibid.*, 2d ed., vol. 1, pp. xxx-xxxii.

ferent characters. Tables of mortality, for example, may be valid for an average, but are false for both indigent and leisured classes, as every insurance company must soon discover.[11] In short, the numerical results obtained by amassing information about a variety of individuals lack the purity and clarity needed for any genuinely scientific theory of political economy. Statistical results, collected regularly and presented coherently, convey an impression of the condition of a people whose value must not be gainsaid, but which cannot provide the basis of a science.

Since Auguste Comte was deeply impressed by the interpretation of science of the *idéologue* physiologists Bichat, Broussais, and Cabanis, and particularly by their reasons for rejecting mathematics in the study of life, it is not entirely fortuitous that his view of statistics should bear a certain resemblance to that of Say. Comte rejected the use of mathematics in social science, for it was a central truth of positivism that each science must have its own distinctive method and that reduction of sciences is impossible. Beyond that, he was wholly unsympathetic to mathematical probability. Like Destutt de Tracy, he thought Condorcet unjustified, and Laplace and his imitators positively ridiculous, to seek to apply probability to judicial decisions and testimonies. This he regarded as an "abuse . . . of the mathematical spirit," and accordingly he applauded Poinsot's refutation of Poisson which, he thought, relieved him of the obligation to present his own full critique. Comte went so far as to reject the mathematics of probability itself—he was the most prominent scientific writer since d'Alembert to do so. He wrote: "As to the philosophical conception upon which such a doctrine rests, I believe it radically false and tending to lead to the most absurd consequences. . . . It would habitually lead us in practice to reject as numerically improbable (*invraisemblables*) events which will nonetheless take place."[12] Probability had led geometers into "utter delusion," Comte thought, for it "presupposes that the phenomena considered are not subject to law."[13]

Comte's social philosophy was not so incompatible with statistics as

[11] Say, "De l'objet" (n. 3), p. 548. On Say, see Claude Ménard, "Three Forms of Resistance to Statistics: Say, Cournot, Walras," *History of Political Economy*, 12 (1980), 524-541.

[12] Auguste Comte, *Cours de philosophie positive* (6 vols., Paris, 1830-1842), vol. 2 (1835), p. 371; also vol. 4 (1839), pp. 513-515. See also "Appendice, XIV" (1819), in *Revue occidentale philosophique, social, et politique*, 8 (1882), 400-409. For Poinsot rebuttal to Poisson, see *Comptes rendus hebdomadaires des séances de l'Académie des sciences*, 2 (1836), 380, 389.

[13] Comte, *A General View of Positivism* (1848), J. H. Bridges, trans. (New York, 1975), p. 28.

to prevent Quetelet from plagiarizing his general historical conception. Comte had argued from the early 1820s that the object of social science was to learn the natural course of civilization, which, he proposed, could not be redirected by politics, but only smoothed or made more turbulent. Comte thought the secular course of progress to be largely a consequence of advancing knowledge. Quetelet adopted these ideas wholesale, and with them the title of Comte's science, *physique sociale*. Comte was greatly annoyed, observing that "a Belgian savant" had expropriated this designation and misconstrued it as mere statistics, thus obliging him to coin a new title for the same science, sociology.[14] This theft was all the more offensive because Comte was unimpressed by the achievements, actual or potential, of statistics. He meant by social science a study of beliefs and of institutions in history, not the construction of a mythical average man. Comte held that the current state of society involved a confusion of the three fundamental stages—theological, metaphysical, and positive—so that mass statistics could only smear together what ought to be kept distinct. A "genuine and precise statistics of the social body," he thought, was impossible unless based on study of the historical development of human institutions.[15]

There were a few other social scientists during the nineteenth century who entirely rejected this new science of statistics. Paul von Lilienfeld argued in 1873 that the law of large numbers, the foundation of statistical science, was no law at all, since it did not act with necessity, and hence that even the most interesting statistical tables presented not "necessary natural laws" but only "accidental circumstances."[16] This argument, however, was less compelling against statistics as social science than it was against other applications of statistical method, for it could be answered by pointing out that the object of statistics was not the individual, but society. Statistics maintained much of its persuasiveness so long as observers were willing to admit the existence of a coherent social

[14] Comte, *Cours* (n. 12), vol. 4, pp. 7, 252. See Julien Freund, "Quetelet et Auguste Comte," Académie royale de Belgique, *Adolphe Quetelet*, 1796-1874 *(Mémorial)* (4 vols., Brussels, 1974), vol. 4, pp. 46-64.

[15] Comte, "Plan des travaux scientifique nécessaires pour réorganiser la société" (1822), in *Opuscules de philosophie sociale*, 1819-1828 (Paris, 1883), 60-180, p. 130. Comte's disciples were less hostile to statistics, though they did not use them. See G. Wyrouboff, "De la méthode dans la statistique," a review of Quetelet's *Physique sociale*, in *La philosophie positive*, 6 (1870), 23-43.

[16] Paul von Lilienfeld, *Gedanken über die Socialwissenschaft der Zukunft* (1873; 3 vols., Berlin, 1901), vol. 3, p. 10; also vol. 1, pp. 387-388. Lilienfeld conceded the usefulness of averages in political economy and medicine (vol. 1, pp. 101-102).

domain standing above that of individuals. Indeed, the regularities of statistics had helped to define that domain.

The statistical method was much more difficult to defend in relation to medicine, particularly diagnostics and therapeutics, and opposition to it was more in evidence there than in any other field. For physicians, the direct object of concern was quite naturally the sick individual, and it was far from clear that the mean result from some large number of assorted trials in a hospital provided an appropriate basis for treating that individual. Blind use of averages threatened to supplant that *je ne sais quoi*, medical tact, upon which physicians prided themselves. Accordingly, the so-called numerical method in medicine was viewed by many with suspicion, and subjected on occasion to vitriolic attacks.

Medical statistics, under other names, was at least as old as political arithmetic, and the numerical method was applied in a variety of instances even to therapeutics during the late eighteenth century.[17] Philippe Pinel, who admired Condorcet's work, championed the numerical method during the first decades of the nineteenth century,[18] but the collection of numbers in order to guide the physician in his practice on individual patients was largely a product of the 1820s and 1830s. The name with which that method became most strongly associated was that of Pierre Charles Alexandre Louis. Louis' most famous and controversial work was a quantitative study of the effects of bloodletting on pneumonia, published in 1828. His results were equivocal—early bleeding seemed to lessen the duration of the illness slightly, but to have far less effect than had generally been thought—but his work became notorious. This was due partly to the circumstance that his results were interpreted as a challenge to conventional practice. More generally, however, the numerical method was put forward as a systematic technique for improving medical treatment at a time of diagnostic triumph and therapeutic crisis, and hence much more was at stake than any particular remedy.

Louis exalted his numerical method as "simultaneously natural and rigorous."[19] It offered an unprecedented level of precision, and was fully

[17] See Ulrich Tröhler, *Quantification in British Medicine and Surgery, 1750-1830, with Special Reference to its Introduction into Therapeutics* (Ph.D. Dissertation, University of London, 1979).

[18] See Erwin Ackerknecht, *Medicine at the Paris Hospital, 1789-1848* (Baltimore, 1967), pp. 47-48.

[19] P.C.A. Louis, *Recherches sur les effets de la saignée dans quelques maladies inflammatoires et sur l'action de l'émétique et des vésicatoires dans la pneumonie* (Paris, 1835), p. 70.

consistent with accepted medical standards. To those who objected that individual cases were too diverse to be grouped together statistically, Louis responded that diseases can be classified by type, even though no two cases are identical, and that if his opponents chose to deny all resemblance among separate cases, then nothing would be left in medicine but pure individuality. Then, there would be no prospect of attaining general knowledge by any method. In fact, "it is precisely on account of the impossibility of comprehending each case with, as it were, mathematical precision, that it becomes necessary to count; since the errors, the inevitable errors, are the same for two groups of diseases treated by different procedures, these errors are compensated and can be neglected, without sensibly altering the exactitude of the results."[20] The numerical method captures what every physician does when he learns from experience, Louis wrote, and he had been led to it naturally, even involuntarily. Still, the difference between those who count and those who are content saying more or less "is the difference of truth from error, of something clear and genuinely scientific from something vague and without value."[21] Louis acknowledged that his numbers were too small, and appealed for a more systematic collection of them. His student Danvin argued that the only impediment to the attainment of certainty with the numerical method was the unavailability of sufficiently large numbers, and maintained that once the number of cases reaches 500 the resulting proportions are thereafter fixed and exact.[22]

In the view of opponents of the numerical method in medicine, of whom there was no shortage, Louis' strategy was based on a false analogy between medicine, whose facts are complex, variable, and often hidden, and physics, where they are always simple and uniform. Risueño d'Amador, who launched a debate in the Académie Royale de Médecine in 1836 with a memoir on numbers and medicine, denounced the use of probability in therapeutics as anti-scientific. Among his arguments was the assertion made familiar by Say that statistics is mere description, valid only for the fleeting moment. His respect for the mathematics of probability was on the same level as Comte's; he called it skepticism embracing empiricism, and remarked that to resort to prob-

[20] *Ibid.*, p. 76.

[21] *Ibid.*, p. 85; *Recherches anatomiques, pathologiques, et thérapeutiques sur la phthisie* (Paris, 2d ed., 1843), p. xx.

[22] B. Danvin, *De La méthode numérique et de ses avantages dans l'étude de la médecine* (Paris, 1831), pp. 30-31.

ability is to appeal to chance and to give up the possibility of certitude. Amador further claimed that the repetition of events in the past proves nothing about the future. When probability is applied "to real facts in the physical and moral world, it becomes either useless or illusory."[23]

If this inflexible and mechanical calculus were made the basis of medicine, Amador announced, healing would cease to be an art and become a lottery. The numerical method denies the variability of medical facts, which can only be fully appreciated through induction and medical intuition. Since the enumerators lump together disparate cases, their aim is clearly "not to cure this or that disease, but to cure the most possible out of a certain number. This problem is essentially anti-medical." After all, since "the law of the majority has no authority over refractory cases," the physician must either ignore the results of statistics for these variant individuals or condemn them to death.[24] Even strong advocates of the numerical method, like the Englishman William Guy, admitted that its application to individual cases was problematical.[25] Risueño d'Amador argued, somewhat oddly, that it was the task of nature to conserve the species, and that since medicine had no desire to usurp this function, it could not be judged by results *en masse*.[26]

Louis pleaded in defense of his method that medical statistics could demonstrate variability as well as uniformity and that its object was not necessarily "la détermination d'un homme moyen ou imaginaire."[27] The numerical method was by no means banished, although the level of enthusiasm had begun to wane by 1850; the cumulative results were sufficient by 1865 to fill a massive German handbook compiled by Friedrich Oesterlen.[28] Most writers on the numerical method had little understanding of probability, and tiny samples were common. For this they were sharply criticized. In addition, resistance to the very idea of basing medicine on numbers, however large, remained strong. Auguste Comte contended that the numerical method was merely "absolute empiricism, masquerading under frivolous mathematical appearances," and could only lead to a "profound, direct degeneration of the medical

[23] Risueño d'Amador, "Mémoire sur le calcul des probabilités appliqué à la médecine," *Bulletin de l'Académie Royale de Médecine*, 1 (1836), 622-680, p. 624. This was followed by "Discussions sur la statistique médicale," pp. 684-806.

[24] *Ibid.*, pp. 634-635.

[25] William Augustus Guy, "On the Value of the Numerical Method as applied to Science, but especially to Physiology and Medicine," *JRSS*, 2 (1839), 25-47, p. 40.

[26] "Discussions" (n. 23), p. 805.

[27] *Ibid.*, p. 741.

[28] Fr. Oesterlen, *Handbuch der medicinischen Statistik* (Tübingen, 1865).

art"—to the trial of therapeutic procedures at random.[29] During the 1840s and 1850s, an alternative to statistics for improving therapeutics began to gain increasing support. This was experimental physiology, a field in which it was and has remained customary to emphasize control over all aspects of the experiment rather than the averaging out of errors. The reductionist Emil Du Bois-Reymond was forthright and vocal in opposing numerical tables and stressing complete determinism.[30] The most influential critique of medical statistics, however, came from the positivist Claude Bernard.

Bernard did not, like Comte, reject the applicability of mathematics to biological phenomena. The precision of mathematics was always desirable, he wrote, for the discovery of exact relations was the proper aim of all science. But the profitable employment of mathematics presupposed, in his view, that all relevant phenomena were completely understood, and on this basis he maintained that the study of vital phenomena was not yet ready for mathematics. The complexity of vital phenomena constituted the central problem of physiology, but no solution to it could be attained by averaging variety out. Instead, the physician must learn to understand in detail.

Bernard asserted repeatedly that physiological functions are not homogeneous, either over diverse individuals or over time for a single individual. There is, he held, no average pulse; there is only a pulse when resting, or exercising, or eating. It was the height of folly to seek, as some unnamed physiologist allegedly had, the "average European urine" by collecting from a railroad station urinal in a great city. Averages "confuse while aiming to unify and distort while aiming to simplify," and consequently, the physiologist "must never make average descriptions of experiments, because the true relations of phenomena disappear in the average."[31] This was most especially the case in medical therapeutics. "True enough, statistics can tell you if an illness is more serious than another; you can tell your patient that, of every hundred such cases, eighty are cured . . . but that will scarcely move him. What he wants to know is whether he is numbered among those who are cured."[32]

[29] Comte, *Cours* (n. 12), vol. 3 (1838), p. 420.

[30] See Brigitte Lohff, "Emil Du Bois-Reymonds Theorie des Experiments," in Gunther Mann, ed., *Naturwissen und Erkenntnis im 19. Jahrhundert: Emil Du Bois-Reymond* (Hildesheim, 1981), 117-128, p. 122.

[31] Claude Bernard, *An Introduction to the Study of Experimental Medicine* (1865), H. C. Greene, trans. (New York, 1957), p. 135.

[32] Bernard, *Principes de médecine expérimentale*, L. Delhoume, ed. (Paris, 1947), p. 67, quoted in William Coleman, "Neither Empiricism nor Probability: The Experimental Ap-

Bernard's point was not, as the physician Double argued, that statistics "strip man of his individuality,"[33] and he certainly did not agree with Cruveilhier that general results in therapeutics are unattainable "because in medicine there is nothing but individuals."[34] Nor did he favor the doctrine that the imperfect knowledge gleaned through observation should be supplemented with some obscure medical tact. Bernard's modest aim was to find through completely controlled experimental manipulations the general laws that governed vital phenomena deterministically, without exception. He thought it absurd to label statistical results laws, when statistics had been applied precisely because the constituent facts were not fully comparable. The numerical method gives only probabilities, he insisted, whereas the only proper aim of scientific medicine is certainty. It cannot rest content with relative proportions, but must know why, when eighty are cured, twenty still die, so that knowledge will be complete and medicine reliable. Bernard proclaimed again and again the need for experimental determinism; "the absolute principle of experimental science is conscious and necessary determinism in the conditions of phenomena."[35] Physicians, he wrote, "have nothing to do with what is called the law of large numbers, a law which, according to a great mathematician's expression, is always true in general and false in particular. This amounts to saying that the law of large numbers never teaches us anything about any particular case. What a physician needs to know is whether his patient will recover, and only the search for scientific determinism can lead to this knowledge."[36]

Bernard doubtless underestimated what could be accomplished through careful controlled trials performed on large numbers of patients using drugs and procedures whose mode of action was imperfectly understood, if at all, and his denunciation of statistics has become a considerable embarrassment to his latter-day medical admirers. But physiologists, in fact, still have little use for statistics, which they rarely present except out of a sense of duty, and then, quite often, badly. Those

proach," in Michael Heidelberger et al., eds., *Probability since 1800: Interdisciplinary Studies of Scientific Development, Report Wissenschaftsforschung*, 25 (Bielefeld, 1983), 275-286, p. 282.

[33] Quoted in Joseph Schiller, "Claude Bernard et la statistique," *Archives internationales d'histoire des sciences*, 17 (1963), 405-418, p. 408.

[34] Quoted in Paul Delaunay, "Les doctrines médicales au début du XIXe siècle: Louis et la méthode numérique," in E. A. Underwood, ed., *Science, Medicine, and History* (2 vols., London, 1953), vol. 2, pp. 321-330, p. 326.

[35] Bernard, *Introduction* (n. 31), p. 53.

[36] *Ibid.*, p. 138.

who, like Bernard, seek universally valid principles but not, in the first instance, exact quantitative ones, can simply ignore statistics. Bernard's critique should be recognized as among the most insightful commentaries on statistical reasoning up to his time, impressive for its clear formulation of epistemological issues if not for its tolerance of alternative approaches or its understanding of probability mathematics. Bernard understood perfectly that adopting a statistical approach implied acceptance of a considerable domain of ignorance—the lumping together of heterogeneous material so as to establish generalizations that could not be applied with certainty to individuals.

Bernard thought this expedient entirely unnecessary, as did Say and Comte. Certainly none of these positivistically inclined French opponents of statistics believed that the imperfections of that science had any implications for a proper understanding of nature or society. Thus, although their role in the development of statistical thinking was significant, it is still more important to look at those critics of statistics who thought detailed knowledge of individuals was inaccessible and who thus accepted the statistical study of mass phenomena, with all its limitations, as the best available approach to certain sciences. For these writers, the reinterpretation of statistics required a new understanding of scientific law, either as an expression of the possibilities of human knowledge, or perhaps even as a characterization of the real world.

STATISTICS AND FREE WILL

As early as 1819, Thomas Young remarked that the surprising regularity of phenomena like the dead letters in the Paris postal system implied no "mysterious fatality" and that Laplace among others had interpreted them "as implying something approaching more nearly to constancy in the original causes of the events, than there is any just reason for inferring from them."[37] The standard interpretation of statistical regularity, however, remained precisely the one that Young opposed. Quetelet's statistical laws were not only seen as indicating some constancy in underlying causes, but came even to be associated with an ill-defined determinism, if not mysterious fatality, and it was seriously debated for

[37] Thomas Young, "Remarks on the probabilities of error in physical observations, and on the density of the earth, considered especially with regard to the reduction of experiments on the pendulums," *Phil. Trans*, 109 (1819), 70-95, p. 71.

some time whether statistics might be inconsistent with human freedom. Although the great champions of statistics, including Quetelet and Buckle, were aware that its success depended on a strategy of ignoring individual phenomena, it was almost universally associated with an impressive extension of the domain of knowledge, and not with its limitations. The more skeptical view of statistics that gained influence during the 1860s and 1870s was largely a reaction to the extravagance of its pretensions.

The possibility that statistical regularity might be seen as inconsistent with human freedom troubled Quetelet almost from the beginning of his career. Even before he invented his *mécanique sociale*, it appears, he was subjected to sharp criticism for his attempt to quantify the human sciences. Already in 1829 he complained of accusations that his researches were materialistic and that he viewed states as cadavers, or that his observations on the constancy of crime led to a dangerous fatalism. Villermé, evidently, had been similarly reproached. In a letter of 1829 to Quetelet, he observed that natality is an economic phenomenon—the production of people (births) being determined by their expenditure (deaths)—and then remarked: "You see that if you are reproached for viewing societies as cadavers, I can equally be reproached for viewing them as merchandise."[38]

Quetelet's 1848 paper seeking to reconcile statistical regularity with free will was criticized in the Belgian Academy by two conservative scholars. P. De Decker argued that statistical regularity must be attributed to divine will, not laws of probability, while M. Van Meenen maintained that moral statistics is a contradiction in terms, since nothing can be known by men about the human soul.[39] Tom Gradgrind's remarks in *Hard Times*, quoted at the head of Part One, reveal Charles Dickens's view of the effects of such teachings on young minds. From William Whewell, Quetelet's ideas brought implied criticism in the form of displaced congratulations. No subject, wrote Whewell, could be better suited to reveal the relation between moral statistics and free will than the one Quetelet had chosen for his 1848 paper: marriage. Whewell clearly was more skeptical of the conclusions that might be drawn

[38] Letter, 25 April 1829, Villermé to Quetelet, cahier 2560, *AQP*; also Quetelet, "Recherches statistiques sur le Royaume des Pays-Bas," *NMAB*, 5 (1829), pp. v, 33.

[39] "De l'influence de libre arbitre de l'homme sur les faits sociaux," *NMAB*, 21 (1848), by De Decker, pp. 69-92, and Van Meenen, pp. 93-112. See also a review by Moritz Wilhelm Drobisch, "Moralische Statistik," *Leipziger Repertorium der deutschen und ausländischen Literatur*, Jg. 1849, vol. 2, pp. 28-39.

from the statistics of crime. Only external influences, not the internal process of the mind, can be given an exact numerical form through statistics, wrote Whewell. "Your statistical results are highly valuable to the legislator, but they cannot guide the moralist. A crime is no less a crime, because it is committed at the age of greater criminality, or in the month of more frequent transgressions."[40]

Objections like these, however, were not often heard. Until 1857, most readers found the regularity of moral statistics somewhat surprising in view of the evident freedom of the human will, but only slightly, if at all, threatening. Thus George Boole, who followed Scottish Common Sense in regarding metaphysical freedom as a self-evident fact of consciousness, wrote: "The consideration of human free-agency would seem at first sight to preclude the idea that the movements of the social system should ever manifest that character of orderly evolution which we are prepared to expect under the reign of a physical necessity. Yet already do the researches of the statist reveal to us facts at variance with such an anticipation."[41]

The forceful arguments in Buckle's 1847 *History of Civilization* destroyed this spirit of statistical moderation. The reception of that book should suffice to dissolve a modern reader's residual inclination to suppose that the expansion of statistics was associated in the first instance with an increasing tendency to philosophical probabilism. On the contrary, statistics was intended to expand the domain of exact scientific certainty to include man and society as well as physical and organic nature. Buckle's book was an enormous success, reaching a popular as well as an intellectual audience. The fear he provoked that a new and all-embracing determinism had at last succeeded in excluding the possibility of divine or human freedom extended from America and Britain to Germany and even to Dostoevsky in Russia, whose underground man complains about statistics and then about Buckle.[42] It is far from clear that Darwin or Comte was discussed with greater urgency during the 1860s and 1870s.

Friends of statistics were obliged to take note of the controversy in which their science was suddenly engulfed, for the charges directed at them threatened to undermine the legitimacy of their science. At the

[40] Whewell to Quetelet, 7 Oct. 1847, cahier 2644, AQP.

[41] Boole, *An Investigation of the Laws of Thought* (1854; New York repr., 1958), p. 20.

[42] See Fyodor Dostoevsky, *Notes from Underground* (1864), Mirra Ginsburg, trans. (New York, 1981), pp. 23, 25.

1860 session of the International Statistical Congress, held in London, these accusations were summarized in their most extreme form by the Prince Consort, a former pupil and continuing correspondent of Quetelet's, and a strong advocate of statistics, who served as president of the Congress. It had been alleged, he explained, that the numerical study of society

> leads necessarily to Pantheism and the destruction of true religion, as it deprives, in man's estimation the Almighty of His power of free self-determination, making His world a mere machine, working according to a general, pre-arranged scheme, the parts of which are capable of mathematical measurement, and the scheme itself of numerical expression, that it leads to fatalism, and therefore deprives man of his dignity, of his virtue and morality, as it would prove him to be a mere wheel in this machine, incapable of exercising a free choice of action, but predestined to fulfil a given task and to run a prescribed course, whether for good or for evil.[43]

In the wake of this challenge to accepted views, it became necessary to think more critically about the character and implications of statistical laws. The urgency of the task was depicted with unmatched melodrama by William Cyples, author of an essay in *Cornhill Magazine* on the "Morality of the Doctrine of Averages." Statistics "brings the human heart to a standstill," he wrote.

> When the choice lies betwixt this wholesale ruin of the human world, and concluding that some few persons, led away by an enthusiasm for statistics, have applied logic to a matter outside the limits of proof, an appeal lies to common sense; and a protest may safely be entered against this modern superstition of arithmetic, which, if acquiesced in, would seem to threaten mankind with a later and worse blight than any it has yet suffered,—that not so much of a fixed destiny, as of a fate expressive in decimal fractions, falling upon us, not personally, but in averages.

Various possibilities were put forward to deflect this argument from statistics against human freedom and responsibility. One of the more creative was suggested by the same William Cyples. "If a prisoner standing in the dock pleaded in bar of punishment, that the commission of his

[43] "The Address of the Prince Consort on opening as President the Fourth Session of the International Statistical Congress," *JRSS*, 23 (1860), 280.

crime was necessary for maintenance of the statistics, it would be perfectly logical . . . for the judge to urge that, in the same way, it was requisite for his ten years' penal servitude to be ordered, to prevent an irregularity in the annual tables of sentences."[44] Several of Buckle's reviewers challenged the supposition that free will ought to express itself in the form of disorder. One, at least, maintained that the lawlike regularities of statistics do not contradict freedom, but involve merely "an observed uniformity of succession" in the phenomena whose cause remains unexplained.[45]

British and American critics of Buckle also found it attractive to stress the negative side of what had appeared in Quetelet and Buckle as the principal justification for adopting a statistical approach. Statistics embodied the long view; it was the expedient through which the confusion and unpredictability of individual phenomena gave way to splendid regularity and order in the mass. But a mass regularity composed wholly of individual idiosyncrasies could hardly stand as proof of some iron necessity regulating the events whose diversity had been evaded by the very act of averaging. Lord Acton, who opposed Buckle with the claim that history was really about individuals, and had nothing to do with statistics, argued that "it is only to men as persons that free-will belongs: look at them in masses, and they become machines; with their personality you abstract their freedom."[46] T. C. Sanders, more sympathetically, observed that Buckle's statistical viewpoint was fully appropriate for social history, but could not justify any conclusions about the freedom of individuals.[47]

These issues were treated more clearly and comprehensively by Fitzjames Stephen, writing in the *Edinburgh Review* in 1958. According to Stephen, Buckle had erred in assuming that free will must act irregularly, or that the role of divine providence must be to cause deviations from natural laws. At the same time he asserted that laws of nature pertain only to the minds which conceive them, not to the "facts which they are supposed to govern."[48] Most pertinently, he insisted that the

[44] [William Cyples], "Morality of the Doctrine of Averages," *Cornhill Magazine*, 10 (1864), 218-224, p. 224.

[45] Anon., "Statistical Averages and Human Actions," *Temple Bar*, 15 (1865), 495-504.

[46] Lord Acton, "Mr. Buckle's Thesis and Method" (1858), in William McNeill, ed., *Essays in the Liberal Interpretation of History* (Chicago, 1967), 3-21, p. 8.

[47] T. C. Sanders, "Buckle's History of Civilization in England," *Fraser's Magazine*, 56 (1857), 409-424; Sanders, *Saturday Review*, 4 (1857), 38-40.

[48] [James Fitzjames Stephen], "Buckle's History of Civilization in England," *Edinburgh Review*, 107 (1858), 465-512, p. 483. See also his "The Study of History," *Cornhill Magazine*, 3 (1861), 666-680; 4 (1861), 25-41, p. 27.

general fact of statistical regularity could in no way justify claims about the behavior of any constituent individual. He wrote:

> If free will exists at all, it cannot on any hypothesis introduce more confusion into statistical calculations than any other cause of action, of the operation and nature of which we are ignorant, but it is the very object of the science to which Mr. Buckle refers to enable us to make general assertions about the effects of such causes, and it is the strangest perversion of its doctrines to infer from them that unknown causes do not exist. If the question whether one man should or should not murder another had to be decided by a throw of the dice, the uncertainty whether the murder would take place, would be quite as great as it could be if the question depended on free will. With respect to the dice, we can foretel to a nicety how many sixes and aces will be thrown in ten thousand throws, but we are absolutely unable to foretel what any particular throw will be, nor does our certainty as to the general result help us in the least degree to a conclusion as to the particular one. This is surely an exact parallel to the case of human action. We can foretel its aggregate, but we cannot foretel its individual results.[49]

Buckle's counterposition of free will and statistics was less controversial in France and Belgium, perhaps because his book attracted far less attention there. Quetelet's grand pronouncements on *l'homme moyen* and *physique sociale* had also been received more coolly in France than in Great Britain. Quetelet himself was pleased at the publicity he received from Buckle's book, and he quoted at length from the relevant chapter in his book *Physique sociale* of 1869. The statistical determinism did not offend him. When John Herschel complained to him of the discredit cast on statistics by Buckle's unjustified remarks on free will, Quetelet responded placidly that his good friend had analyzed these ideas "*trop de rigueur.*" [50]

By contrast, in Germany, where Quetelet had been little noticed except by specialists before 1860, Buckle's book elevated both men to prominence and inspired a controversy over free will and statistics at least as vehement as that in Britain. Oddly enough, Buckle was translated by Arnold Ruge, the aging Young Hegelian, by then an exile in

[49] Stephen, "Buckle's History" (n. 48), p. 473.

[50] Quetelet, "Notice sur Sir John Frédéric William Herschel," *AOB*, 39 (1872), 153-197, pp. 185-186; also cahier 1289, *AOP*.

England, who had decided to make the translation and exposition of Buckle to his countrymen his life work. This implied no rejection of his former commitments, he explained; Buckle's history captured the true spirit of Hegelian philosophy. Its author, he went on, should not be too harshly judged for his inability to comprehend the true concept of freedom, and it would be a mistake to speak "as if Buckle were a materialist, when he is really only an Englishman."[51] Buckle's remarkable success in Germany was likely due to the fortunate appearance of his book during the ascendancy of liberalism there—for which, according to one critic, it helped set the tone. Buckle was received with less enthusiasm by academic historians, like Droysen, who tended to find his book most noteworthy for its *Dilettantismus* and unphilosophical empiricism.[52] Nevertheless, it was principally he who made statistics a philosophical problem, in Germany as in England, and his history inspired a number of scholars to go back and investigate the writings of Quetelet.

Among the earliest and most prominent of these was Adolph Wagner, who was also among the few to express an unreservedly favorable opinion of Quetelet's and Buckle's contributions. Wagner was the son of a holistically inclined Göttingen biologist, but though he dedicated his book on statistics to his father, he preferred to interpret science in terms of rigid natural laws. His field was economics, and as a young man he embraced the deductive approach to that science along with free-market liberalism. Similarly, he took Mill's doctrine of universal causation as the point of departure for statistics, observing that while this postulate might seem self-evident "its significance immediately becomes apparent once we recollect that we so often explain our own actions and events in our lives in a diametrically opposite manner."[53]

However great his admiration of statistics, Wagner was unwilling to allow that the regularities revealed by statistics were themselves natural laws. He challenged the coherence of the law of large numbers, insisting

[51] Arnold Ruge, "Ueber Heinrich Thomas Buckle, und zur zweiten Auflage" (1864), in H. T. Buckle, *Geschichte der Civilisation in England* (2 vols., Leipzig, 7th ed., 1901), pp. xvi-xvii. On Ruge's ambition to translate Buckle, see the letters of another young Hegelian in English exile, Karl Marx to Friedrich Engels, 18 June 1862, in *Briefwechsel, Marx-Engels Gesamtausgabe (MEGA)*, vol. 3, sec. III ("Glashütten in Taunus," 1970), p. 78.

[52] Joh. Droysen, "Die Erhebung der Geschichte zum Rang einer Wissenschaft," *Historische Zeitschrift*, 9 (1863), 1-22. On German liberalism, see James J. Sheehan, *German Liberalism in the Nineteenth Century* (Chicago, 1978); on Buckle and the German liberals, see G. F. Knapp, *Aus der Jugend eines deutschen Gelehrten* (Stuttgart, 1927), p. 155.

[53] Adolph Wagner, "Statistik," in J. C. Bluntschli and K. Brater, eds., *Deutsches Staats-Wörterbuch* (11 vols., Stuttgart and Leipzig, 1867), vol. 10, 400-481, p. 457.

that no law can apply to the mass which does not also pertain to the individuals. Wagner was nonetheless persuaded that mass regularity indicated the existence of genuine laws acting on every individual, and he accounted for their failure to express themselves uniformly by pointing to the disturbing forces which often cancel their effects. He believed that statisticians should not rest content with the demonstration of regularities, but decompose and recombine data in order to demonstrate law-likeness by finding actual causes. The fluctuation of crimes, suicides, and marriages by season or with grain prices seemed to Wagner the best statistical argument against freedom of the will, though he thought that their annual regularity by itself was sufficient to reject with high probability the idea that these events could be produced by free will, since the will is inherently diverse when unconditioned by the constant causes acting in society. Wagner attracted the most attention with his fable of a land where an autocrat decreed at the beginning of every year the number of marriages in the different age groups, suicide by age, sex, profession, and weapon, and crimes of all sorts, so that virtually every aspect of life was controlled. It is obvious, he pointed out, that no state has the power to accomplish all this—yet what "could never be effected artificially through human will and authority is executed by itself in a remarkable manner due to the natural organization of human society."[54]

Despite his enthusiasm for Buckle and Quetelet, Wagner reached no firm conclusion as to the relations between statistics and free will. He acknowledged the existence of subjective feelings of moral responsibility, and while he felt that Quetelet and others had not been successful in their efforts to get beneath this apparent contradiction with the lawlike regularity of phenomena, Wagner thought a successful reconciliation might yet be possible. Others were less circumspect. The materialist J. C. Fischer published a tract against free will in 1859, including all the time-honored disproofs of human freedom and attributing special significance to the results of statistics. The argument that society can be ruled by laws even as its members retain a certain measure of freedom he dismissed with contempt. One might as plausibly suppose that a congregation of blind men could form a society with vision, or that atoms

[54] Wagner, *Die Gesetzmässigkeit in den scheinbar willkührlichen menschlichen Handlungen vom Standpunkt der Statistik* (Hamburg, 1864), p. 46. See also his note 22 on p. 54, the essay on pp. 63-80, and two reviews he wrote for ZGSW, 21 (1865), 273-291, p. 280, and 36 (1880), 189-203, p. 192, in the latter of which he retreats somewhat from this position.

are unaffected by force even though macroscopic bodies follow rigidly the laws of gravitational attraction.[55]

No argument against free will was likely to win unreserved popular acclaim, and the one based on statistics was no exception. Its influence was predominantly negative, for it was routinely considered, and then rejected, in late nineteenth-century German discussions of free will and determinism. The argument was, to say the least, not without its flaws, but the very imprecision of its logic served to hinder the emergence of an accepted, sharply defined refutation. The most common opposing argument had already been implied in an odd passage by Quetelet, when he noted that the enlightened will tended to combat all perturbing forces and thus to maintain a state of equilibrium more perfect even than that of nature. Wagner's title highlighted the inconsistency of statistics with caprice (*Willkühr*) rather than will (*Wille*). In the Kantian tradition, the enlightened will was unambiguously a source of order, and the Leipzig philosopher Moritz Wilhelm Drobisch placed great emphasis on this aspect in his influential book on moral statistics and human freedom. Gustav Schmoller maintained that the stability of the will provided the only possible explanation for statistical regularity, since material factors could never exert a causative influence in the domain of spirit. Carl Göring argued that the systematic variation of marriages with grain prices was no argument for the powerlessness of will, but the opposite; it shows that human rationality can sometimes overcome blind impulse. Hermann Siebeck made the point that willful acts do not cancel to yield the results of moral statistics, but sum to them.[56]

That mass regularity need not imply causal necessity was also a widespread argument among German statistical critics, who developed it much more fully than any other group. One variant of this approach was introduced by the philosopher Hermann Lotze, who sought to refute what he took to be the core of Buckle's position, that a fixed quantity of evil is present in society, and must somehow be expressed. If this constancy of evil were genuine, Lotze thought, it would indeed contradict

[55] J. C. Fischer, *Die Freiheit des menschlichen Willens und die Einheit der Naturgesetze* (Leipzig, 2d ed., 1871), pp. 235-236.

[56] See M. W. Drobisch, *Die moralische Statistik und die menschliche Willensfreiheit: Eine Untersuchung* (Leipzig, 1867); Gustav Schmoller, "Die neueren Ansichten über Bevölkerungs- und Moralstatistik" (1869) in *Zur Litteraturgeschichte der Staats- und Sozialwissenschaften* (Leipzig, 1888), pp. 172-203; Carl Göring, *Ueber die menschliche Freiheit und Zurechnungsfähigkeit. Eine kritische Untersuchung* (Leipzig, 1876), p. 128; Hermann Siebeck, "Das Verhältniss des Einzelwillens zur Gesammtheit im Lichte der Moralstatistik," *Jbb*, 33 (1879), 347-370. See also C. Schaarschmidt, "Zur Widerlegung des Determinismus," *Philosophische Monatsschrift*, 20 (1884), 193-218, pp. 213-214.

free will, implying that a thief, for example, "is not free in reference to his decision to steal, but only in whether to do so on horseback or on foot."[57] Lotze held, however, that crimes of the same legal category are of very different ethical worth, so that the regularity obtained by lumping together hundreds of such acts can indicate nothing about the measure of evil in a society, and hence cannot be explained in terms of any constancy of causes, but remains a great mystery.

This view, which would seem to deny that enumeration of moral acts has any value or legitimacy, was not widely adopted by German statisticians. They did, however, insist on a fundamental distinction between those who committed and those who abstained from crime. That is, they rejected Quetelet's ploy of converting crime from a series of discrete individual acts into a continuous social phenomenon through the assignment of penchants. The Austrian Leopold Neumann reflected that the wide diversity of responses to reasonably uniform circumstances—the fact that some placed under given conditions commit crimes, and others do not—confirmed the existence of a large measure of personal autonomy.[58] Drobisch held that criminal statistics do not reveal tendencies shared by a whole population, but only those of a small minority, so that no inference is possible from an all-embracing average to the traits of individuals. He wrote: "Only through a great failure of understanding can the mathematical fiction of an average man . . . be elaborated as if all individuals . . . possess a real part of whatever obtains for this average person."[59]

THE SCIENCE OF DIVERSITY

Drobisch's argument against determinism, however, opens up a much wider statistical issue, the extent to which statistics presupposes homo-

[57] Hermann Lotze, *Mikrokosmus: Ideen zur Naturgeschichte und Geschichte der Menschheit: Versuch einer Anthropologie* (3 vols., Leipzig, 1856-58-64), vol. 3, p. 78; also *Grundzüge der praktischen Philosophie: Diktate aus den Vorlesungen* (Leipzig, 1884), pp. 24-28. This argument was repeated approvingly by Friedrich Albert Lange, *Geschichte des Materialismus* (2 vols., Iserlohn, 1866), vol. 2, pp. 479-480. It was challenged on the ground that deviations of moral worth would average out by Johannes Wahn, "Kritik der Lehre Lotzes von der menschlichen Wahlfreiheit," *Zeitschrift für Philosophie und philosophische Kritik*, 94 (1888), 88-141, pp. 111ff.

[58] L. N. [Leopold Neumann], "Zur Moralstatistik," *Preussische Jahrbücher*, 27 (1871), 223-247, p. 245; also Richard Wahle, "Eine Verteidigung der Willensfreiheit," *Zeitschrift für Philosophie und philosophische Kritik*, 92 (1888), 1-64, pp. 35-36.

[59] Drobisch, *Die moralische Statistik* (n. 56), p. 18; also "Moralische Statistik" (n. 39).

geneity in the phenomena counted. This issue was widely discussed, and not exclusively in response to Buckle's claims or even in reference to the matter of human freedom. As early as 1831, an appreciative reviewer of Quetelet's first papers on *mécanique sociale* cautioned that a mean value taken by itself could not provide an adequate representation of any given trait over a whole society. One must also know the extremes, the critic observed, pointing to the great moral and intellectual benefits bestowed on society as a result of the accumulation of wealth by a few privileged individuals. The mean might suffice for most problems in physics, but the most important questions of social mechanics required more detailed knowledge.[60]

The coherence of the average man concept was also sharply challenged during its early childhood. In 1843 Cournot pointed out that the mean taken for each side from a great number of right triangles could in no way represent the type of a right triangle, since it almost certainly would not be a right triangle at all. For the same reason, the mean of all organs and limbs for a particular species of animal would likely fail even to be a viable organism. Hence *l'homme moyen* might well prove to be simply *l'homme impossible.*[61] The Prussian pastor and astronomer J.W.H. Lehmann showed that the *Schuss*, or spurt of growth accompanying adolescence, disappeared completely when a curve of mean height according to age was plotted, since the *Schuss* occurred at different ages in different men.[62] Quetelet did not accede to these objections. He held up his subsequent demonstration of the applicability of the error law to humanity as proof that the average man was a genuine type, notwithstanding Cournot, and he answered Lehmann by asserting that the growth spurt was no normal characteristic of human growth, but merely an artifact of contemporary civilization.[63] That these rebuttals persuaded anyone, however, is unclear.

The value and usefulness of averages was challenged in a more practical way by the uncongenial results that too frequently presented themselves as the outcomes of direct and simple-minded studies. Reformist

[60] See *Bulletin universel des sciences et de l'industrie*, 6th sec., *Bulletin des sciences géographiques*, 28 (1831), 113-136.

[61] Cournot, p. 214.

[62] Jacob Wilhelm Heinrich Lehmann, "Bemerkungen bei Gelegenheit der Abhandlung von Quetelet: Über den Menschen und die Gesetze seiner Entwicklung, in diesem Jahrbuche, Jahrgang 1839," in H. C. Schumacher, ed., *Jahrbuch für* 1841, 137-219, p. 139; *Jahrbuch für* 1843, 146-230, pp. 146-147.

[63] Quetelet, *Du système social et des lois qui le régissent* (Paris, 1848), pp. 23ff.

statisticians received a shock in 1828 when A. Balbi, A. M. Guerry, and Benoiston de Châteauneuf showed that crime and education were not, at least on their face, inversely correlated, but that, on the contrary, areas where education was most widely distributed tended also to have more crimes. Unwilling to forsake their campaign for public education, statisticians were obliged to reinterpret their science as a subtle and complex affair. One could no longer suppose, as had the devout educational reformer Lucas, that the influence of moral factors on crimes against persons could be established directly by comparing a table giving the state of civilization based on instruction and leisure with a table of the pertinent crimes. Balbi, after learning of the preliminary results, decided it was essential to consider a multitude of factors—extremes of wealth and poverty, density of population, the proximity of borders or of seacoast, cities, and the presence or absence of released criminals, as well as ignorance and superstition—before the numbers could be interpreted. Guerry, similarly, examined a wide range of factors that might influence the level of crime, and concluded somewhat vaguely that differences of culture and mores were the most important factors.[64]

The issue of crime and education inspired recognition that in statistics the effects of given causes are often masked, and can easily be confused with other factors not explicitly considered. Alphonse De Candolle argued that the positive relationship between crime and instruction derived from no direct effect of the latter, but only reflected the tendency for education to be most widespread in prosperous districts where differentials of wealth were most pronounced. He emphasized that in social science, even more than in physics, it was essential to sort out the effects of perturbing causes before drawing conclusions.[65] In England, Charles Morgan and others found reason to dismiss Guerry's results as a kind of false correlation, inadequate to justify a firm conclusion on the relation of crime to education.[66] Quetelet, too, accepted De Candolle's argu-

[64] A. Balbi and A. M. Guerry, *Statistique comparée de l'état de l'instruction et du nombre des crimes dans les divers Arrondissements des Académies et des Cours de France* (Paris, 1828), a large printed table; M. Lucas, "Influence morale de l'instruction et de la civilisation en général sur la diminution des délits et des crimes," *Bulletin universel des sciences et de l'industrie*, 6th sec., *Bulletin des sciences géographiques*, 15 (1828), 105-117; also E. Héreau's review of the Balbi-Guerry chart, *ibid.*, 16 (1829), 6-10; A. Balbi, "Rapport du nombre des crimes à l'état de l'instruction publique en France," *ibid.*, 20 (1829), 252-264; A. M. Guerry, *Essai sur la statistique morale de la France* (Paris, 1833), p. 40.

[65] Alphonse De Candolle, "Considérations sur la statistique des délits," *Bibliothèque universelle des sciences, belles-lettres et arts*, 104 (1830), *Littérature*, pp. 159-186.

[66] Charles Morgan, review of Guerry in *Athenaeum*, no. 303 (1833).

ment, though he added with his customary retrospective prescience that he had never expected the alleviation of crime to result from mere instruction, but only from moral education.[67] The bewildering complexity of statistics was invariably rediscovered whenever a result arose that contradicted expectations or prejudices. The relatively high suicide rate among Protestants troubled German scholars later in the nineteenth century, and writers like the Göttingen economist Helferich found reason to suppose that Protestantism could not itself be a cause of suicide, but rather was entangled with other factors.[68] Problems of heterogeneity were invoked by William Lucas Sargant to criticize Buckle's conclusions on crime and free will, and by others to point out the deficiencies of sanitary statistics.[69]

A coherent tradition of critical thinking about the character of statistics only developed in the wake of Buckle's discomfitting pronouncements. It was predominantly German; the French showed a minimum of interest, and the British, being less professionalized, tended to be more impulsive and to consign their contributions to the relatively ephemeral critical reviews. This was, however, by no means universally the case. The remarks of W. S. Jevons and, especially, John Venn on this issue were studied and cited for years afterward.

Venn's *Logic of Chance*, which first appeared in 1866, had received a considerable impetus from Buckle's *History*. Venn was no militant on free will. He confessed himself sympathetic to a limited doctrine of necessity, which allowed a prominent role for the will but insisted that identical antecedents must produce the very same consequents. On the other hand, he rejected as a dangerous fallacy the doctrine, which he labeled fatalism, that events occur quite independently of men's motives, and he categorically denied that even his less extreme version of philosophical necessity received the slightest measure of support from the existence of regularities in statistics. Universal causality could be established only from systematic covariation, not from broad uniformities; "no amount of regularity seems to me to bring us nearer to proving that

[67] Quetelet, "Recherches sur le penchant au crime aux différens ages," *NMAB*, 7 (1832), separate pagination, pp. 26ff., esp. pp. 43-44.

[68] Helferich, review of Wagner in *Göttingische gelehrte Anzeigen*, 1865, vol. 1, 486-506, p. 500.

[69] William Lucas Sargant, "Lies of Statistics," in *Essays of a Birmingham Manufacturer* (4 vols., London, 1869), vol. 2, 56-182, pp. 166-167; Anon., "The Fallacies and Shortcomings of our Sanitary Statistics," *The Social Science Review and Journal of the Sciences*, n.s. 4 (1865), 234-250, 358-363, 403-414, 481-495; 5 (1866), 21-43, 310-321, 436-447; 7 (1866), 98-110.

each separate event comprised in the statistics has its invariable and unconditional antecedents."[70]

Venn, we may recall, held that probability statements apply only to series and not to individual events. The use of probability presupposes that the events can be regarded as homogeneous for certain purposes, and Venn rejected its application to testimonies, judicial decisions, and some aspects of medicine on the ground that too much is generally known about the peculiarities of the individual cases for an assumption of homogeneity to be acceptable. Venn understood that the use of probability implies also a measure of ignorance about individuals, proceeding from their intricate and complex differences. He held that in statistics—though not, for practical purposes, in games of chance—there is also irregularity over the very long run, reflecting the gradual evolution of customs, beliefs, and laws. Probability, then, applies to that limited extent of order that characterizes in an imperfect way the middle term.[71]

These mass regularities constituted, for Venn, the fundamental data of probability. He opposed Quetelet's view that they provided a sufficient justification for inferring the existence of an average man, representing the type of a race or people. This error, he thought, embodied a deep subjectivism, and was analogous to the defective argumentation of Laplace and De Morgan, for it assumed a fixed, pre-existent probability regulating individual trials, from which laws of the mass were to be derived. He wrote: "We are concerned only with averages, or with the single event as deduced from an average and conceived to form one of a series. We start with the assumption, grounded on experience, that there is uniformity in this average, and, so long as this is secured to us, we can afford to be perfectly indifferent to the fate, as regards causation, of the individuals which compose the average."[72] That is, probability neither presupposes nor demonstrates causality.

William Stanley Jevons stood diametrically opposite Venn on the proper interpretation of probability, but concurred with him on the implications of mass regularity for the understanding of individual events or prediction of the future. Jevons is best known to posterity as a pioneer of the marginal approach to economics and, to a lesser extent, as a logician and philosopher of science, but he began his career as a meteorologist in Australia and he took a leading role in the Manchester Sta-

[70] John Venn, *The Logic of Chance* (London and Cambridge, Eng., 1866), p. 335.

[71] *Ibid.*, pp. 20-22, 234-236.

[72] *Ibid.*, p. 330; see also p. 43.

tistical Society during the 1860s and 1870s. Like Venn, he received a considerable impetus from Buckle, and he discussed the nature of statistics at some length in the context of his wide-ranging remarks on probability. Jevons set out from a classical view of probability derived from the instruction of De Morgan and the writings of Laplace and Poisson. Our knowledge is always uncertain, he explained, so that probability is essential for scientific investigation. "Events come out like balls from the vast ballot-box of nature,"[73] he remarked, and indeed the world might have been governed by chance. "Happily the Universe in which we dwell is not the result of chance, and where chance seems to work it is our own deficient faculties which prevents us from recognizing the operation of Law and Design."[74] Jevons marveled that natural order extended even to errors and deviations, a result he attributed to Quetelet. This remarkable order provided assurance that the same principles of scientific method which had proved so fruitful in the study of nature could be applied with like success to man. "Little allusion need be made in this work to the fact that man in his economic, sanitary, intellectual, aesthetic, or moral relations may become the subject of sciences, the highest and most useful of all sciences. Every one who is engaged in statistical inquiry must acknowledge the possibility of natural laws governing such statistical facts."[75]

Despite his unqualified acceptance of the subjective logic of probability, and not obviously reconcilable with it, Jevons emphasized the interest of natural variety and the appropriateness of probability theory for dealing with it. "Number is but another name for diversity,"[76] he announced, and he rejected the "fallacious impression . . . that the theory of probabilities necessitates uniformity in the happening of events," as Buckle had "superficially" believed.[77] On the contrary, he argued, the theory of probability positively mandates runs of luck of a limited extent. Jevons remarked that in statistics "quantities which are called errors in one case, may really be most important and interesting phenomena in another investigation. When we speak of eliminating error we really mean disentangling the complicated phenomena of nature."[78] He emphasized repeatedly that real events, especially the phenomena of mind

[73] W. S. Jevons, *The Principles of Science* (1874, London, 1924), p. 239; see also p. 149.
[74] *Ibid.*, p. 2
[75] *Ibid.*, p. 334. See also preface and pp. 359ff., esp. p. 374.
[76] *Ibid.*, p. 156.
[77] *Ibid.*, p. 655.
[78] *Ibid.*, p. 339.

and of society, are always complicated, so that "laws and explanations are in a certain sense hypothetical, and apply exactly to nothing which we can know to exist."[79] Science proceeds through abstraction, by substituting imaginary objects for those subsisting in the real world. Hence "those who speak of the uniformity of nature, and the reign of law, misinterpret the meaning involved in those expressions. Law is not inconsistent with extreme diversity."[80]

Like Laplace, Quetelet, and De Morgan, Jevons thought the concept of pure, irreducible chance unthinkable. He was, however, opposed to the idea that the events of the world are regulated in all their details by blind, purposeless laws of the mechanical sort. He noted that by varying the initial conditions, the same laws could be made to yield quite different effects, observing that on this account one could still think in terms of purpose and design even after the triumph of Darwinism. Jevons ridiculed Comte and Buckle, arguing that the inaccessibility of laws of individuals ruled out the discovery of determinate social laws, and he most emphatically rejected the possibility of a science of history, maintaining that society was unstable, and that the slightest disturbance could magnify itself until everything was altered. With regard to free will, moral responsibility, and the efficacy of prayer, Jevons denied they could be contradicted by statistics. "Laws of nature are uniformities observed to exist in the action of certain material agents, but it is logically impossible to show that all other agents must behave as these do. The too exclusive study of particular branches of physical science seems to generate an over-confident and dogmatic spirit."[81] Science should respect the teeming diversity of the world, and recognize its own imperfection. Jevons wished to make clear "that atheism and materialism are no necessary results of scientific method."[82]

Statistik: BETWEEN NATURE AND HISTORY

For practitioners of the academicized German version of statistics, the idea of a comprehensive mean value uniting the whir of appearances into a lawlike totality had very limited appeal. The only prominent ex-

[79] *Ibid.*, p. 458.
[80] *Ibid.*, p. 750.
[81] *Ibid.*, p. 737; see also pp. 759-761, 764-765.
[82] *Ibid.*, p. 766.

ception to this generalization is Alexander von Oettingen, the conservative Lutheran theologian and statistician, whose influential and successful tome on moral statistics and social ethics appeared in 1866. Even Oettingen did not subscribe to this view unreservedly, for he was deeply concerned about the alleged inconsistency between statistics and free will, and hence was reluctant to speak without qualification of law. Oettingen sought to find the middle way between collectivism and individualism. On the one hand, he opposed what he viewed as the social determinism of Adolph Wagner and Quetelet, and even persuaded Wagner during the latter's brief tenure at Dorpat that his ideas on this matter were "exaggerated."[83] At the same time he rejected the individualism of his critic, the criminologist Emil Wilhelm Wahlberg, and he dismissed Drobisch as a Pelagian and "proud pharisee" for his claim that most individuals were uninfluenced by the urge to crime.[84] The union of free individuals into a higher collective whole seemed to Oettingen essential to the very concept of a society; he even imputed Germany's victory in the Franco-Prussian War to the contrast between her deep sense of community and French atomism.[85] Oettingen's aim was to exemplify this social dimension, not to find the causes of individual behavior, and for this purpose there was no need to break the numbers down. The regularity of mean values was sufficiently inspiring by itself; "if all individuals were free to conduct their lives as autonomous and unfettered selves, how could this constancy in the ethical activity of the whole have arisen and how could it be explained?"[86] Still, Oettingen denied that this regularity was "so undeviating, that we can infer from it the existence of a necessary law of nature."[87]

The Göttingen professor J. E. Wappäus is perhaps also to be excluded from the general claim that German statisticians were comparatively unimpressed by the virtues of wide mean values. He announced that the truths of statistics "are only valid for the totality of a population, considered as a whole, or, as Quetelet puts it, for the average man (*l'homme*

[83] On Wagner, see Oettingen, *Die Moralstatistik in ihrer Bedeutung für eine Socialethik* (Erlangen, 3d ed., 1882), p. 19; and Wagner, ZGSW, 36 (1880), 159-203, p. 192.

[84] Alexander von Oettingen, *Die Moralstatistik und die Christliche Sittenlehre: Versuch einer Socialethik auf empirischer Grundlage*, vol. 2, *Die Christliche Sittenlehre: Deductive Entwicklung der Gesetze Christlichen Heilslebens im Organismus der Menschheit* (Erlangen, 1873), on Wahlberg, pp. 8-9, on Drobisch p. 33; see also Wahlberg's review of Oettingen, *Die Moralstatistik in ihrer Bedeutung* (n. 83), (1866) 1st ed., in ZGSW, 26 (1870), 567-576.

[85] Oettingen, *Moralstatistik und Christliche Sittenlehre* (n. 84), p. 3.

[86] Oettingen, *Die Moralstatistik in ihrer Bedeutung* (n. 83), p. 37.

[87] Oettingen, *Ueber akuten und chronischen Selbstmord: Ein Zeitbild* (Dorpat, 1881), p. 12.

moyen) of a nation."[88] Wappäus, however, was a statistician of the old school, writing in 1859 when a new tradition of university statistics based on numbers and dedicated to the solution of social problems was just beginning to take hold. The new generation, educated principally in history though partly in economics, was motivated by an entirely different set of concerns. They were far more closely attuned to issues of methodology, and had definite preconceptions about the character of social science. Buckle's book appeared at just the right time to form a key part of their collective education. Still more important, he directed them to Quetelet, whose interpretation of statistical science seemed to them at once inspiring and deeply flawed. Quetelet's exaltation of mean values and dismissal of variation appeared to them as a central deficiency of his approach.

The idea of a science built on the regularities of mean values stirred opposition for a variety of reasons. One objection, whose appeal was wide both within and outside of Germany, simply took note of the circumstance that the regularities of statistics were not causal. In Germany, the Tübingen statistician Fallati and the prominent Berlin public-health writer J. L. Casper both remarked as early as the 1840s that the observed constancy of statistical aggregates could not by itself justify the term "law," since there was no basis for certainty that the same regularity would persist into the future.[89] The most compelling evidence of this deficiency was the impossibility of applying those so-called laws to individuals. A critic of Adolph Wagner's book observed that a law must provide a determinate and necessary relation between cause and effect, and hence must apply not only to an abstract whole, but also to the parts. Since statistical regularities do not prevail for small numbers, they cannot apply with necessity to large numbers either. Hence the law of large numbers is a misnomer, and "statistical laws" are at best numerically expressed properties of lands or nations.[90]

From the 1850s, German numerical statisticians stressed the search for systematic covariation, and not mere regularities, as the primary aim

[88] J. E. Wappäus, *Allgemeine Bevölkerungsstatistik: Vorlesungen* (2 vols., Leipzig, 1859), vol. 1, p. 17. Franz Vorländer also argued that statistical principles apply only to mean values; see "Die moralische Statistik und die sittliche Freiheit," *ZGSW*, 22 (1866), 477-511, p. 483.

[89] Fallati, *Einleitung in die Wissenschaft der Statistik* (Tübingen, 1843), p. 54; J. L. Casper, *Über die wahrscheinliche Lebensdauer des Menschen* (Berlin, 1843), pp. 29-30. I treat the German discussion of statistical heterogeneity in greater detail in "Lawless Society: Social Science and the Reinterpretation of Statistics in Germany, 1850-1880," in *Prob Rev*.

[90] Anonymous review in *Jbb*, 4 (1865), 286-301.

of their science. Ernst Engel, founder of the Berlin statistical seminar in which so many German statisticians received their training, and to all appearances one of the most fervent admirers of Quetelet, argued in his first essay on statistics, in 1851, that regularities amounted only to empirical laws, not causal ones.[91] Twenty-five years later, in his *éloge* for Quetelet for the International Statistical Congress meeting in Budapest, Engel announced that their leader's infatuation with mean values had led the congress in the wrong direction, and that "not the largest, but relatively local averages constitute what is really worth knowing."[92] Wagner aimed his book at the discovery of partial causes through the presentation of tables giving suicide rates in relation to every possible variable—climate, weather, time of day and year, sex, age, religion, profession, education, political and economic conditions, and so forth. He felt that suicide would be explained, or reduced to law, when all variation over time and place was in this way accounted for. That the search for causes of variation constituted the highest aim of statistics was similarly endorsed by Bruno Hildebrand, Adolf Held, Leopold Neumann, and Gustav Schmoller, among others.[93]

These practical reasons for stressing the study of variation, however, were supported by, and often subordinate to, an organic conception of state and society widely shared among academic liberal reformers who opposed both socialism and laissez-faire liberalism. Statisticians of the generation that came to maturity during the 1860s derived much of their philosophical outlook from the German idealist tradition; but to the extent they adopted the viewpoint of their predecessors they did so because it seemed to answer the needs of their own day. Social science in Germany was closely bound to the so-called worker question. The idea that society was a dynamic and powerful entity, requiring a science of its own which the state could ignore at its peril, had been introduced to Germany by the radical Lorenz von Stein in an 1843 study of socialism and communism in France. This one beneficial result of the "ill-formed so-

[91] Ernst Engel, "Mein Standpunkt der Frage gegenüber ob die Statistik eine selbständige Wissenschaft oder nur eine Methode sei" (1851), repr. in Engel, "Das statistische Seminar und das Studium der Statistik überhaupt," *Zeitschrift des königl. preussischen statistischen Bureaus*, 11 (1871), 188-194, p. 189. On Engel see Ian Hacking, "Prussian Numbers," in *Prob Rev*.

[92] Engel, "L.A.J. Quetelet: Ein Gedächtnisrede," *ibid.*, 16 (1876), 207-220, p. 217.

[93] Wagner, *Gesetzmässigkeit* (n. 4), part 2; Bruno Hildebrand, "Die wissenschaftliche Aufgabe der Statistik," *Jbb*, 6 (1866), 1-11; Adolf Held's review of Quetelet in *Jbb*, 14 (1870), 81-95; Leopold Neumann, "Zur Moralstatistik" (n. 58); Gustav Schmoller, "Die neueren Ansichten" (n. 56).

cialist and barbaric communist doctrines and intrigues"[94] of the time was forceably brought to public attention by the events of 1848, and then purged of its radical implications by the Heidelberg political scientist Robert von Mohl, who substituted for "the false social doctrine" of class conflict a "true science of society" based on organic diversity and "communities of interest."[95] The explosive growth of industry beginning in the 1850s produced disruptions that made the worker question seem all the more pressing. Numerical study of society was undertaken in large measure to facilitate the discovery and implementation of a moderate solution to it.

Virtually all of the influential German statisticians of the 1860s and 1870s were or became members of the Verein für Sozialpolitik, which was formed in 1871. Most, including Engel and Schmoller, favored a solution to the problems of workers through the organization of voluntary cooperatives,[96] although Adolph Wagner, who renounced his free-market orientation in a "Damascus experience" about 1870, championed state socialism and became increasingly influential after Bismarck's change of economic policy in 1878.[97] In any event, all these "academic socialists" (*Kathedersozialisten*) were opposed to rigid laissez-faire economics, and especially to the supposition that economic phenomena are wholly governed by natural laws, before which governments are powerless. Schmoller, Lujo Brentano, and G. F. Knapp, among others, were greatly vexed by the so-called "iron law of wages," which held that the "wage fund" in any economy at a given time was fixed, so that collective bargaining could do no more than shift wages from one class of workers to another.[98] The Sozialpolitikers were persuaded that this so-

[94] Anon., "Neuere deutsche Leistungen auf dem Gebiete der Staatswissenschaften," *Deutsche Vierteljahrsschrift* (1854), no. 3, 1-78, p. 12.

[95] Robert von Mohl, "Gesellschafts-Wissenschaften und Staats-Wissenschaften," ZGSW, 7 (1851), p. 25 and *passim*. See also Erich Angermann, *Robert von Mohl, 1799-1875: Leben und Werk eines altliberalen Staatsgelehrten* (Neuwied, 1962), and Eckart Pankoke, *Sociale Bewegung—Sociale Frage—Sociale Politik: Grundfragen der deutschen "Socialwissenschaft" im 19 Jahrhundert* (Stuttgart, 1970).

[96] See Gustav Schmoller, "Die Arbeiterfrage," *Preussische Jahrbücher*, 14 (1864), 393-424, 523-547; 15 (1865), 32-63.

[97] See Wagner reviews in ZGSW, 34 (1878), 199-233, p. 211, and Lujo Brentano, *Mein Leben im Kampf um die soziale Entwicklung Deutschlands* (Jena, 1931), pp. 63-76.

[98] See Schmoller, "Arbeiterfrage" (n. 96), p. 413; James Sheehan, *The Career of Lujo Brentano: A Study of Liberalism and Social Reform in Imperial Germany* (Chicago, 1966), esp. p. 21. Even Jevons, English founder of the mathematical and deductive approach to economics—marginalism—called in 1876 for a new branch of political and statistical science to investigate the limitations of laissez faire. See T. W. Hutchison, "Economists and Economic Policy in Britain after 1870," *History of Political Economy*, 1 (1969), 231-255, p. 236.

called law was inconsistent with empirical evidence of an actual rise in wages, that it had survived because it supported the presumed interests of myopic manufacturers, and that in the end adherence to it would only promote a more radical and violent process of social change.

Classical economics, in fact, was rejected almost entirely in Wilhelmian Germany. It seemed to sanctify greed as the expression of natural law and to deny the possibility of conscious reform. It made society a machine rather than an organism, an aggregation rather than a community, and though it was conceded to have introduced economic dynamism by liberating the energy of individuals, the free market was seen as having since degenerated to financial speculation and mindless *Manchestertum*. According to the historical view, the liberal hypertrophy of atomistic individualism and rigid deduction embodied traits characteristic of England and France. The time had come to subordinate these to the essentially German ideas of community and history. German economics in the late nineteenth century meant historical economics, and the new enthusiasm for numerical statistics reflected a commitment to empiricism rather than abstract deduction as well as a heightened concern about social unrest. The German reinterpretation of statistics was parallel to that of economics.[99] The historical economists opposed the idea of statistical law for the same reasons they objected to timeless natural laws of economics. The physics of the average man defined society in terms of the similarities of individuals, whereas the key to social science was seen by historicist Germans as the harmonious interactions of diverse social groups. Quetelet's statistical regularities offered a subject matter which could justify the establishment of statistics as a university discipline, but only if statistics were redefined to accord with acceptable German social and political ideas.

That redefinition began with the work of Gustav Rümelin. Rümelin was trained at Tübingen in theology, and served as a pastor and then teacher before his political activity led to his election as a Frankfurt parliamentarian. Subsequently he became director of the department of churches and schools in the Württemberg ministry of culture, during which time he presided over a controversy regarding government treatment of Catholic schools that amounted to a minor *Kulturkampf*. His position had already become untenable by 1861, and he was appointed

[99] On the historical school, see Pankoke, *Sociale Bewegung* (n. 95); Sheehan, *Brentano* (n. 98); Ulla G. Schäfer, *Historische Nationalökonomie und Sozialstatistik als Gesellschaftswissenschaften* (Cologne and Vienna, 1971).

head of the statistical-topographical office primarily because the post became open at a convenient time. He was two decades older than most German statisticians of this new generation, and was virtually the only one without formal academic training in statistics or some other branch of *Staatswissenschaft* (Ernst Engel, a graduate of the mining school at Freiberg, was also an exception on both counts). Nevertheless, he was a notable success as a statistician, and wrote two highly influential papers on the theory of statistics while an administrator that justified his assumption of a university post at Tübingen in 1867. His writings, even more than Wagner's, formed the point of departure for the "younger historical school" in statistics, much as the "older historical school" of Wilhelm Roscher, Bruno Hildebrand, and Carl Knies defined their orientation on economic matters.[100]

In his first paper on statistics, written in 1863, Rümelin addressed himself to the longstanding debate in Germany as to whether statistics was properly a science or a method. His answer, conveniently, was that it was both, although he made the methodological definition primary. Statistics, Rümelin argued, is intrinsically a technique for the observation and study of "mass phenomena." As such, it is applicable precisely to those composite phenomena made up of thoroughly heterogeneous individuals. Invoking a distinction set out by the French statistician Dufau in 1840,[101] Rümelin explained that in the sciences of nature, the individual is wholly or largely typical, so that a single well-recorded fact is sufficient to justify an induction. Society, by contrast, is the domain of diversity, and each person, though governed by law, is yet subject to so many perturbing causes that the action of particular laws cannot be inferred from single cases. This dichotomy, he explained, is really a matter of degree; animals are less uniform than plants, people than apes, moderns than ancients, adults than children, Caucasians than Negroes, men than women, and the educated than the unschooled. The gulf between man and nature, however, was so wide that the human sciences "could never have raised themselves above infancy, where they have been for some generations and where they in part remain, if there were no means of observation through which the inadequacy of individual-

[100] See the biography by Rümelin's much younger brother-in-law, Gustav Schmoller, in *Charakterbilder* (Munich and Leipzig, 1913), pp. 140-188. On the older historical school, see Gottfried Eisermann, *Die Grundlagen des Historismus in der deutschen Nationalökonomie* (Stuttgart, 1956).

[101] P. A. Dufau, *Traité de statistique, ou théorie de l'étude des lois d'après lesquelles se développent les faits sociaux* (Paris, 1840), p. 24.

ized and idiosyncratic experience can be alleviated, and our experience grasped as a whole."[102]

Although statistics by its very nature dealt with diverse objects, and hence could not reduce the phenomena to a few simple laws, neither should the statistician rest content with a general impression of the whole. The lawlikeness of statistical aggregates, shown by Quetelet, was merely a starting point. "In the end," Rümelin wrote, "the interest of moral statistics lies not at all in the demonstration of these regularities in willful acts of man, but rather in the perpetual movement and alteration that these numbers undergo."[103] Indeed, statistical uniformity is mostly superficial, attained only when one considers a great mass and smears together the variety of the phenomena. An average by itself, he observed, is completely inadequate, for a given mean of wealth, housing space, age of death, or price of grain could arise from completely different distributions.[104] More significantly, the disposition to emphasize a single aggregate figure from a large, miscellaneous population must inhibit rather than promote the search for causes. Hence, according to Rümelin, the proper statistical procedure is to fracture the population into tiny pieces, and then regroup these in various ways. For example, by finding the conditions under which crime and suicide increase or decrease, and the subgroups for which they are most common, the statistician goes beyond the mere assertion of regularity and attains propositions of real worth and interest.

Even so, the tendencies unveiled through this procedure would pertain not to any particular individual, but to "the collective whole of greater or lesser groups of persons or processes."[105] Like Drobisch, he denied that statistical conclusions could be used as a basis to infer psychological states; "What I say of the forest does not hold for the separate trees."[106] Because statistical generalizations do not permit prediction of individual cases, Rümelin held that phrases like "statistical law" and

[102] Gustav Rümelin, "Zur Theorie der Statistik, I" (1863), in *Reden und Aufsätze* (Freiburg, 1875), 208-264, pp. 218-219.

[103] Rümelin, "Moralstatistik und Willensfreiheit," in *ibid.*, 370-377, p. 375.

[104] Rümelin, "Zahl und Arten der Haushaltungen in Württemberg nach dem Stand der Zählung vom 3 Dec. 1864," in *Württembergische Jahrbücher für Statistik und Landeskunde*, Jg. 1865, 162-217, pp. 173, 185; "Ergebnisse der Zählung der ortsanwesenden Bevölkerung nach dem Stande vom 3 December 1867," in *ibid.*, Jg. 1867, 174-226, p. 192. The same point was made by Fr. J. Neumann, "Unsere Kenntniss von den socialen Zustände um uns," *Jbb*, 18 (1872), 279-341, p. 288.

[105] Rümelin, "Statistik," in Gustav Schönberg, ed., *Handbuch der politischen Oekonomie, Finanzwissenschaft und Verwaltungslehre* (3 vols., Tübingen, 3d ed., 1891), vol. 3, 803-822, p. 814.

[106] Rümelin, "Zur Theorie der Statistik, II" (1874), in *Reden* (n. 102), 265-284, p. 270.

"law of large numbers" reflect a fundamental misunderstanding. He argued in 1867 that the term "law" should be reserved for expressing a relation of cause and effect that is elementary, constant, and recognizable in every case. The peculiar function of statistics, in contrast, is to facilitate the acquisition of knowledge about domains like society where a complex multitude of forces acts simultaneously. Statistical analysis might lead to laws, but once the law is found, the need for statistics must vanish, since every case can then be explained. A genuine law of sex at birth, for example, would not belong to statistics but to physiology, and would apply with certainty to every individual case.[107]

Although Rümelin remained a wholehearted proponent of statistics, he became increasingly impressed in subsequent years by its limitations. In a lecture on the laws of history, delivered in 1878, he reflected that nearly two decades of intensive statistical research had yielded him nothing that could properly be called a social law. For this reason, he explained, he had been led to consider how different in character physical phenomena were from psychical, and how unreasonable to suppose that the same concept of law applied to each. He had decided that where the idea of freedom is involved, the method and ideal of knowledge must change.[108] In view of these considerations, the uncertainty of statistical knowledge ceased to be a defect, as it had seemed to the French positivists, and was transformed into a virtue, an accurate reflection of the reality it was meant to describe. Rümelin became persuaded that statistics provided the appropriate method for dealing with the collective behavior of highly diverse individuals—social science—precisely because it did not require the discovery of fixed and timeless laws.

Rümelin's definition of statistics as a method of mass observation, and his characterization of its principal object, society, in terms of fundamental diversity won wide and almost immediate acceptance in the German statistical community. Georg Mayr of the University of Munich and Bavarian statistical office held that statistics was the only suitable method for studying human communities, and Etienne Laspeyres exalted it as a new category of scientific inference, the key to research in those areas where "all other things" cannot be set equal.[109] In this guise, statistics won a prominent place in logic textbooks, such as Christoph

[107] Rümelin, "Ueber den Begriff eines socialen Gesetzes" (1867), *ibid.*, 1-31, pp. 8-10.

[108] Rümelin, "Ueber Geseze der Geschichte," *Reden und Aufsätze. Neue Folge* (Freiburg and Tübingen, 1881), 118-148, p. 139.

[109] Georg Mayr, *Die Gesetzmässigkeit im Gesellschaftsleben: Statistische Studien* (Munich, 1877), part 1; Etienne Laspeyres, "Die Kathedersocialisten und dic statistische Congresse," *Deutsche Zeit- und Streitfragen*, Jg. 4 (1875), Heft 51, 137-184, p. 164.

Sigwart's.[110] On the matter of the actual existence of social laws, irrespective of the fitness of statistics to discover them, Rümelin did not lead, but followed younger statisticians. Reformers and moralists like Schmoller, Knapp, and Wilhelm Lexis invested in the idea of the *Gemeinschaft*, or community, their hopes for the preservation of order through shared values and mutual assistance—the latter to be guided by professors of economy. As Knapp indicated most explicitly, to argue that humanity was subject to uniform and unvarying laws was to erase the essential distinction between man and nature and to deny the importance of unique cultural values. The societies that subsumed human individuals were not mechanical aggregates ruled by necessary laws, but diverse communities drawing strength from common feeling and defined by a unique history rooted in freedom.

Georg Friedrich Knapp was one of the most outspoken opponents of the idea of statistical law, and more generally of the idea that societies are subject to fixed natural laws of any sort. He was, for the same reason, a sharp critic of Quetelet, although his was a criticism born of a deep consciousness of intellectual debts, and the consequent need to sift out what was valuable and true from what was misleading or untenable. Knapp was among the few German statisticians in the school of historical economics and Sozialpolitik who knew much mathematics, and he took his mathematics seriously. He had no interest in mere works of statistical compilation, and was disgusted while at Göttingen by the old-fashioned and uninspired writings of the aging statistician Wappäus. Hence he fully approved of Quetelet's devotion to mathematics, though Quetelet's actual work impressed him less than the mathematical models of population and mortality published by Joseph Fourier and Ludwig Moser. Knapp's intellectual orientation was devoutly historicist, but this did not stifle his lively interest in abstract and deductive models of population change.[111]

Knapp, the son of Justus Liebig's sister and her husband, one of Liebig's early students, was born in Giessen. His autobiography records an early infatuation with French culture, but it also reveals that his mother placed flowers on the grave of the nationalist economist Friedrich List while the family was on vacation in the Tirol. He began his university

[110] Christoph Sigwart, *Logik* (2 vols., Freiburg, 2d ed., 1893), vol. 2, sec. 101.

[111] See Georg Friedrich Knapp, *Die Sterblichkeit in Sachsen nach amtlichen Quellen dargestellt* (Leipzig, 1869), and especially, *Theorie des Bevölkerungs-Wechsels: Abhandlungen zur Angewandten Mathematik* (Brunswick, 1874).

studies in 1861 at Munich, where he studied political economy with the head of the Bavarian statistical office, Hermann. He records that in 1862 he was deeply impressed by the "Enlightenment" viewpoint in David Friedrich Strauss's *Leben Jesu*, but his politics and social philosophy began to change at about that time when he moved to the University of Berlin. There the Prussian parliament was deadlocked in its struggles with Bismarck, at a time when a revolution seemed imminent, and he began to feel that liberalism was sterile and ineffectual. The sermon of a Straussian liberal preacher at this time impressed him as vapid, even contemptible, and though he still regarded himself as a "nominal liberal" he found the argument about the powerlessness of liberalism in Ferdinand Lassalle's *Was Nun* striking and persuasive. He took his doctorate from the classical economist Helferich at Göttingen, whom he held in high esteem, but by the time he finished he was persuaded that "dogmatic" liberal political economy was a useless *Gymnastik*, and that the solution to real problems could never be discovered through the abstractions of formal theory. He sought out a more practical education at Engel's statistical seminar in Berlin, and his first real position was as a statistical administrator. He set up the city statistical office in Leipzig in 1867, and was called to a post at the University of Leipzig two years later. From 1874 until the end of the First World War he held a professorship at Strasbourg, although by 1875 his interests had moved away from statistics.[112]

In Leipzig, Knapp was able to seek out Drobisch, whose remarks on moral statistics he found highly admirable. He also benefited from a paper on "Adam Smith and Quetelet" by his close friend Adolf Held, published in 1867. Held expressed measured admiration for both these authors, but he dismissed Quetelet's idea of a social physics and criticized the Belgian astronomer for his lack of interest in the autonomy and freedom of individuals. Quetelet's shallow treatment of mankind as a "homogeneous mass," according to the historicist Held, was obviously the outcome of his experience of advancing democracy and leveling. Held also denied that the statistical regularities uncovered by Quetelet could properly be regarded as laws, or that they contradicted free will.[113]

[112] Knapp, *Aus der Jugend* (n. 52); also Knapp, "Ernst Engel, Erinnerungen aus den Jahren 1865-1866," in *Einführung in einige Hauptgebiete der Nationalökonomie* (Munich and Leipzig, 1925), pp. 322-327.

[113] Adolf Held, "Adam Smith und Quetelet," *Jbb*, 9 (1867), 249-279, p. 271; see also his review of Quetelet's *Physique sociale* in *Jbb*, 14 (1870), 81-95.

Knapp included in his first book on demographic models a chapter "Are there Laws of Mortality?" in which he determined that unless death was a fixed function of age, independent of time and place, which it was not, it could not be justified to speak of laws in this regard. He also argued that there was no reason to be astonished by the regularities of statistics.[114] During 1871 and 1872 Knapp published a series of papers on Quetelet and other modern statistical writers in which he developed more fully his interpretation of statistics. He attributed to Quetelet two interconnected sets of errors, the social determinism of physics and the atomistic individualism of anthropology, both inconsistent with true social science. The first was the work of Quetelet the physicist, and was the source of the doctrine of "natural laws of society," that, when exaggerated by Buckle and Wagner, had inspired the supposed conflict between statistics and human freedom. Knapp defined the "astronomical conception of society" as the view "that forces act on society which, as we recognize from the regularity of their effects, seem to be independent of those affecting individual events and actions, and that therefore must be conceived as external forces."[115] This, incidentally, was the definition that guided Wilhelm Lexis's work on dispersion, which was designed as a refutation of Quetelet and his followers. Quetelet's other great error, according to Knapp, arose from his confusion of social science with the natural science of man, anthropology. Knapp's objection here was to the theory of the average man, which explained social phenomena in terms of a set of penchants that were, at least to a first order of approximation, common to all members of society. This view made society a mere sum of individuals. Both the anthropological and the astronomical errors, Knapp thought, were characteristic of Enlightenment liberalism and of French thought generally, and indeed were responsible for the habitual instability of French political life, particularly "the frightful catastrophe of the present," the Paris commune.[116]

The proper understanding of statistics, according to Knapp, was inconsistent with both these conceptions, yet also derived from the "unphilosophical head" of Adolphe Quetelet, and even embodied the proper lesson of "true *Queteletismus*." Recognition of statistical regu-

[114] Knapp, *Sterblichkeit* (n. 111), pp. 95-101.

[115] Knapp, "Bericht über die Schriften Quetelet's zur Sozialstatistik und Anthropologie," *Jbb*, 17 (1871), 167-174, 342-358, 427-445, pp. 438-439. Quetelet almost certainly would not have endorsed this statement, though Buckle might and Oettingen did (see n. 86). See also review by Knapp in *Jbb*, 16 (1871), 182-186.

[116] Knapp, "Die neueren Ansichten über Moralstatistik," *Jbb*, 16 (1871), 237-250, p. 250.

larity really did provide the key to social science, but not if it was explained in terms of some mysterious and diffuse force that negated the freedom of individuals, or of properties held in common by all members of society. Society was not a mere aggregation of like individuals, but a union of free persons imbedded in a common culture that depended as much on their differences as on their similarities. For this reason, Knapp thought Quetelet's ideas on the error law was justified for the anthropological domain but useless to social science, and he went on to deny that probabilistic error theory could contribute significantly to statistics. A statistician cannot use the methods of error analysis for the very simple reason that every individual is genuinely different. There are no errors, but only variation, and hence there is no "true" value beneath the diversity of phenomena awaiting discovery.[117]

Knapp's emphasis on the intrinsic importance of human diversity was associated with his belief that the observed regularity of various statistical series constituted the most secure knowledge that was accessible through the empirical study of society. Unlike Wagner and the early Rümelin, Knapp refused to believe that true laws of society could be uncovered through the study of variation. According to Knapp, it was the great shortcoming of Quetelet's French contemporary A. M. Guerry to believe that "social-historical constants" such as a timeless relation between crime and education could be discovered through statistical manipulation. In this respect, Quetelet's viewpoint on crime seemed to him altogether more satisfactory, for he read Quetelet as being content to interpret crime wholly in terms of the society in which it occurred. Society, Knapp believed, is so complex an entity that a negative relation between crime and education in one culture could in no way be taken as definitive for another, and he implied that statisticians could do little more than affirm the close relationship between individual actions and social milieux. Since he was equally persuaded that deductive principles of population or economy could not be applied directly to real social entities, his viewpoint amounted to a denial of the possibility of social laws of any sort.[118]

Knapp's wide-ranging critique of Quetelet was widely read and exerted an influence comparable to those of Wagner, Rümelin, Oettingen, and Drobisch. His positive ideas about statistical science, along

[117] Knapp, "A. Quetelet als Theoretiker," *Jbb*, 18 (1872), 89-124; also his review of Quetelet's *Anthropométrie* in *Jbb*, 17 (1871), 160-167.

[118] Knapp, "Quetelet als Theoretiker" (n. 117), pp. 98-100.

with those of Rümelin, were enlisted in the campaign to make statistics the basis for the emerging science of sociology. That campaign was conducted less effectively in Germany, where the academic social science of statistics was increasingly giving way to bureaucratic compilation by 1880,[119] than in Austria, where academic statisticians maintained their theoretical and ideological interests for at least a decade longer. In Vienna, as in Berlin, a statistical seminar had been instituted under the joint direction of the university and the government statistical office during the early 1860s. The head of Austrian statistics, Baron Czoernig, respected the organization of Belgian statistics sufficiently to adopt a similar model, and his successor, Adolf Ficker, strongly approved of Quetelet's attempt to transform this descriptive discipline into an inductive science, even if he was disturbed by the Belgian's ostensible materialism and his tendency to dissolve the individual in society. The occupant of the statistics chair at the University of Vienna, Leopold Neumann, had already in 1865 expressed skepticism that statistical regularities could amount to laws of society.[120] The Austrians, evidently, were sensitive to the same issues as their German neighbors.

During the late 1870s, leadership of Viennese statistics passed to F. X. von Neumann-Spallart and K. T. von Inama-Sternegg. These writers developed the interpretation of statistics as a science of mass phenomena, derived from Rümelin and Wilhelm Lexis, into an argument for the importance and appropriateness of statistical sociology. They held that an alternative was needed to the deductive sociology of Herbert Spencer, the American H. C. Carey, the Russian P. F. Lilienfeld, and the German Albert Schäffle. They criticized the tendency to rely on biological or physical analogies and argued that neither Carey's insistence on analyzing society down to its most elementary parts nor Schäffle's unwillingness to view society as anything but a seamless web represented a viable approach to social science.

One error was common to both these writers, according to the Austrians—an insistence on perfect certainty in social science. Neumann-Spallart and Inama-Sternegg deemed this unattainable. Along with their colleague Gustav Adolf Schimmer they stressed the importance of

[119] See Anthony Oberschall, *Empirical Social Research in Germany*, 1848-1914 (Paris, 1965).

[120] See Adolf Ficker, Quetelet Nekrolog, *Statistische Monatschrift*, 1 (1875), 6-14; Leopold Neumann, "Über Theorie der Statistik," *Oesterreichische Vierteljahresschrift für Rechts- und Staatswissenschaft*, 16 (1865), 40-62.

social heterogeneity, and insisted that a strategy of mass observations was indispensable if reliable knowledge about society was to be gained. Their aim was not to find laws of society, nor to pursue international comparison. Following Knapp, Inama-Sternegg maintained that since each nation is an intricate, highly differentiated community, possessing a distinctive individual character, the search for cross-cultural constants was likely to be less fruitful than the alliance of statistics with history. His aim was not to dissolve sociology in traditional political history, but to create a social history founded on the quantitative study of mass phenomena. That statistics lacked the power to give strict proof of causal connections did not vitiate, but confirmed, its appropriateness as the method of social science.[121]

Notwithstanding Durkheim's celebrated study of suicide, the battle to make sociology statistical was lost, at least in the short term. Naúm Reichesberg, a Swiss academic and biographer of Quetelet, maintained the faith during the 1890s, and his book of 1893 on statistics and social science is perhaps the most comprehensive summary of statistical ideas developed in Germany and Austria during the preceding three decades, but academic sociology was unmoved.[122] The Danish social mathematician Harald Westergaard tried without luck to revive the old science of statistics and to infuse it with greater mathematical sophistication.[123] To be sure, the statistical study of society did not by any means disappear. Ferdinand Tönnies wrote on the methods of statistical social analysis into the twentieth century.[124] The historian Karl Lamprecht sought to revive Buckle's approach while purifying it of atomistic materialism

[121] See F. X. v. Neumann-Spallart, "Sociologie und Statistik," *Statistische Monatschrift*, 4 (1878), 1-18, 57-72; Karl Theodor von Inama-Sternegg, "Vom Wesen und den Wegen der Socialwissenschaft," *ibid.*, 7 (1881), 481-488; idem, "Geschichte und Statistik," *ibid.*, 8 (1882), 3-15; idem, "Zur Kritik der Moralstatistik," *Jbb*, N.F. 7 (1883), 505-525; idem, "Die Quellen der historischen Bevölkerungsstatistik," *Statistische Monatschrift*, 12 (1886), 387-408; idem, "Die Quellen der historischen Preisstatistik," *ibid.*, pp. 579-594; idem, "Neue Beiträge zur allgemeinen Methodenlehre der Statistik," *ibid.*, 16 (1890), 101-110; idem, "Geographie und Statistik," *ibid.*, 17 (1891), 375-385; Gustav Adolf Schimmer, "Die Statistik in ihren Beziehungen zur Anthropologie und Ethnographie," *ibid.*, 10 (1884), 262-267.

[122] See Naúm Reichesberg, *Die Statistik und die Gesellschaftswissenschaft* (Stuttgart, 1893); idem, "Was ist Statistik," *Zeitschrift für schweizerische Statistik*, 33 (1897), 269-275; idem, "Adolf Quetelet als Moralstatistiker," *ibid.*, 29 (1893), 490-498; idem, "Der berühmte Statistiker Adolf Quetelet, sein Leben und sein Wirken: Ein biographische Skizze," *ibid.*, 32 (1896), 418-460.

[123] Harald Westergaard, "Zur Theorie der Statistik," *Jbb*, N.F. 10 (1885), 1-23.

[124] Ferdinand Tönnies, "Eine neue Methode der Vergleichung statistischer Reihen," *Jahrbuch für Gesetzgebung, Verwaltung, und Volkswirtschaft im Deutschen Reiche*, 33 (1909), 699-720.

by adopting the statistical viewpoint of Rümelin, Knapp, and Lexis and wedding it to Wilhelm Wundt's nondeterministic psychical causality.[125] The Italians became increasingly interested in mathematical social statistics near the end of the century,[126] and the concrete statistical work in Germany around 1900 appears enormously more sophisticated than that of Quetelet's generation. Heinz Maus concludes that statistics deserves the credit for the establishment of a bond between sociology and empirical social research at the end of the nineteenth century.[127]

The most important effects of these discussions about the nature of statistical knowledge, however, were felt by other disciplines. Robert Campbell and Wilhelm Lexis, the first writers to compare the dispersion of actual distributions with expected values from a combinatorial model, were both inspired by their opposition to certain ideas of Quetelet and Buckle. Their important mathematical contributions, discussed in chapter 8, belonged to the traditions discussed here. Norton Wise has suggested that these German ideas, and particularly an idea of "statistical causality" developed by Karl Lamprecht, may have contributed to the radical new views on causality that arose in German quantum physics during ensuing decades.[128] Even before the rise of quantum physics, a close and significant relationship between social statistics and the origins of probabilism in physics is apparent. It pertains to the last third of the nineteenth century.

[125] See M. Norton Wise, "How Do Sums Count? On the Cultural Origins of Statistical Causality," in *Prob Rev*.

[126] See Luigi Perozzo, "Nuove applicazione del calcolo delle probabilità allo studio dei fenomeni statistici . . . ," *Atti della R. Accademia dei Lincei, Memorie della classe di scienze morali, storiche, e filologiche*, 10 (1882), 473-503; Antonio Gabaglio, *Storia e teoria generale della statistica* (Milan, 1880).

[127] Heinz Maus, "Zur Vorgeschichte der empirischen Sozialforschung," in René König, ed., *Handbuch der empirischen Sozialforschung*, vol. 1, *Geschichte und Grundprobleme* (Stuttgart, 1967), 21-56, p. 33. To appreciate the increased sophistication in Germany one need only look at the *Archiv für soziale Gesetzgebung und Statistik*, edited by Heinrich Braun from 1888-1903, or its continuation, edited by Werner Sombart, Max Weber, and Edgar Jaffé, *Archiv für Sozialwissenschaft und Sozialpolitik*.

[128] Wise, "How Do Sums Count?" (n. 125).

Chapter Seven

TIME'S ARROW AND STATISTICAL UNCERTAINTY IN PHYSICS AND PHILOSOPHY

German economists and statisticians of the historical school viewed the idea of social or statistical law as the product of confusion between spirit and matter or, equivalently, between history and nature. Their sense of inconsistency between mechanical law and progressive history was in most cases uncorrupted by any actual knowledge of physics, but the contradiction they identified proved more pervasive than they realized. That the laws of Newtonian mechanics are fully time-symmetric and hence can be equally run backwards or forwards could not easily be reconciled with the commonplace observation that heat always flows from warmer to cooler bodies. This discrepancy became for a time one of the deepest theoretical problems of the dynamical—or, as it came to be regarded, statistical—gas theory. James Clerk Maxwell, responding to the apparent threat to the doctrine of free will posed by thermodynamics and statistics, pointed out that the second law of thermodynamics was only probable, and that heat could be made to flow from a cold body to a warm one by a being sufficiently quick and perceptive. Ludwig Boltzmann resisted this incursion of probabilism into physics but in the end he was obliged, largely as a result of difficulties presented by the issue of mechanical reversibility, to admit at least the theoretical possibility of chance effects in thermodynamics. The American philosopher and physicist C. S. Peirce determined that progress—the production of heterogeneity out of homogeneity—could never flow from rigid mechanical laws, but demanded the existence of objective chance throughout the universe. His position can by no means be represented as the consensus view at the end of the nineteenth century, but neither was it a radical departure from all contemporary discussion.

BUCKLE'S LAWS AND MAXWELL'S DEMON

Statistical law had been presented to the world by Quetelet and Buckle as proof that disorder and chance were epiphenomenal. The application by Maxwell of that most remarkable law of unreason, Quetelet's error curve, to the otherwise intractable problem of molecular velocities implied no vindication of what was still almost universally regarded as an unthinkable and self-contradictory illusion of untrained minds, objective chance, but an impressive extension of the domain of scientific order. The combinatorial operations it made possible were both rigorous and elegant, and it was no more to be anticipated that the reduction of thermodynamic propositions to mechanical ones would introduce a new element of uncertainty into the subject than that the application of number would have such an effect on social science. The idea that macroscopic regularities such as the second law of thermodynamics are only probable was manifestly a development of certain ideas associated with the statistical approach, but it involved at the same time a repudiation of the statistical viewpoint presented by Quetelet and Buckle.

That Maxwell should play a major role in the reinterpretation of statistical reasoning was fully consistent with his character and commitments. He was, of course, a physicist of exceptional creativity, revealed in work on electricity and magnetism that began in 1854 no less clearly than in his papers on gas physics. He was always eager to look at old truths from new perspectives, and in his inaugural lectures at Aberdeen in 1856 and King's College, London, in 1860 he warned his students against "assuming that the higher laws which we do not know are capable of being stated in the same forms as the lower ones which we do know."[1] A deeply religious man, Maxwell was sensitive to the limitations of natural science. His philosophy professor at Edinburgh, Sir William Hamilton, had observed that since the Deity could not be subject to necessity in the material universe, the study of nature could never attain more than "probable certainty,"[2] and although Maxwell's enthu-

[1] Maxwell, "Inaugural Lecture at Aberdeen, 2 Nov. 1856," *Notes and Records of the Royal Society of London*, 28 (1973), 69-81, p. 78; also "James Clerk Maxwell's Inaugural Lecture at King's College, London, 1860," *American Journal of Physics*, 47 (1979), 928-933, p. 930.

[2] Sir William Hamilton, *Discussions on Philosophy and Literature* (New York, 1856), p. 41; also pp. 275, 297. On Maxwell and Scottish Common Sense, see George Elder Davie, *The Democratic Intellect: Scotland and her Universities in the Nineteenth Century* (Edinburgh, 1961); Richard Olson, *Scottish Philosophy and British Physics, 1750-1880* (Princeton, 1975).

siasm for Hamilton waned after he left Edinburgh for Cambridge in 1850, he always took care to avoid implicating the scientific temperament as an obstacle to religious belief. "I have endeavoured to show that it is the peculiar function of physical science to lead us to the confines of the incomprehensible, and to bid us behold and receive it in faith, till such time as the mystery shall open."[3]

Maxwell was not persuaded that the bounds of the incomprehensible were being pushed back quite as rapidly as some of his contemporaries would have it, and he was troubled by the extravagant pronouncements sometimes made in the name of science. Prominent among the writers whose necessitarian claims bothered him was Henry Thomas Buckle. Maxwell had been highly impressed by certain aspects of Buckle's first volume and mentioned Buckle to his friend and subsequent biographer Lewis Campbell within a few months of its appearance: "One night I read 160 pages of Buckle's *History of Civilization*—a bumptious book, strong positivism, emancipation from exploded notions and that style of thing, but a great deal of actually original matter, the true result of fertile study, and not mere brainspinning."[4] Evidently Buckle set him, like so many of his fellows, to thinking about the relation between the remarkable regularities of statistics and free will, for three months later he wrote to his friend R. B. Litchfield:

> Now, I am going to put down something on my own authority, which you must not take for more than it is worth. There are certain men who write books, who assume that whatever things are orderly, certain, and capable of being accurately predicted by men of experience, belong to one category; and whatever things are the result of conscious action, whatever are capricious, contingent, and cannot be foreseen, belong to another category.
>
> All the time I have lived and thought, I have seen more and more reason to disagree with this opinion, and to hold that all want of order, caprice, and unaccountableness results from interference with liberty, which would, if unimpeded, result in order, certainty, and trustworthiness (certainty of success of predicting). Remember I do not say that caprice and disorder are not the result of free will

[3] Maxwell, "Aberdeen Lecture" (n. 1), p. 78. On Maxwell's relationship to Hamilton, and on his life generally, see C.W.F. Everitt, "Maxwell's Scientific Creativity," in Rutherford Aris et al., eds., *Springs of Scientific Creativity* (Minneapolis, 1983), pp. 71-141.

[4] *Maxwell*, pp. 294-295.

(so called), only I say that there is a liberty which is not disorder, and that this is by no means less free than the other, but more.[5]

No less objectionable to devout and conservative men like Maxwell were the public lectures and popular writings of the Victorian scientific naturalists. Maxwell seems to have gotten along well with T. H. Huxley, but there is ample evidence that he found John Tyndall's ideas wrongheaded and disagreeable. In a commentary inspired by some remarks of P. G. Tait, Maxwell implied that Tyndall had "martyred his scientific authority," adding: "If he writes it in a dry manner it is bad enough, but the harm is confined to students. But if he seasons it for the public and the public swallows it, then it is a sad misuse of words to say that this is useful work."[6] More playful, but perhaps equally pointed, was the "Tyndallic Ode" that Maxwell composed to parody Tyndall's very successful lecture style. Maxwell's Tyndallic lecturer began with some showy demonstration experiments, then resolved that "These transient facts / These fugitive impressions / Must be transformed by mental acts / To permanent possessions," and finally proceeded to construct a metaphysical tower of Babel upon his experimental sand:

> Go to! prepare your mental bricks,
> Fetch them from every quarter,
> Firm on the sand your basement fix
> With best sensation mortar.
> The top shall rise to heaven on high—
> Or such an elevation,
> That the swift whirl with which we fly
> Shall conquer gravitation.[7]

Scientific naturalism sprang up in Britain in defense of Darwin's theory of evolution, but its prominent spokesmen included the physicist Tyndall and the mathematician W. K. Clifford as well as biologists like Huxley, and its rhetoric was saturated with concepts and terminology from mechanics. Naturalism was in part an instrument of professionalism, and biologists felt far less threatened by the scientific imperialism of physics than by the Anglican clergymen who insisted on bringing biology within the domain of natural theology and, worse, who occupied

[5] *Ibid.*, pp. 305-306.
[6] See Maxwell comment on galleys of Tait rebuttal to Tyndall in *JCMP* 7655 III d/5.
[7] *Maxwell*, pp. 635-636.

scarce scientific posts when dedicated dissenting scientists had difficulty finding suitable positions. The appeal to physics embodied a drive to sever ties with natural theology, to remove biology from the empire of teleology. The object, as Huxley put it, was to "reduce all scientific problems, except those which are purely mathematical, to questions of molecular physics."[8] Huxley was appalled by the audacity of those who claimed authority to judge scientific theories such as Darwin's on the basis of what amounted to "sacerdotal pretensions,"[9] and Tyndall, displaying equal sensitivity to the purity of science and the sanctity of the scholar's turf, boldly announced the "impregnable position of science" as follows: "We claim, and we shall wrest from theology the entire domain of cosmological theory. All schemes and systems which thus infringe upon the domain of science must, in so far as they do this, submit to its control, and relinquish all thought of controlling it."[10]

If natural science was to implement these ambitious claims, it could afford to leave no gaps, and the idea of chance in science was viewed with skepticism by Huxley and Tyndall. One reason for opposition to Darwin's theory was that it replaced teleological purpose by chance variation—what John Herschel was reputed to have called "the law of higgledy-piggledy."[11] By "chance," Darwin of course meant undirected, not uncaused, but his defenders were nonetheless disturbed by the absence of a law of biological variation. Thus Huxley wrote to Hooker in 1861: "Because no law has yet been made out, Darwin is obliged to speak of variation as if it were spontaneous or a matter of chance, so that the bishops and superior clergy generally (the only real atheists and believers in chance left in the world) gird at him as if he were another Lucretius."[12]

This was by no means the last mention in late Victorian scientific discussion of the ancient atomists or of issues involving chance and mechanical determinism with which they were habitually associated. John

[8] T. H. Huxley, "The Scientific Aspects of Positivism," in *Lay Sermons, Addresses, and Reviews* (London, 1895), p. 144.

[9] Huxley, "On the Hypothesis that Animals are Automata, and its History" (1874), in *Collected Essays* (9 vols., New York, 1968), vol. 1, p. 249. See also Frank M. Turner, "The Victorian Conflict Between Science and Religion: A Professional Dimension," *Isis*, 69 (1978), 356-376.

[10] John Tyndall, "The Belfast Address" (1874), in George Bassala et al., eds., *Victorian Science* (New York, 1970), 436-478, pp. 474-475.

[11] David L. Hull, *Darwin and his Critics: The Reception of Darwin's Theory of Evolution by the Scientific Community* (Cambridge, Mass., 1973), pp. 7, 61.

[12] Leonard Huxley, *Life and Letters of Thomas H. Huxley* (2 vols., New York, 1901), vol. 1, p. 245.

Tyndall found it possible to enlist ancient atomism in the naturalists' cause by sidestepping Lucretius, whose *De rerum natura* contained the troubling doctrine of the swerve, and invoking directly his predecessor Democritus, who had thoughtfully refrained from burdening the popular historian with any surviving works. In Tyndall's controversial 1874 address to the British Association in Belfast, he listed six fundamental scientific propositions of Democritus and pronounced the first five of them "a fair general statement of the atomic philosophy as now held." These were the following:

> 1. From nothing comes nothing. Nothing that exists can be destroyed. All changes are due to the combination and separation of molecules. 2. Nothing happens by chance. Every occurrence has its cause from which it follows by necessity. 3. The only existing things are the atoms and empty space, all else is mere opinion. 4. The atoms are infinite in number and infinitely various in form; they strike together, and the lateral motions and whirlings which thus arise are the beginnings of worlds. 5. The varieties of all things depend upon the varieties of their atoms, in number, size, and aggregation.[13]

Tyndall's line of thinking was more troubling to Maxwell even than Buckle's since it was precisely one of Maxwell's specialties, atomism, that was dressed up to support this deterministic and materialistic doctrine. Maxwell had long held that dynamical explanation represented a scientific ideal, and he had written just before the storm broke, in 1856, that "if we know what is at any assigned point of space at any assigned instant of time, we may be said to know all the events in Nature. We cannot conceive any other thing which it would be necessary to know."[14] It was not a little discomfiting to see these ideas applied vigorously to life and mind, in explicit opposition to religion. Maxwell preferred to cast his lot with Lucretius:

> When Lucretius wishes us to form a mental representation of the motion of atoms, he tells us to look at a sunbeam shining through a darkened room (the same instrument of research by which Dr Tyndall makes visible to us the dust we breathe), and to observe the

[13] Tyndall, "Belfast Address" (n. 10), p. 443. See also Frank M. Turner, "Lucretius among the Victorians," *Victorian Studies*, 16 (1972-73), 329-348; and Turner, *The Greek Heritage in Victorian Britain* (New Haven, 1981).

[14] *Maxwell*, p. 238.

motes which chase each other in all directions through it. This motion of the visible motes, he tells us, is but a result of the far more complicated motion of the invisible atoms, which knock the motes about. In his dream of nature, as Tennyson tells us he

> "Saw the flaring atom streams
> And torrents of her myriad universe
> Running along the illimitable inane,
> Fly on to clash together again, and make
> Another and another frame of things
> For ever."

And it is no wonder that he should have attempted to burst the bonds of Fate by making his atoms deviate from their courses at quite uncertain times and places, thus attributing to them a kind of irrational free will, which on his materialistic theory is the only explanation of that power of which we ourselves are conscious.[15]

Maxwell certainly did not wish to imply that natural science should be constrained to support certain religious doctrines. "The rate of change of scientific hypotheses is naturally much more rapid than that of Biblical interpretations," he told a clergyman in 1876, "so that if an interpretation is founded on such an hypothesis, it may help to keep the hypothesis above ground long after it ought to be buried and forgotten."[16] He thought it appropriate for Christians to seek to harmonize their science with their faith, but insisted that the results of this effort were valid only for the individual involved, and only for a limited time.[17] Arguments about the inconsistency of contemporary science with religion seemed to him to violate this dictum, proving, as his friends P. G. Tait and Balfour Stewart explained, that the opponents of faith, and not religious men, are the true dogmatists.[18] "No mind ever delighted more in speculation," observed Lewis Campbell of Maxwell, "and yet none was ever more jealous of the practical application or the popular dissemination of what appeared to him as crude and half-baked theories about the highest subjects."[19] Karl Pearson came away from a

[15] Maxwell, "Molecules," in *SP*, vol. 2, p. 373.

[16] *Maxwell*, p. 394.

[17] *Ibid.*, p. 405.

[18] Peter Guthrie Tait and Balfour Stewart, *The Unseen Universe, or Physical Speculations on a Future State* (London, 1875), p. v.

[19] *Maxwell*, p. 322.

Cambridge examination by Maxwell for the Smith's prize with a less charitable view:

> The conversation turned on Darwinian evolution; I can't say how it came about, but I spoke disrespectfully of Noah's flood. Clerk Maxwell was instantly aroused to the highest pitch of anger, reproving me for want of faith in the Bible. I had no idea at the time that he had retained the rigid faith of his childhood, and was, if possible, a firmer believer than Gladstone in the accuracy of Genesis.[20]

It was in this context that Maxwell developed his ideas about the statistical character of the second law of thermodynamics, and more generally about the inescapable imperfection of human knowledge. Maxwell first pointed out that his dynamical theory of gases implied the possibility of violating the second law in 1867, in a playful letter to P. G. Tait. There he introduced the "very observant and neat-figured being," later dubbed by William Thomson Maxwell's "demon," whose mission was to "pick a hole" in the second law. The demon needed merely to be set to guard a small hole in the elastic wall between gases at different temperatures and it could—by allowing only the most energetic molecules to pass from the cold to the hot side, and only the least energetic in the opposite direction—cause heat to flow from a cold gas to a warm one. Notwithstanding Thomson's name for this fictional creature, which Maxwell found objectionable, no supernatural powers were required, but only an exaggerated level of ordinary ones.[21]

Maxwell's invention implied that some physical principles, among them the second law, were really as much attributes of human perceptions as of nature itself. He wrote a decade later in the *Encyclopaedia Britannica* under the heading "Diffusion":

> The idea of dissipation of energy depends on the extent of our knowledge. Available energy is energy which we can direct into any desired channel. Dissipated energy is energy which we cannot lay hold of and direct at pleasure, such as the energy of the confused agitation of molecules which we call heat. Now, confusion, like

[20] Karl Pearson, "Old Tripos Days at Cambridge, as seen from another Viewpoint," *Mathematical Gazette*, 20 (1936), 27-36.

[21] The original letter is printed in C. G. Knott, *Life and Scientific Work of Peter Guthrie Tait* (Cambridge, 1911), pp. 213-214. See also the undated letter of Maxwell to Tait in *ibid.*, pp. 214-215.

> the correlative term order, is not a property of material things in themselves, but only in relation to the mind that perceives them. A memorandum-book does not, provided it is neatly written, appear confused to an illiterate person, or the owner who understands it thoroughly, but to any other person able to read it appears to be inextricably confused. Similarly the notion of dissipated energy could not occur to a being who could not turn any of the energies of nature to his own account, or to one who could trace the motion of every molecule and seize it at the right moment. It is only to a being in the intermediate stage, who can lay hold of some forms of energy while others elude his grasp, that energy appears to be passing from the available to the dissipated state.[22]

Maxwell first adumbrated a connection between the indeterminacy of certain thermodynamic principles and their statistical character in 1868, when he compared the tendency for gas molecules to assume the normal velocity distribution to the mixing of black and white balls in a box. Two years later he wrote to the young physicist Rayleigh that the second law had "the same degree of truth as the statement that if you throw a tumblerful of water into the sea you cannot get the same tumblerful of water out again."[23] Maxwell began to develop an argument about the wider implications of the statistical method in physics in the inaugural lecture he delivered in 1871 upon becoming head of the new Cavendish Laboratory at Cambridge. There he remarked, explicitly and publicly, that predictions based on statistical knowledge were inherently uncertain. The gas laws evidently were of a different character from dynamical principles, yielding a form of knowledge whose implications he thought had wide interest. Maxwell pointed immediately to the vexed question of human freedom, arguing:

> . . . the statistical method . . . , which in the present state of our knowledge is the only available method of studying the properties of real bodies, involves an abandonment of strict dynamical principles, and an adoption of the mathematical methods belonging to

[22] Maxwell, "Diffusion," in *SP*, vol. 2, p. 646.

[23] R. J. Strutt, Fourth Baron Rayleigh, *Life of John William Strutt, Third Baron Rayleigh* (Madison, Wisc., augmented ed., 1968), p. 47. See also Stephen Brush, "Randomness and Irreversibility," in *The Kind of Motion We Call Heat* (2 vols., Amsterdam, 1976), 543-654, p. 590, *passim*; also Brush, "Irreversibility and Indeterminism," chap. 2 of his *Statistical Physics and the Atomic Theory of Matter from Boyle and Newton to Landau and Onsager* (Princeton, 1981).

> the theory of probability. It is probable that important results will be obtained by the application of this method, which is as yet little known and is not familiar to our minds. If the actual history of Science had been different, and if the scientific doctrines most familiar to us had been those which must be expressed in this way, it is possible that we might have considered the existence of a certain kind of contingency a self-evident truth, and treated the doctrine of philosophical necessity as a mere sophism.[24]

By 1873 Maxwell had developed this line of reasoning into a full argument against the increasingly threatening doctrine of mechanical determinism. His aim was not to use physics to demonstrate the existence of human freedom, for he adhered to the Common Sense tenet that belief in free will arises naturally from reflection on the mind's own activity. He carried out his examination of physical knowledge precisely in order to show that known scientific principles in fact proved nothing on this vital issue. Maxwell did not, however, deny natural science all relevance to metaphysical issues. He suggested that by showing what freedom could not be, physics offered valuable guidance in identifying what it was. The inherent interest of this project, as well as the need to demonstrate the possibility of free action that did not violate any laws of physics, inspired Maxwell to devote considerable attention to the physical interpretation of free will.

Already in 1862 Maxwell had observed to Lewis Campbell that the conservation of energy permitted the soul to be switchman, but not mover of the body. "There is action and reaction between body and soul," he wrote, "but it is not of a kind in which energy passes from one to the other." The direction which the soul could give to the energy of the body was comparable, he thought, to the relation between trigger and gun, or pointsman and train. Maxwell disclaimed any pretense to having solved this issue—"It is well that it will go, and that we remain in possession, though we do not understand it,"[25]—but he was not long willing to rest secure in ignorance. The frequent invocation of conservation laws against human freedom convinced the defenders of metaphysical freedom of their obligation to demonstrate the possibility that the will could operate without any expenditure of energy.[26] The most-

[24] Maxwell, "Introductory Lecture on Experimental Physics," in *SP*, vol. 2, p. 253.

[25] *Maxwell*, p. 336.

[26] As C. S. Peirce noted of Simon Newcomb, defenders of free will were strangely indifferent to violations of Newton's third law of motion. See Charles Sanders Peirce, "Variety and

cited physicist's defense of free will was published anonymously in the *North British Review* in 1868 by Fleeming Jenkin, who evidently sent Maxwell the proofs in advance.[27] His argument, cast as a defense of the Lucretian doctrine of the swerve, was virtually the same as Maxwell's; Jenkin wrote that "if mind or will deflects matter as it moves, it may produce all the consequences claimed by the Wilful school, and yet it will add neither energy not matter to the universe."[28]

The main incentive for Maxwell's reflections on the limits of statistical knowledge was his desire to show that freedom was not inconsistent with the laws of nature properly known by contemporary science. His fundamental argument was that the statistical method, the only means by which humans can attain general knowledge of a molecular universe, yields only generalizations about the mass of molecules and provides no information about individuals. He wrote in his 1871 textbook on heat:

> It is therefore possible that we may arrive at results which, though they fairly represent the facts as long as we are supposed to deal with a gas in mass, would cease to be applicable if our faculties and instruments were so sharpened that we could detect and lay hold of each molecule and trace it through all its course.
>
> For the same reason, a theory of the effects of education deduced from a study of the returns of registrars, in which no names of individuals are given, might be found not to be applicable to the experience of a schoolmaster who is able to trace the progress of each individual pupil.[29]

Hence the need to resort to statistics guaranteed not only that nothing was known of the particular circumstances of individual molecules but that even the laws of their motion might bear no determinate relation to the observable regularities of the mass.

Maxwell developed these arguments most fully in a paper that he read early in 1873 for Eranus, a club of past Cambridge Apostles, titled "Does the Progress of Physical Science tend to give any advantage to the

---

Uniformity" (1903), in Charles Hartshorne et al., eds., *Collected Papers of Charles Sanders Peirce* (8 vols., Cambridge, Mass., 1931-1958), vol. 6, 67-85, p. 70.

[27] See Jenkin's letter to Maxwell, 10 July 1868, in *JCMP*, 7655/II, Box 1.

[28] [Fleeming Jenkin], "The Atomic Theory of Lucretius," *North British Review*, 48 (1868), 111-128, p. 118. This paper was widely circulated, and quoted by, among others, William Thomson, "The Structure of Matter and the Unity of Science," in Bassala, *Victorian Science* (n. 10), 101-128, p. 109, and Tait and Stewart, *Unseen Universe* (n. 18), pp. 181-182.

[29] Maxwell, *Theory of Heat* (1871; London, 1904), pp. 315-316.

opinion of Necessity (or Determinism) over that of the Contingency of Events and the Freedom of the Will?" Here atomism, thermodynamics, and statistics, a trio of sources for those deterministic arguments seen as most compelling in mid-Victorian Britain, were turned on their heads and reinterpreted as evidence for the possibility of human freedom. Maxwell began by noting that the statistical method, which "has Laplace for its most scientific and Buckle for its most popular expounder," was by its nature an imperfect one, applicable precisely when the course of individual events cannot be charted or explained. Its use, he observed depends on an assumption "that the effects of widespread causes, though very different in each individual, will produce an average result on the whole nation, from a study of which we may estimate the character and propensities of an imaginary being called the Mean Man." Far from being unique to social science, however, this form of reasoning constituted the basis for "all our knowledge of matter," assuming that the molecular hypothesis was true. "A constituent molecule of a body has properties very different from those of the body to which it belongs," he continued, and "those uniformities which we observe in our experiments with quantities of matter containing millions of millions of molecules are uniformities of the same kind as those explained by Laplace and wondered at by Buckle arising from the slumping together of multitudes of causes each of which is by no means uniform with the others."[30]

This, evidently, left open a certain space of ignorance, within which the will could operate without its direct effects being perceived. In this respect "our free will is at best like that of Lucretius's atoms," extending only over an infinitesimal range. Under certain circumstances, however, these limitations might be transcended. Maxwell wrote: "It has been well pointed out by Professor Balfour Stewart that physical stability is the characteristic of those systems from the contemplation of which determinists draw their arguments and physical instability that of those living bodies, and moral instability that of those developable souls, which furnish to consciousness the conviction of free will." The mind, he indicated, was a system for generating and regulating instabilities, through which human freedom and moral responsibility are expressed. "In the course of this our mortal life, we more or less frequently find ourselves on a physical or moral watershed, where an imperceptible de-

[30] *Maxwell*, 434-444, pp. 438-439.

viation is sufficient to determine into which of two valleys we shall descend. The doctrine of free will asserts that in some cases the Ego alone is the determining cause."[31] The statistical character of knowledge insured that no finite observer could be in a position to refute this possibility.

A few years later, Maxwell encountered a refinement to this solution to the free-will problem that seemed to him more satisfactory. His source was a group of French and Belgian Catholic scientists who were similarly repelled by the deterministic materialism of Tyndall, and also of Comte's disciple Emile Littré and the American John Draper. As Mary Jo Nye has shown, Ignace Carbonelle, Joseph Delsaulx, and Julien Thirion had, like Maxwell, found in the kinetic gas theory justification for the belief that some measure of uncertainty and unpredictability characterized scientific laws. They were especially interested in the problem of Brownian motion, the irregular movements of tiny but visible particles suspended in fluids, which was first linked to the kinetic theory by Delsaulx. Maxwell found his inspiration in the work on fluid mechanics of another member of this group, Joseph Boussinesq, who showed in 1878 that under certain conditions the differential equations regulating a mechanical system should have multiple solutions at certain points of singularity, and hence that determinate forces might produce no uniquely determined motion. In that event, it would be possible for a "directing principle" such as the will to determine which possible solution actually occurred.[32]

Oddly enough, Maxwell developed this line of thinking most fully in a letter to Francis Galton, with whom he did not regularly correspond and whom he once characterized as a man "whose mission it seems to be to ride other men's hobbies to death."[33] Maxwell's comments, which appeared as a postscript to an otherwise brief and routine letter renewing his subscription to the Philosophical Club, were as follows:

> Do you have any interest in Fixt Fate, Free Will, &c. If so Boussinesq [of hydrodynamic reputation] "Conciliation du veritable determinisme mécanique avec l'existence de la vie et de la liberté mo-

[31] *Ibid.*, pp. 440-441.

[32] See Mary Jo Nye, "The Moral Freedom of Man and the Determinism of Nature: The Catholic Synthesis of Science and History in the *Revue des questions scientifiques*," *BJHS*, 9 (1976), 274-292, pp. 277-281. The idea of a directing principle derives from A. A. Cournot and was revived by Boussinesq's mentor, St. Venant. Nye also mentions Claude Bernard in this connection.

[33] *Maxwell*, p. 390.

rale" (Paris, 1878) does the whole business by the theory of the singular solutions of the differential equations of motion. Two other Frenchmen have been working on the same or similar tracks. Cournot (now dead) and de St. Venant [of elastic reputation Torsion of Prism &c.].

Another, also in the engineering line of research. Philippe Breton seems to me to be somewhat like minded with these.

There are certain cases in which a material system, when it comes to a phase in which the particular path which it is describing coincides with the envelope of all such paths may either continue in the particular path or take to the envelope (which in these cases is also a possible path) and which course it takes is not determined by the forces of the system (which are the same for both cases) but when the bifurcation of path occurs, the system, ipso facto, invokes some determining principle which is extra physical (but not extra natural) to determine which of the two paths it is to follow.

When it is on the enveloping path it may at any instant, at its own sweet will, without exerting any force or spending any energy, go off along that one of the particular paths which happens to coincide with the actual condition of the system at that instant.

In most of the former methods Dr. Balfour Stewart's, &c. there was a certain small but finite amount of travail decrochant or trigger-work for the will to do. Boussinesq has managed to reduce this to mathematical zero, but at the expense of having to restrict certain of the arbitrary constants of the motion to mathematically definite values, and this I think will be found in the long run very expensive. But I think Boussinesq's method is a very powerful one against metaphysical arguments about cause and effect and much better than the insinuation that there is something loose about the laws of nature, not of sensible magnitude but enough to bring her round in time.[34]

That Maxwell's argument about the imperfection of physical knowledge and even the possible incompleteness of mechanical laws should have developed out of his thoughts on the implications of statistics for physics seems completely plausible—perhaps too plausible—in retrospect. To be sure, Maxwell himself came to see this connection as a natural one. He often noted how much looseness there was in statistical

[34] Maxwell to Galton, 26 Feb. 1879, *FGP*, folder 191. I have silently closed a parenthesis in the first paragraph. The material inside brackets is Maxwell's.

generalizations, if not in the laws of nature themselves. Using one of Galton's favorite similes, he remarked in an unpublished manuscript that "the population of a watering-place, considered as a mere number, varies in the same way whether its visitors return to it season after season or whether the annual flock consists each year of fresh individuals."[35] He was right, of course, but it should not be forgotten that these ideas were new, that the imperfection of statistical knowledge had received scarcely any attention before 1857 except by those who rejected it entirely, and that Maxwell began pursuing this line of thought only when he was provoked by the vehement expression of statistical and mechanical determinism by Buckle and Tyndall.

The persuasive force of Maxwell's conclusions on statistical knowledge is evident from the circumstance that Boltzmann, who was completely unsympathetic to probabilism in thermodynamics, was compelled by problems arising within statistical gas theory to move a long way in the same direction. Still, statistics did not require the embrace of non-mechanical causation, much less of indeterminism, for it was always possible to believe the underlying phenomena to be mechanically determined, and only the large-scale numerical regularities probabilistic. Francis Galton saw no reason to move away from determinism on account of the statistical character of knowledge, as is in some way indicated by his unresponsive response to Maxwell's unsolicited missive, where he noted that in his own introspective psychological experiments he had been "almost frightened to find how distinctly cause and effect seem to govern everything."[36] Boltzmann eventually accepted the im-

[35] Undated manuscript, *JCMP*, 7655/V f/11.

[36] Galton to Maxwell, 27 Feb. 1879 in *JCMP*, 7655/II (Box I). The relevant text of Galton's reply is as follows:

> Very many thanks for the free will &c.—After all, does the question not coincide in final principle with that of unstable equilibrium?—I am rather busy just now with experiments on the workings of my own mind, and am almost frightened to find how distinctly cause & effect seem to govern everything.—If you happen to see the forthcoming "Nineteenth Century" (March) & care to look at a short paper in it by me "Psychometric facts"—it will as you will see describe something of what I mean, & how I continue to drag ideas that are almost out of the reach of consciousness into full light.
>
> I look forward with infinite interest to—moreover—to hear Sir W. Thomson's exposition of your sorting demon at the Royal Institution. I have little doubt that some will go to see art illustrations of the medieval Devil.

In "Free Will—Observations and Inferences," *Mind*, 9 (1884), 406-413, p. 412, Galton concluded even more decisively that "man is little more than a conscious machine," and that "we must understand the word 'spontaneity' in the same sense that a scientific man understands the word 'chance.' He thereby affirms his ignorance of the precise causes of an event, but he does not in any way deny the possibility of determining them."

plications of his statistical approach, but always emphasized the positive content and power of its accomplishments, not its limitations.

BOLTZMANN, STATISTICS, AND IRREVERSIBILITY

The kinetic gas theory constituted Boltzmann's life work, and the language and concepts of probability theory were central to his research in this field from the beginning. He was never comfortable, however, with the idea that the form of his mathematics implied any indeterminacy in the resulting laws. Boltzmann would have derived no satisfaction from "picking a hole" in the second law of thermodynamics, as Maxwell had. He always stressed instead the certainty of science:

> A precondition of all scientific knowledge is the principle of the complete (*eindeutig*) determination of all natural processes, or, as applied to mechanics, the complete determination of all movements. This principle declares, that the movements of a body do not occur purely accidentally, going sometimes here, sometimes there, but that they are completely determined by the circumstances to which the body is subject.[37]

As we have seen, Boltzmann invoked the analogy of gas theory with social statistics in order to bolster, not undermine, the certainty of his scientific conclusions. Nevertheless, he did not glide over difficulties—at least not for long—but faced them with resilience and creativity. His faith in molecular models was so deep as to be virtually a matter of principle—he thought continuity was meaningful even in mathematics only as the limit of finite differentials[38]—and he held to atomism even when he found himself compelled by it to concede that certain macroscopic laws, such as the second law of thermodynamics, were only statements of high probability.

Boltzmann was an Austrian, the son of a Viennese tax official. He was born in 1844, about a decade later than most of the holistic historical

[37] Boltzmann, "Über die Grundprinzipien und Grundgleichungen der Mechanik" (1899), in *PS*, pp. 276-277.

[38] It may be noted in this connection that, according to his friend and Vienna colleague Franz Clemens Brentano, Boltzmann thought Joseph Bertrand's paradox of probability (see n. 44, chap. 3, above) would be cleared up once the fallacy of assuming a continuum was recognized. See Brentano, "Von der Unmöglichkeit absoluten Zufalls" (1916) in *Versuch über die Erkenntnis* (Hamburg, 1970), p. 141.

economists from Germany and Austria who sought to reinterpret statistical reasoning so as to purge it of all connection with atomism, mechanical determinism, and natural law. Boltzmann grew up under the ascendancy of liberalism, and such evidence as we have suggests that he was a steadfast liberal throughout his life. That his political ideas did not change seriously under the pressure of the "worker question," the German unification, and the depression that began in 1873 may perhaps be attributed in part to the greater remove from politics that is possible for physicists than for economists, although it may be significant that in Austria, deductive, liberal economics held sway even when economists in the new *Reich* had moved almost unanimously to historicism. Boltzmann, it may be observed, applauded the precision and efficiency of a free economy, defending capitalism against those who derided it as worship of Mammon.[39] His loyalty to a culture of *Recht und Wissenschaft* can be seen in his opposition to the radical pan-Germans and anti-Semites who became increasingly troublesome for the University of Vienna, and indeed the whole Austrian empire, during the 1880s and thereafter. Boltzmann's love for Austria was shown in his dedication to a cosmopolitan empire, and his lament about 1866 as the "year of misfortune" can be seen as reaffirmation of his liberal faith.[40] It is also worth noting his persistent fascination with the United States, which he visited three times.

In science, Boltzmann applauded mechanical explanation in every domain. He called Darwin's theory of evolution a mechanical one, and he argued that human thought, whether expressed in scientific theory or in fondness for music, was "mechanically necessary." As a young physicist, he worked for a time in the lab of Hermann von Helmholtz, the great champion of the ideas of the *Naturwissenschaften* against the *Geisteswissenschaften*. Boltzmann's consistent adherence to atomism transgressed the guiding metaphysics not only of historical economists, but also of late-century energeticists and Machian positivists like Wilhelm Ostwald and Ernst Zermelo, who held that the transformations of energy perceptible in experiment and observation must be accepted as

[39] Boltzmann, "Über die Principien der Mechanik" (1900), in *PS*, p. 322.

[40] See Engelbert Broda, *Ludwig Boltzmann: Mensch, Physiker, Philosoph* (Vienna, 1955), p. 7; also Boltzmann, "Josef Stefan" (1895), in *PS*, p. 102. On fin-de-siècle Austrian liberalism and the revolts against it, see Allen Janik and Stephen Toulmin, *Wittgenstein's Vienna* (New York, 1973); William J. McGrath, *Dionysian Art and Populist Politics in Austria* (New Haven, 1974), and, especially, Carl Schorske, *Fin-de-Siècle Vienna: Politics and Culture* (New York, 1980).

primary, not explained as interactions of purely hypothetical molecules. During his last years—before he fulfilled his statistical destiny by taking his own life—Boltzmann came to feel scientifically isolated. Faced with a wide resistance to the reduction of communities or of perceptible phenomena to individuals, whether human or atomic, he began to call himself a "reactionary," one who has been left behind, "the last one remaining of those who embraced the old [mechanical scientific picture] with their full souls." But he held to his principles, vowing "so much as it lies in my power, to work out in as clear and logically-ordered a way as possible the results of the old classical theory in order that the many good and eternally useful doctrines which I am persuaded are yet contained in it will not have to be discovered a second time."[41]

Boltzmann's earliest paper on the kinetic theory made frequent reference to probability, and used probability mathematics extensively, yet it contained, as Lorenz Krüger has pointed out, not a trace of indeterminacy. Boltzmann there treated probabilities as perfectly interchangeable with frequencies. That is, he moved from probabilities expressed as the fraction of time which an arbitrary molecule would, in the long run, spend in any given state to the actual proportion of molecules in that state at a given time without evincing the least awareness that this might be problematical. Like Clausius, he invoked probability largely in order to justify the introduction of simplifying assumptions so as to rise above the chaos of molecular motions and to attain simple, analytic expressions. Boltzmann believed then that he had given a rigorous analytical proof of the second law, reducing it to a proposition in mechanics.[42] Although his methods had become much more refined by 1872, when he introduced his H-theorem, that derivation also flowed from the assumption that frequencies were interchangeable with probabilities. Boltzmann presupposed, as indeed had Maxwell in his technical papers, that collisions of every sort occurred in exact proportion to certain sta-

[41] Boltzmann, "Über die Entwicklung der Methoden der theoretischen Physik in neuerer Zeit" (1899), in *PS*, p. 205. More generally, see Erwin N. Hiebert, "The Energetics Controversy and the New Thermodynamics," in Duane H. D. Roller, ed., *Perspectives in the History of Science and Technology* (Norman, Oklahoma, 1971), pp. 67-86; also Daniel Gasman, *The Scientific Origins of National Socialism* (New York, 1971); and, for a French perspective, chap. 1 of Mary Jo Nye, *Molecular Reality* (New York, 1972).

[42] Boltzmann, "Über die mechanische Bedeutung des zweiten Hauptsatzes der Wärmetheorie," in *WA*, vol. 1, pp. 9-35; see also Lorenz Krüger, "Reduction as a Problem: Some Remarks on the History of Statistical Mechanics from a Philosophical Point of View," in Jaakko Hintikka et al., eds., *Probabilistic Thinking, Thermodynamics and the Interaction of the History and Philosophy of Science* (Dordrecht, 1981), 147-174, pp. 152-154.

tistical parameters, such as the product of frequencies of molecules of specified types. He made no allowance for variation around some mean value, and his conclusion admitted no exceptions. He wrote in 1872: "It is accordingly rigorously proved that, whatever the initial distribution of kinetic energy may have been, it must always necessarily approach the Maxwellian form after a very long time has elapsed."[43]

To be sure, Boltzmann had acknowledged as early as 1868 that some initial configurations, among them the one in which all molecules move in the same direction with the same velocity, perpendicular to the walls with which they collide, would not converge to the Maxwell distribution.[44] The possibility of exceptions to his H-theorem only became interesting to him, however, when he was confronted with the so-called "reversibility paradox" by his colleague and former teacher at Vienna, Josef Loschmidt. That the second law could be violated simply by reversing the velocities of all particles in a closed system, causing time, in effect, to run backwards, had been discussed privately in Britain since 1867, when it was noted by William Thomson on the letter in which Maxwell had first mentioned his sorting demon to their mutual friend P. G. Tait.[45] Loschmidt, an atomist like Boltzmann, was inspired to present this argument, which he invented independently of Thomson and Maxwell, by his concern over the "heat death" that would necessarily ensue in time if, as the second law indicated, entropy must increase monotonically until all the energy in the universe has been converted into the disordered motion of heat. Boltzmann quickly recognized that he was confronted with a deep paradox, the problem of reconciling the flow of heat, which manifestly depends on the direction of time, with the laws of mechanics, which do not. Elastic collisions in a mechanical system are perfectly reversible, but heat is always observed to flow from warm to cold bodies.

Boltzmann responded to this objection with some of his best technical work, but he required more than a decade to formulate a satisfactory answer to it. During the interim he offered several contradictory resolutions of the paradox, without ever admitting that his opinions had altered in the least. He conceded immediately what he now held up as

[43] Boltzmann, "Weitere Studien über das Wärmegleichgewicht unter Gasmolekülen," in WS, vol. 1, p. 345; translation from Martin Klein, "The Development of Boltzmann's Statistical Ideas," *Acta Physica Austrica*, Suppl. X (1973), 53-106, p. 73.

[44] Boltzmann, "Studien über das Gleichgewicht der lebendigen Kraft zwischen bewegten materiellen Punkten" (1868), in WA, vol. 1, p. 96.

[45] See Klein, "Boltzmann's Statistical Ideas" (n. 43), p. 75.

self-evident—that a mechanical derivation of the second law could not possibly apply with absolute certainty, since every distribution of velocities is at least possible. "Mathematical probability itself teaches this."[46] But, he noted, there are infinitely many more uniform than nonuniform distributions of heat, so that "the number of states which lead to a uniform distribution after a time interval t_1 must be much larger than the number of those which lead to a non-uniform distribution."[47] It was in this context that Boltzmann resolved to perform the calculation that led to his important combinatorial paper of 1877, discussed above, which gave not just the probability that a given molecule would occupy a particular velocity differential, but the probability of any given distribution of velocities in the entire system. He solved the problem, we may recall, by using combinatorial mathematics to find an expression for the probability of any given distribution of velocities, then showing that to maximize this probability was equivalent to maximizing entropy, and hence to minimizing the H-function of his 1872 paper.

This was, by any standard, an impressive achievement, but as a solution to Loschmidt's reversibility paradox it had certain deficiencies. The characteristic property of the H-function was to decline uninterruptedly until it reached its minimum, which corresponded to the Maxwell distribution. Hence it appeared that through the use of sophisticated probability, the laws of mechanics had somehow been forced to yield directionality, even though they were themselves completely reversible. Boltzmann explained: "The initial state will in most cases be a highly improbable one; from it, the system will hasten to ever more probable states until it finally reaches the most probable of all, i.e. thermal equilibrium."[48] Boltzmann evidently found this resolution attractive for a considerable time, since he wrote a decade later that "a given system of bodies can never by itself pass into an absolutely equally probable state, but instead passes always into a more probable one."[49] As late as 1897 he referred to the tendency for a system to evolve towards ever

[46] Boltzmann, "Bemerkungen über einige Probleme der mechanischen Wärmetheorie" (1877), in *WA*, vol. 2, p. 120.

[47] *Ibid.*, pp. 120-121; translation from Thomas Kuhn, *Black-Body Theory and the Quantum Discontinuity* (Oxford, 1978), p. 47.

[48] Boltzmann, "Über die Beziehung zwischen dem zweiten Hauptsatze der mechanischen Wärmegleichgewicht und der Wahrscheinlichkeitsrechnung . . . ," in *WA*, vol. 2, p. 165; translation from Kuhn, *Black-Body Theory* (n. 47), p. 58.

[49] Boltzmann, "Der zweite Hauptsatz der mechanischen Wärmetheorie" (1886), in *PS*, p. 48; translation from Brush, *Heat* (n. 23), p. 612.

more probable states as "almost tautologous."[50] Loschmidt's objection implied, however, that for every entropy-increasing process there is an entropy-decreasing process, namely the one that runs backwards from the final to the initial state. Clearly it is tautologous that a system at a given time is much more likely to be in a probable than an improbable state, but neither the laws of mechanics nor those of probability furnish grounds for preferring a transition from an improbable to a probable state over the reverse, unless exceptional initial conditions are assumed.

Boltzmann's treatment of the uncertainty implied by his statistical approach was not fully free of contradiction even in the *Lectures on Gas Theory* that he published in 1896 and 1898. In Part One of the *Lectures*, he published his H-theorem in what might be regarded as its deterministic form, which here depended on the assumption of "molecular disorder"—that is, he excluded in advance all those special configurations, such as Loschmidt's, that lead to a massive decrease of entropy. These, he argued, were artificial products of calculation, chosen "so as to violate intentionally the laws of probability."[51] Clearly, however, he did not at this time still view the H-theorem as fully deterministic, for only a few pages later he spoke of fluctuations in H, noting that from time to time it must rise slightly, though with a very high probability that it will fall quickly back to its minimum.[52] The H-theorem in its deterministic form does not permit H to do anything but fall until the minimum is reached, and it is evident that Boltzmann conceptualized his H-theorem in a way that was not expressed in his mathematics. He presumably viewed his exclusion of molecular-ordered states as a useful assumption, not strictly valid in theory but quite reliable enough in practice, by which he could be exempted from writing exceptions into the mathematics of the H-theorem.

In fact, Boltzmann had come to a satisfactory resolution of the reversibility paradox by this time. Pursuant to some 1894 discussions at a meeting of the British Association, where this difficulty was raised and discussed by Edward Culverwell, S. H. Burbury, George Bryan, and Joseph Larmor, Boltzmann published a paper in English in which he explained his position directly. According to the laws of probability, he

[50] Boltzmann, "Über die Unentbehrlichkeit der Atomistik in der Naturwissenschaft" (1897), in *PS*, p. 154.

[51] Boltzmann, *Lectures on Gas Theory* (1896-98), Stephen Brush, trans. (Berkeley, 1964), part 1, p. 41.

[52] *Ibid.*, p. 59.

wrote, it is, for an arbitrary system at a given time "extremely probable that H is very near to its minimum value; if it is greater, it may increase or decrease, but the probability that it decreases is always greater."[53] Although any given deviation from the Maxwell distribution has a finite probability, that probability declines precipitously as the extent of the deviation increases. The decline is so steep, in fact, that if one choose some arbitrary H-value significantly above the minimum, and observes all cases in which the system reaches that deviation in a very great time interval, the chosen level will almost always be a local maximum. The system will exceed that level by only an insignificant amount—Boltzmann, always disdainful of differentials, said not at all—and then decline back to its minimum.

The observation that perceptible fluctuations from the minimum are exceedingly rare provided Boltzmann with his answer to another so-called paradox that arose in the mid-1890s, this one called the recurrence paradox. It was formulated by Ernst Zermelo, an assistant in Max Planck's laboratory, who argued from a recurrence theorem by Poincaré that, given enough time, any system of molecules must eventually return to a configuration arbitrarily close to its initial state. Since no such recurrences have been observed, Zermelo argued, the molecular hypothesis must be false. Boltzmann answered straightforwardly that these recurrences must be quite rare, and that, for example, a small volume of gas could be expected to separate spontaneously into nitrogen and oxygen only once in $10^{10^{10}}$ years. He wrote:

> One may recognize that this is practically equivalent to *never*, if one recalls that in this length of time, according to the laws of probability, there will have been many years in which every inhabitant of a large country committed suicide, purely by accident, on the same day, or every building burned down at the same time—yet the insurance companies get along quite well by ignoring the possibility of such events. If a much smaller probability than this is not practically equivalent to impossibility, then no one can be sure that today will be followed by a night and then a day.[54]

[53] Boltzmann, "On Certain Questions of the Theory of Gases" (1895), in WA, vol. 3, p. 541. I have silently corrected the spelling in quotes from this essay.

[54] Boltzmann, *Lectures* (n. 51), part 2, p. 444.

The second law of thermodynamics is one of those principles that "have, theoretically, only the character of propositions in probability," but are "practically equivalent to natural laws."[55]

By the standard of his insurance simile, the real universe appeared to be a highly improbable place indeed, and Boltzmann began to feel the need for caution in applying statistical thermodynamics to the real world. His attitude moved toward that enunicated most clearly at the beginning of the new century by the American Josiah Willard Gibbs, who introduced statistical mechanics as a purely deductive science and observed that "there can be no mistake in regard to the agreement of the hypothesis with the facts of nature, for nothing is assumed in that respect. The only error into which one can fall, is the want of agreement between the premises and the conclusions, and this, with care, one may hope, in the main, to avoid."[56] Boltzmann did not go quite this far, but he did disclaim all obligation to account for the actual universe. The kinetic gas theory was a form of rational mechanics, whose object was not the actual universe but the laws governing the molecular configurations of an arbitrary gaseous system with certain assumed properties.

Nevertheless, Boltzmann did offer some opinions as to the place of the second law of thermodynamics in the observable universe. Two possibilities immediately presented themselves. One was that the universe may simply have begun in a highly improbable state, from which it was evolving gradually towards thermal equilibrium. It was also conceivable, however, that the region of space neighboring the earth represented a monumental fluctuation of the entropy curve, whose observation by man was not wholly fortuitous, since without this fluctuation there could be no life to perceive the surrounding circumstances. Already in 1877, Boltzmann had mentioned "a peculiar consequence of Loschmidt's principle" that the laws of thermodynamics, as deduced from mechanics, lead to the prediction of a primeval equilibrium of heat with no less certainty than they predict a heat death in the indefinite future.[57] By 1895 he was prepared to announce to his British colleagues a hypothesis, which he attributed to his former laboratory assistant, Dr.

[55] Boltzmann, "Entgegnung auf die wärmetheoretischen Betrachtungen des Hrn. Zermelo" (1896), in *WA*, vol. 3, p. 578.

[56] Josiah Willard Gibbs, *Elementary Principles in Statistical Mechanics* (New York, 1902), p. x.

[57] Boltzmann, "Bemerkungen" (n. 46), pp. 121-122.

Schuetz, that deviations from thermal equilibrium such as that observed from earth represented normal fluctuations in an inconceivably vast universe. He wrote:

> We assume that the whole universe is, and rests for ever, in thermal equilibrium. The probability that one (only one) part of the universe is in a certain state, is the smaller the further this state is from thermal equilibrium; but this probability is greater, the greater the universe itself is. If we assume the universe great enough we can make the probability of one relatively small part being in any given state (however far from the state of thermal equilibrium), as great as we please. We can also make the probability great that, though the whole universe is in thermal equilibrium, our world is in its present state. It may be said that the world is so far from thermal equilibrium that we cannot imagine the improbability of such a state. But can we imagine, on the other hand, how small a part of the universe this world is? Assuming the universe great enough, the probability that such a small part of it as our world should be in its present state, is no longer small.
>
> If this assumption were correct, our world would return more and more to thermal equilibrium; but because the whole universe is so great, it might be probable that at some future time some other world might deviate as far from thermal equilibrium as our world does at present. Then the aforementioned H-curve would form a representation of what takes place in the universe. The summits of the curve would represent the worlds where visible motion and life exist.[58]

The passage also illustrates aptly the difference between Maxwell's treatment of the uncertainty of the second law and that of Boltzmann. Whereas Maxwell stressed the imperfection of statistics and hence of most knowledge in physics, emphasizing the possible existence of instabilities and singularities, Boltzmann took the exceptional improbability of the visible universe and sought to fit it into a pattern of equilibrium conditioned by statistical regularity. Maxwell, while never advocating positive indeterminism or acausality, wished to establish the possibility of a nonphysical causality that depended on the action of the will. Boltzmann, never comfortable with the dependence of science on probabil-

[58] Boltzmann, "On Certain Questions" (n. 53), pp. 543-544.

ity, except in terms of stable frequencies, refused to countenance the idea that the most fundamental phenomena of nature could be other than mechanically determined. He came to regard the statistical character of knowledge as consistent with this outlook, even if, as he thought possible, probability was shown to apply even to molecular motions. He wrote in 1898:

> Since today it is popular to look forward to the time when our view of nature will have been completely changed, I will mention the possibility that the fundamental equations for the motion of individual molecules will turn out to be only approximate formulas which give average values, resulting according to the probability calculus from the interactions of many independent moving entities forming the surrounding medium—as for example in meteorology the laws are valid only for average values obtained by long series of observations using the probability calculus. These entities must of course be so numerous and must act so rapidly that the correct average values are attained in millionths of a second.[59]

By the 1890s, when Boltzmann's neglected work at last began attracting the attention it deserved, probability had also begun to enter another area of physics, the study of the newly discovered phenomenon of radioactive decay, which proved to conform to the Poisson distribution. Quantum physics was from the beginning statistical. Max Planck applied Boltzmann's combinatorial mathematics to the radiation of black bodies, to which he was led, ironically, in pursuit of a reconciliation of mechanics and irreversibility through the assumption of a continuum. Einstein's route to the quantum discontinuity began with some important work on statistical mechanics. By 1918 Marian Smoluchowski, who with Einstein had derived from the kinetic theory predictions for Brownian motion that enabled Claude Perrin at last to demonstrate conclusively the reality of molecules, could write that probability had become central to most of modern physics. "From this trend," he observed, "only Lorentz's equations, electron theory, the energy [conservation] law, and the principle of relativity have remained unaffected, but it is quite possible that in the course of time exact laws may even here be replaced by statistical regularities."[60]

[59] Boltzmann, *Lectures* (n. 51), part 2, p. 449.

[60] Marian Smoluchowski, "Ueber den Begriff des Zufalls und den Ursprung der Wahrscheinlichkeitsgesetz in der Physik" (1918), in *Oeuvres* (3 vols., Cracow, 1924-1928), vol. 3,

Boltzmann's prophecy that even molecules might be governed by statistical principles proved too cautious. Probability was applied during the 1920s to the most fundamental particles known, and as the hypothesis of hidden variables, through which a deterministic interpretation of these phenomena could be salvaged, became increasingly untenable, it became evident that the work of Max Born, Niels Bohr, and Erwin Schrödinger required an acausal physics. Paul Forman had argued that these developments reflected the defensiveness of German physicists in the face of attacks by the likes of Oswald Spengler during the Weimar period,[61] and it may be noted that various physicists in Britain and Germany applauded this work because it seemed to provide a niche for God and human freedom.

The statistical approach did not mandate a rejection of determinism. Albert Einstein's late remark that God does not play dice reflected the view he had presupposed from the beginning, and inherited from the nineteenth century, that statistical laws were based on causal assumptions and reflected a causal reality.[62] It was statistics, nevertheless, that made this rejection possible. The case of physics reinforces the evidence from nineteenth-century social thought that it was the success of statistics, as much as its limitations, that set in motion what Ian Hacking calls the "erosion of determinism"—one of the noteworthy intellectual developments of our time.[63] Statistics provided grounds for believing that large-scale order could be explained without presupposing a determining cause for each minute event and, at the same time, brought forth extravagant claims about the impossibility of divine or human freedom that inspired a critical examination of its own teachings. Physics was not alone in following this path all the way to a rejection of determinism. Indeed, it was not even first. It did, however, contribute significantly to

---

87-110, p. 87. E. Amaldi has emphasized the discovery of radioactivity as a source of the new indeterminism, but J. van Brakel argues convincingly that its influence was limited; see van Brakel, "The Possible Influence of the Discovery of Radio-active Decay on the Concept of Physical Probability," *Archive*, 31 (1984-85), 369-385.

[61] Paul Forman, "Weimar Culture, Causality, and Quantum Theory: Adaptation of German Physicists and Mathematicians to a Hostile Intellectual Environment," *Historical Studies in the Physical Sciences*, 3 (1971), 1-115.

[62] See Patrick H. Byrne, "Statistical and Causal Concepts in Einstein's Early Thought," *Annals of Science*, 37 (1980), 215-228.

[63] See Ian Hacking, "From the Emergence of Probability to the Erosion of Determinism," in Hintikka, *Probabilistic Thinking* (n. 42), pp. 105-123; "Nineteenth-Century Cracks in the Concept of Determinism," *Journal for the History of Ideas*, 44 (1983), 455-475; "The Autonomy of Statistical Law," in N. Rescher, ed., *Pittsburgh Lectures in the Philosophy of Science, 1981* (Berkeley, 1983), pp. 3-20.

the movement away from determinism by a few philosophical thinkers even before the twentieth century began. The most compelling and unqualified such defense of "contingency" in nature was written by the American physicist and philosopher Charles Sanders Peirce.

PEIRCE'S REJECTION OF NECESSITY

It was in reference to his belief in panpsychism, and related eccentricities, that Peirce composed the following fragment of autobiography:

> I may mention for those who are curious in studying mental biographies that I was born and raised in the neighborhood of Concord—I mean in Cambridge—at the time when Emerson, Hedge, and their friends were disseminating the ideas that they had caught from Schelling, and Schelling from Plotinus, from Boehm, or from God knows what minds stricken with the monstrous mysticism of the East. But the atmosphere of Cambridge held many an antiseptic against Concord transcendentalism; and I am not conscious of having contracted any of that virus. Nevertheless it is probable that some cultured bacilli, some benignant form of the disease was implanted in my soul, unawares, and that now, after long incubation, it comes to the surface, modified by mathematical conceptions and by training in physical investigations.[64]

Another bacillus contracted by Cambridge intellectuals was statistics. Peirce was exposed to it during his youth, both as an object of metaphysical curiosity and as an instrument for reducing observations. Ralph Waldo Emerson learned of the regularities of statistics from Quetelet even before Buckle published his *History of Civilization*, and in his 1845 lecture on Swedenborg he spoke of the "terrible tabulation of the French statists" that "brings every piece of whim and humor to be reducible also to exact numerical ratios. If one man in twenty thousand, or thirty thousand, eats shoes or marries his grandmother, then in every twenty thousand or thirty thousand is found one man who eats shoes or marries his grandmother."[65] Statistics also had a prominent place in his 1860 essay "Fate," where he held it a statistical principle that "Punch

[64] Peirce, "The Law of Mind" (1892), in *Papers* (n. 26), vol. 6, 86-113, p. 86.

[65] Ralph Waldo Emerson, "Swedenborg," in *Representative Men* (1850; Boston and New York, 1903), 93-146, p. 109.

makes exactly one capital joke a week," and drew comfort from the supposed truth that the same fate which requires one in twenty thousand to eat shoes ineluctably brings forth heroic genius with a like regularity.[66]

A more prosaic, but more easily demonstrated intellectual connection is between Peirce and his father, the distinguished Harvard astronomer and mathematician Benjamin Peirce. The father made routine use of error theory in his work, and even published a noteworthy contribution to it, a probabilistic criterion for the rejection of "doubtful observations," or outliers.[67] C. S. Peirce, too, devoted much of his career to the sciences of observation and measurement of Quetelet and Herschel, which had grown highly sophisticated by the late nineteenth century. For about thirty years, beginning in 1861 when he graduated from Harvard and saw little immediate opportunity for a scientific career except by becoming his father's assistant, he drew a salary from the United States Coast Survey. In the field of measurement he achieved some distinction, especially through his work on the use of pendulums to measure the force of gravity, and thereby to estimate the degree of ellipticity of the earth. He was also a strong advocate of the "new," or Wundtian, psychology, for he believed, like Quetelet, that number and measurement were essential attributes of science.[68] Peirce, however, never deluded himself that perfect exactitude was possible. On the contrary, he invoked his scientific experience to buttress his assertion that errors cannot be made to disappear by making observations maximally sophisticated and painstaking. Peirce held this truth to have important philosophical consequences. He wrote in 1892:

> For the essence of the necessitarian position is that certain continuous quantities have certain exact values. Now, how can observation determine the value of such a quantity with a probable error absolutely nil? To one who is behind the scenes, and knows that the most refined comparison of masses, lengths, and angles, far surpassing in precision all other measurements, yet fall behind the accuracy of bank accounts, and that the ordinary determinations of

[66] Emerson, "Fate," in *The Conduct of Life* (1860; Boston and New York, 1904), 1-49, pp. 17-19. I am grateful to Barbara Packer for calling my attention to these essays.

[67] Benjamin Peirce, "Criterion for the Rejection of Doubtful Observations" *Astronomical Journal*, 2 (1852), 161-163.

[68] On Peirce and natural science, see Victor F. Lenzen, "Charles S. Peirce as Astronomer," in E. C. More and R. S. Robin, eds., *Studies in the Philosophy of Charles Sanders Peirce* (Amherst, 1964), pp. 33-50, and articles by Lenzen, Thomas Cadwallader, and Thomas Manning in *Transactions of the Charles S. Peirce Society*, 11 (1975), 160-194.

physical constants, such as appear from month to month in the journals, are about on a par with an upholsterer's measurements of carpets and curtains, the idea of mathematical exactitude being demonstrated in the laboratory will appear simply ridiculous.[69]

Peirce began forming his ideas on the philosophy of probability quite early. Already in 1866 he was arguing for the lawlikeness of chance, supporting his claim in the customary manner on the evidence of statistics. Indeed, he wished to make statistical method central to scientific reasoning. He wrote: "If we count the number of suicides in Massachusetts for ten years and calculate the ratio of that number to the population, it is well known that we can predict with certainty how many suicides will take place the next year. This is induction and corresponds to a special uniformity in the world. It is the type of uniformity and all induction. *Statistics*; that is induction."[70] At the same time he spoke in favor of the frequency interpretation of probability, as set out in the first edition of Mill's *Logic*. He found the logical interpretation of chance in John Venn's book, which he saw shortly afterwards, even more congenial, but there already was more to his enthusiasm for the idea of local irregularity and mass uniformity than logic. Peirce argued in 1868 that the imperfection of knowledge renders each individual like an insurance company, whose soundness depends on the absence of any single interest overbalancing the rest, and hence that truly logical conduct is posible only when motivated by adherence to a social principle.[71] A decade later he argued more clearly, if not altogether persuasively, that because reliable uniformities emerge only in the mass, it is impossible for the individual to pursue his narrow self-interest with strict logicality. To choose the highest probability might be an erroneous procedure in any finite number of cases. The only recourse is to identify private interests "with those of an unlimited community," and to hope that it will last beyond any assignable date. All else is illogical.[72]

One more important Cambridge connection was Peirce's close and enduring friendship with William James, who shared his interest in the question of scientific determinism. James was deeply moved by the free

[69] Peirce, "The Doctrine of Necessity Examined," in *Papers* (n. 26), vol. 6, 28-45, p. 35.

[70] Peirce, "Lowell Lectures, 1866," in Max H. Fisch, ed., *Writings of Charles S. Peirce: A Chronological Edition* (Bloomington, Indiana, 1983 et seq.), vol. 1, p. 423.

[71] *Ibid.*, p. 396; "Grounds of Validity of the Laws of Logic: Further Consequences of Four Incapacities" (1868), in *Papers* (n. 26), vol. 5, 190-222, pp. 220-221.

[72] Peirce, "The Doctrine of Chances" (1878), in *Papers* (n. 26), vol. 2, 389-414, pp. 397-399.

will issue, and he argued passionately that it was at least intellectually respectable to act as if free will existed, since science was in no position to refute this doctrine. Lacking, as we do, exact knowledge about particulars, the claim that every trivial human decision is predetermined is, according to James simply a *Machtspruch*—"a mere conception fulminated as a dogma and based on no insight into details."[73] For James, the defense of freedom was equivalent to the defense of pluralism, of the capacity of individuals or small groups to act autonomously, without being dominated completely by the whole. James was able to conceive, and willing to allow, the existence of genuine chance in the action of individuals, and he noted that local spontaneity was fully consistent with mass order. Conversely, he insisted that scientific demonstration of the lawlikeness of aggregate phenomena did not prejudice the issue of whether freedom, and with it moral responsibility, subsisted at the level of individuals.

James, in turn, drew much of his inspiration from several French philosophers, particularly Charles Renouvier, Alfred Fouillée, and Joseph Delboeuf who, like Boussinesq and St. Venant, were concerned about the alleged inconsistency between modern science and freedom of the will. Peirce also read these philosophers, and he maintained that his doctrine of "Tychism," or belief in objective chance, was closely related to their theories.[74] Fouillée was concerned with an older psychological and philosophical argument for determinism of the will rather than the new debates arising from physics, physiology, and statistics, and was in any event more interested in conciliation than in refutation.[75] Renouvier and Delboeuf, however, did indeed introduce ideas on chance that Peirce subsequently developed.

Renouvier began writing against determinism during the 1850s. In the large essay on logic that made up the first two volumes of his *Essai de critique générale*, he reversed the customary interpretation of probability. Laplace's argument, he maintained, could not easily be reconciled with determinism, and he held the law of large numbers to be specifically dependent on an assumption of *imprédétermination* regarding the individual events.[76] He went on to argue, in a subsequent essay on

[73] William James, "The Dilemma of Determinism" (1884), in *The Will to Believe and Other Essays in Popular Philosophy* (New York, 1915), 145-183, pp. 156-157.

[74] See Peirce, "Answers to Questions Concerning my Belief in God," in *Papers* (n. 26), vol. 6, 340-355, p. 350.

[75] Alfred Fouillée, *La liberté et le déterminisme* (Paris, 1872).

[76] Charles Bernard Renouvier, *Essais de critique générale* (5 vols., 1854-1864; Paris, 2d ed., 1912), 1st essay, vol. 2, p. 146.

psychology, that the fit between probability and the events of moral statistics constituted no vindication of scientific determinism, but a challenge, perhaps an insuperable one, to it. Liberty, according to Renouvier, is highly similar to chance, and the applicability of mathematical probability to moral events established that they behave as if they were genuinely free. This did not preclude order, of course: "I see no reason not to conclude with Buckle that acts, virtuous as well as vicious, and even neutral ones, if such exist, are, no less than crimes, governed by general laws, but I will add, governed in their mean values, and governed approximately."[77] Renouvier did not see freedom as identical to chance, since the former required moral deliberation as well as the absence of law, but he believed that determinists were obliged by logic to abjure probability, as indeed some French writers had.

By 1869, when he published his *Science de la morale*, Renouvier had retreated somewhat from this strong argument on the implications of mathematical probability. Now, evidently, he recognized that a more intricate causal structure might underlie the mass regularities, and he was content in his discussion of statistics to maintain that it lacked compelling force as a positive argument for determinism. He gave his argument for free will in the form of a denial of "what one might call the determinism of chance." This was the viewpoint presented by Cournot, Windelband, and many others and derived from Aristotle—that chance, at least in its objective guise, refers to the intersection of independent causal chains, each fully deterministic. Such an intersection was commonly illustrated by the calamitous meeting of a falling brick and a pedestrian, both arriving at the fatal spot for a well-determined reason, but colliding for no good reason, so that the injury may be characterized as by chance, even though it was fully determined.[78] Renouvier thought that every causal chain must have some point of beginning, itself not caused, and hence that there was no reason to suppose all events along the way fully determined either.[79]

Peirce's first argument for indeterminism, or "Tychism," was yet more closely related to one given by Joseph Delboeuf in 1882. Delboeuf set out to improve on Boussinesq, whose reconcilation of physics and freedom seemed to him defective on two counts—first, because there

[77] *Ibid.*, 2d essay, vol. 1, p. 329; see also chap. 13, *passim*, and 2d essay, vol. 2, p. 341.

[78] See C. C. Gillispie, "Intellectual Factors in the Background of Analysis by Probabilities," in A. C. Crombie, ed., *Scientific Change* (New York, 1963), 431-453, pp. 440-443; also Wilhelm Windelband, *Die Lehren vom Zufall* (Berlin, 1870).

[79] Renouvier, *Science de la morale* (2 vols., 1869; Paris, 2d ed., 1908), vol. 2, pp. 366-373.

are in nature no real points of discontinuity, as required (Delboeuf thought) by Boussinesq's equations; and second, because his argument represented the will as forceless, whereas it in fact is based on reasons. Delboeuf proposed that the solution was to be found in recognizing the unknown, and that the error of the determinists was their supposition that what we don't know doesn't exist. If causes are really equal to their effects, Delboeuf argued, it would be possible for nature to run backwards, for old men to grow progressively younger, for trees to shrink down to acorns, then leap into the air, and so on. But this, he declared, is unimaginable; one need only look at the law proved by Sadi Carnot and William Thomson that heat passes exclusively from warm to cold objects. His conclusions were, first, that our knowledge is significantly incomplete, and second, that usable force, rather than the conserved quantity, absolute force, is the key physical variable. Human freedom, he explained, is imbedded in time. The function of the will is "to transform the accumulated potential forces in the organism into work" at a chosen moment, in order to produce the desired effect.[80] Just how this was to be accomplished without some infinitesimal incursion of mind into the domain of matter remained unspecified. The argument against mechanical determinism, however, did not depend on the design of a trigger device, but on the incompatibility between mechanical law and temporal directionality.

Peirce's first explicit argument against mechanical determinism, published in an essay on "Science and Immortality" in 1887, was, like Delboeuf's, based on the direction of time. Peirce was aware of the problem of irreversibility in thermodynamics, and he illustrated the statistical approach by citing the work of Clausius and Maxwell almost as often as he did so in terms of errors of measurement, but he seems never to have mentioned the ideas of Boltzmann or Loschmidt. In the context of his refutation of determinism, he opposed the inherent reversibility of the laws of force, and particularly energy conservation, to what could be observed in organic nature, and not to the motion of heat. "It is sufficient to go out into the air and open one's eyes to see that the world is not governed altogether by mechanism," he wrote; "the essential of growth is that it takes place in one determinate direction, which is *not* reversed.

[80] Joseph Rémi Leopold Delboeuf, "Determinisme et liberté: la liberté demontrée par la mécanique," *Revue philosophique de la France et de l'étranger*, 13 (1882), 453-480, 608-638; 14 (1882), 156-189, p. 156.

. . . As well as I can read the signs of the times, the doom of necessitarian metaphysics is sealed."[81]

Peirce took up the matter in greater detail during the early 1890s, pointing out the inadequacies of existing scientific arguments for determinism in a way which to a twentieth century reader evokes images of the guileless lad in the fable of the emperor's new clothes. The common nineteenth-century argument that determinism stands as a postulate of scientific reasoning seemed to Peirce no argument at all; it was equivalent to "postulating" a loan when asked for security by a bank. The unavoidable fluctuations shown by repeated measurements barred any claim that the exactitude of laws could be established in the laboratory. The order of nature, moreover, could equally well be accounted for in terms of a universe of chance.

Peirce's positive reasons for belief in chance are perhaps less compelling, and the cosmological system into which he incorporated his doctrine of Tychism was by no means in conformity with the common sense either of his own day or of ours. Peirce, whose ideas were more nearly Spencerian than he liked to think, believed devoutly in evolution not only of life, but of matter as well, and even of the laws that govern matter. As we have seen, however, he denied that growth and evolution could be the product of mechanical laws, such as Herbert Spencer's postulated tendency of matter to evolve from the homogeneous to the heterogeneous. Growth required departure from law—that is, random variation—along with some mechanism for its fixation. Peirce maintained that everything in the universe was subject to spontaneous fluctuations, which in turn were regulated and organized by a principle of association analogous to that in psychology. By this process, Peirce thought, the universe had evolved from pure chaos to ever greater order.

These ideas formed the core of a cosmology that Peirce published between 1891 and 1893 in a series of five papers written for *The Monist*, a Chicago philosophical journal with strong German ties. It may be noted that the chief mission of that journal, the establishment of a monistic or unified science through the reconciliation of matter, life, and mind, and with them of science and religion, concurred broadly with Peirce's own aims, although his belief in irreducible chance was wildly heretical and provoked a detailed refutation by the editor, Paul Carus. Peirce's principle that growth was only possible as a consequence of random

[81] Peirce, "Science and Immortality," in *Papers* (n. 26), vol. 6, 370-374, pp. 372-373.

variation gave new meaning to the proposition that to err is human, for error now appeared not as a defect of mental law, but its essence. In this way, Peirce provided for human freedom, which was indeed important to him, though not quite so integral to his belief in objective chance as to similar arguments by Maxwell and Renouvier. Taken in conjunction with his belief in cosmic evolution, and his idea that measurement fluctuations reflect real changes in the objects measured, Peirce's belief that error was identical to consciousness implied that even matter was partly conscious. He supposed that matter had lost most of its consciousness as its properties became standardized, through a process of associative fixation acting on what had begun as a primordial swamp of intense consciousness bound to unrestrained randomness. Like many *Monist* authors, he thought some form of panpsychism the only possible solution to the mind-body problem.

Peirce's evolutionary philosophy might seem to suggest that Darwinism had contradicted Huxley's expectations and become a pillar of support for a doctrine of pure chance. Indeed, Peirce, and also William James,[82] had come to see evolution as the vindication of chance, though many French biologists had opposed natural selection for precisely this reason. Peirce always regarded Darwin's theory as preeminently statistical, much like the kinetic gas theory. At least by the 1890s, however, this insistently scientific philosopher had become a passionate opponent of natural selection, which he thought a mere extension of the dominant nineteenth-century spirit of economism into the study of nature. Announcing his preference for the "Gospel of Christ" over the "Gospel of Greed," for sentiment and the merging of individualities over self-interest, Peirce identified his view of evolution with Lamarck's, and replaced heartless selection with habit (association) as the regulating force in cosmic and organic evolution. In conjunction with his "Synechism," or belief in absolute continuity, this implied that evolution was fundamentally communal, and thus this lonely and difficult logician contrived to characterize the world in terms of yet another neologism, "Agapasm," indicating that it was composed of pure "Evolutionary Love."[83]

Peirce, clearly enough, was moved by philosophical aims that were not necessarily shared by other probability writers, and he explicitly subordinated his Tychism to Synechism on the ground that the existence of

[82] See Reba N. Soffer, *Ethics and Society in England: The Revolution of the Social Sciences, 1870-1914* (Berkeley, 1978).

[83] Peirce, "Evolutionary Love" (1893), in *Papers* (n. 26), vol. 6, pp. 190-215.

a perfect continuum of possible values for the so-called constants of nature rendered utter exactitude, whether in our measurements or the phenomena themselves, quite inconceivable. Still, it should be clear that Peirce's views on probability were not the pure product of isolated, idiosyncratic genius. He made regular use of error theory, and was much impressed with the exploitation of the error curve in social statistics, biometry, and, especially the kinetic gas theory.[84] He drew on Mill and Venn for his interpretation of probability and, like Maxwell, Venn, Renouvier, and Rümelin, was moved to entertain the possiblity of incomplete physical determinism partly by opposition to the mechanical world view and partly by a faith, deriving ultimately from the arguments of Quetelet and Buckle, that statistics was quite enough to account for the observed order in the world without a postulate of an intricately detailed determinism. Peirce can hardly be made an archetype for what became known in this century, under the influence of popularizing physicists, as the abandonment of the deterministic world picture, for his faith in continuity was directly opposed to the great innovation of quantum physics. No more can his idiosyncratic views on chance and cosmic evolution be seen as characteristic for his own century. His work, however, illustrates again the centrality of statistics to debates about determinism.

Hence the disavowal of determinism was, ironically but essentially, a development of the statistical tradition. It represents one of the most interesting outcomes of the introduction of statistical thinking to science, social thought, and philosophy in the nineteenth century. It also played a role in the creation of a mathematical field of statistics—a coherent set of methods and theories that could be applied to a wide range of sciences, not only to establish regularities, but also to learn about underlying causal structures, without presupposing complete knowledge of constituent individuals. The movement of opposition to the determinism in Buckle's concept of statistical law was responsible for some important early contributions to the new mathematical statistics, and recognition of the crucial significance of heterogeneity in its social and biological objects was absolutely central to it.

[84] *Ibid.*, pp. 197-198; "The Doctrine of Chances" (n. 72), p. 391.

PART FOUR

Few intellectual pleasures are more keen than those enjoyed by a person who, while he is occupied in some special inquiry, suddenly perceives that it admits of a wide generalization, and that his results hold good in previously-unsuspected directions. —FRANCIS GALTON (1890)

For whatever may be said about the importance of aiming at depth rather than width in our studies, and however the demand of the present age may be for specialists, there will always be work, not only for those who build up particular sciences and write monographs on them, but for those who open up such communications between the different groups of builders as will facilitate a healthy interaction between them.

—JAMES CLERK MAXWELL (1878)

Statistics is the premier inexact science.

—EDMOND and JULES GONCOURT (1877)

POLYMATHY AND DISCIPLINE

Statistical mathematics was developed through disciplinary interactions. The mathematical field of statistics that in the twentieth century has become an indispensable resource for all sciences that depend on large numbers of observations or on the study of populations rather than of individuals was in fact dependent on what now appear as applications of it for its very creation. Mathematical statistics became possible only when statistical mathematics and statistical thinking had become sufficiently sophisticated in a diverse range of particular sciences that a coherent branch of mathematics embracing these various techniques and approaches could be envisioned concretely.

"Pure" mathematics was invented during the early nineteenth century and pure probability not until its second half, with the work of Chebyshev and his students on the central limit theorem.[1] The parts of probability most useful for mathematical statistics were developed primarily in the context of studies of insurance, political arithmetic, rational belief, and astronomy. The insights of Laplace and Condorcet that probability could provide a key to the mathematization of the social and biological sciences has been vindicated to a remarkable extent, but the abstract mathematical expression of the laws of probability played a relatively minor role in this process, at least during the nineteenth century. Those statists, physicians, and natural historians who endeavored most successfully to fulfill the Laplacian desideratum looked not to mathematics directly, but to other social or natural sciences that had successfully incorporated probability either as an element of theory or as a means for reducing observations. The diffusion of statistical thinking was thus contingent on the recognition of analogies between the objects of diverse disciplines. More than diffusion was involved here, however, for these analogies were almost never perfect. The application of established ideas and techniques to new objects often led to a wider understanding of their import and limitations, and consequently to the development of new methods and approaches better suited to the new subject matter.

[1] It was still viewed skeptically by many mathematicians as late as the 1930s; see Mark Kac, *Enigmas of Chance: An Autobiography* (New York, 1985), chap. 3, esp. p. 54.

The stimulus for a successful mathematical statistics came less from the theory of errors of observation than from the use of probability distribution formulas to model real events in nature and society. Its beginnings are to be found in the last quarter of the nineteenth century, partly in Germany, where probabilistic analysis was first applied constructively to statistical social science between 1875 and 1880, but primarily in Great Britain. The economist Francis Edgeworth went a long way toward establishing the basis for a general mathematical statistics during the 1880s and 1890s. Still more important was Francis Galton's biometric work, leading him to a general method of correlation that he pubpublished in 1889 and that became the inspiration for a host of social and biological investigations. Karl Pearson, Galton's disciple in this respect, combined superior mathematical skill with a similar breadth of interests and a genius for academic entrepreneurship to establish, both socially and intellectually, a mathematical field of statistics.

Chapter Eight

THE MATHEMATICS OF STATISTICS

Quetelet, Herschel, and the quantitative natural historians of their generation were already moved by a strong sense of the analogies that bound together statistics, observational astronomy, geodesy, meteorology, "tidology," and related fields. The analogy was first of all one of procedure—that is counting and measuring—inspired by a conviction that enumeration and careful analysis must lead to exact laws. These, it was generally anticipated, would be phenomenological, and not reductive. The widespread expectation among early nineteenth-century insurance writers that the true "law of mortality" must soon yield to their exact methods of research is entirely typical. This standard methodology is more impressively revealed by Charles Babbage's notable essay of 1832 calling for measurement of "constants of nature and art" ranging from astronomical distances and atomic weights to mean labor power, average duration of the reigns of sovereigns, and the frequency of occurrence of letters of the alphabet in various languages.[1] Statistical reformers were sufficiently impressed by these ideas to seek standard units of disease and mortality that would enable the sanitary scientist to identify abnormally unhealthful situations.[2]

Numerical social science itself provided a model for this kind of thinking, thereby undermining the official view of the council of the London Statistical Society that statistics was a science and not a method. For example, the economist Poulett Scrope proposed a technique for classifying and dating geological formations based on a statistical analysis of fossils, which Charles Lyell later elaborated in terms of an explicit

[1] Charles Babbage, "On the Advantage of a Collection of Numbers to be Entitled the Constants of Nature and Art," reprinted from *Edinburgh Journal of Science*, n.s. 12 (1832). See also Babbage, *A Comparative View of Various Institutions for the Assurance of Lives* (London, 1826).

[2] See John Eyler, *Victorian Social Medicine. The Ideas and Methods of William Farr* (Baltimore, 1979).

demographic analogy.[3] The extent of the influence of statistics on analogous sciences is attested by the fact that it soon was compelled to give up exclusive possession of its name. That transition, however, was resisted by the most dedicated statistical reformers as well as by continental practitioners of the old descriptive statistics, and was debated, at least in Britain and Germany, until nearly the end of the nineteenth century. In fact, those debates only ended when statistics was supplanted as the general science of society by sociology.

As we have seen, however, the idea that mere collections of facts could constitute a science had been challenged much earlier. The Germans conceded that statistics was, at least implicitly, a method of wide applicability as soon as they admitted quantification into its definition. In Britain, the threat of social unrest diminished after 1850, so that the call for reform which had inspired the initial burst of statistical activity no longer seemed so urgent. During the 1850s, the London Statistical Society entered a new phase, characterized by less passionate studies on a greater range of topics. By the time London hosted the International Statistical Congress, in 1860, many writers, including the Prince Consort and Nassau Senior, were prepared to concede that statistics was a process for accumulating and verifying facts, applicable "to all phenomena that can be counted and recorded."[4]

The central role that probability might play in the various sciences of observation and measurement was by this time widely recognized, if rarely put into effect. Quetelet had long championed this view, which he expressed most fully in his *Letters on Probability* of 1846, and the penetrating mathematical work of the French civil servant I. J. Bienaymé, beginning in 1838, was aimed at a general theory for the analysis of observations of all sorts.[5] Cournot's book on probability of 1843 treated a variety of subjects from a broadly statistical perspective, which

[3] See Martin J. S. Rudwick, "Poulett Scrope on the Volcanoes of Auvergne: Lyellian Time and Political Economy," *BJHS*, 7 (1974), 205-242, and "Charles Lyell's Dream of a Statistical Palaeontology," *Palaeontology*, 21 (1975), 225-244.

[4] "Opening Address of Nassau W. Senior, Esq., as President of Section F . . . ," *JRSS*, 23 (1860), 359. See also papers by Jevons (who argued that statistics was a proper science, and that its use of tables was a "mere accident"), J. Ingram, and Wynnard Hooper, in *JRSS*, 33 (1870), 309-322; 41 (1878), 602-629; 44 (1881), 31-48; also W.S.B. Woolhouse, "On the Philosophy of Statistics," *Assurance Magazine and Journal of the Institute of Actuaries*, 17 (1873), 37-56; [Frederic Marshall], "Statistics," *Eclectic Magazine*, 86 (1876), 335-348.

[5] See C. C. Heyde and E. Seneta, *I. J. Bienaymé: Statistical Theory Anticipated* (New York, 1977); also Irenée Jules Bienaymé, "Mémoire sur la probabilité des résultats moyens des observations," in *Mémoires presentés par divers savants à l'Académie royale des sciences de l'Institut de France* (1838), pp. 513-558.

he held to be indispensable whenever events are produced by multiple, interwoven chains of causation.

> Statistics (as its etymology indicates) is principally understood as the collection of facts to which the aggregations of men in political society give rise: but for us the word will take a more extended acceptation. We will mean by statistics "the science whose object is to collect and arrange numerous facts of every sort, so as to obtain numerical relations which are effectively independent of the anomalies of chance, and which indicate the existence of regular causes whose action is combined with that of fortuitous causes." . . . In order for statistics to merit the name of science, it should not consist simply of a compilation of facts and figures; it must have its theory, its rules, its principles. Now this theory applies to facts of the physical and natural order as it does to those of the social and political order. In this sense, phenomena which take place in celestial regions can be submitted to the rules and to the investigations of statistics, as can the agitations of the atmosphere, the perturbations of animal economy, and the still more complex facts which arise in the state of society from the interactions of individuals and of peoples.[6]

Like Quetelet, Cournot took as his model the most advanced of these statistical sciences, the celestial one. "In a word, just as observational astronomy is the model for sciences of observation, and theoretical astronomy the model for scientific theories; so likewise should the statistics of stars (if it may be permitted to have recourse to this conjünction of terms) serve some day as the model for all other statistical inquiries."[7]

The idea that statistics should be regarded as an abstract mathematical study appeared later in England. It seems to have been first expressed by J. J. Fox, who argued in 1860 in a paper read before the Statistical Society of London: Statistics cannot really be called a science, since it "has no facts of its own; in so far as it is a science, it belongs to the domain of Mathematics. Its great and inestimable value is, that it is a 'method' for the prosecution of other sciences. It is a 'method of investigation' founded upon the laws of abstract science; founded on the mathematical theory of probabilities; founded upon that which had been happily

[6] Cournot, pp. 181-184.
[7] *Ibid.*, p. 262.

termed the 'logic of large numbers.' "[8] Neither Fox nor his more accomplished French predecessors, however, succeeded in showing concretely how the analytical procedures of error analysis or probability theory could be used to shed light on important issues of biology and meteorology, or to lend rigor to empirical economics. Cournot was an important pioneer of mathematical economics, and an astute theorist and philosopher, yet even he was unable to go beyond the programmatic exhortations of probability writers and actually put the techniques of error analysis to work in social science.

Since error theory was well developed by the time of Laplace, and was described clearly in publications by Fourier and Quetelet, it seems at first surprising that probabilistic estimates of uncertainty were so rare in social statistics during the nineteenth century. There is, however, little mystery. Statists and statisticians were almost unanimously distrustful of sampling, and emphasized at every opportunity the importance of complete enumeration. To rely on an incomplete survey was to become dependent on conjecture, a procedure that statisticians associated with the discredited political arithmetic and an obligation from which they had thankfully been absolved due to the massive expansion of official statistics. As late as 1901, Ladislaus von Bortkiewicz observed, correctly, that the degree of exactitude of inferential calculation had been little investigated in social statistics, "because the decisive progress that has been made in the exhaustive mass observation of social phenomena in the 19th century has gradually led statistics away . . . from these inferential calculations."[9]

The skepticism of statisticians about inference from samples was not wholly unjustified, for in the absence of reliable information about the population as a whole it was difficult to know if a particular sample was adequately representative. Sampling, however, continued to be widely rejected in social statistics even after this obstacle had been largely removed, as is apparent from the unfavorable reception of the first serious

[8] J. J. Fox, "On the Province of the Statistician," *JRSS*, 23 (1860), 331. See also papers by William Augustus Guy in *JRSS*, 2 (1839), 25-47, and 36 (1873), 467-485; by William Newmarch in *JRSS*, 41 (1869), 359-384, and by Jevons in *JRSS*, 41 (1878), 597-599.

[9] Ladislaus von Bortkiewicz, "Anwendungen der Wahrscheinlichkeitsrechnung auf Statistik" (1901), in *Encyclopädie der mathematischen Wissenschaften* (Leipzig, 1900-1904), I.2, 821-851, pp. 825-826. That exactitude is the defining attribute of statistics was emphasized by, among others, Alexandre Moreau de Jonnès, *Eléments de statistique* (Paris, 2d ed., 1856), p. 45. Joseph Garnier, "Statistique," in Ch. Cocquelin and Guillaumin, eds., *Dictionnaire de l'économie politique* (2 vols., Paris, 1853), vol. 2, 653-662, pp. 656-668, stressed the contrast between factual statistics and speculative political arithmetic.

proposal for representative sampling in social statistics by the Norwegian A. N. Kiaer at the 1895 meeting of the International Statistical Institute.[10] Those who had advocated the use of error analysis, like Quetelet and Wilhelm Lexis, proposed to use it not to infer attributes of a population from some subset of it, but to infer the existence of underlying causes from a complete census. Even with respect to this aim, moreover, these authors did not conceive their theoretical problems in terms sufficiently structured that any probabilistic test would hold much interest, and the numbers with which they worked were often so large that little was to be gained by computing a standard error. As we have seen, even Quetelet never did so.

As Bortkiewicz pointed out, the need to make use of the theory of errors was appreciated much more readily where large numbers were difficult or impossible to attain, particularly where each datum was the result of an experiment or trial rather than an easily made observation, as by a census agent. The most noteworthy application of probability to the human sciences during the early nineteenth century, apart from insurance, was in the area of medical statistics. The application of probability to the results of therapeutic trials was simple and straightforward, for the problem was the standard one of estimating the error of the mean in order to determine whether a fixed cause could justifiably be inferred. The necessary formalism could be taken directly from Laplace's demographic investigations of the late eighteenth century, subsequently refined and developed by Poisson, and the concepts were fully consistent with those that had become widely known and accepted as a result of the prominence of the method of least squares in astronomy. The very circumstance that established ideas could so readily be applied to medical statistics perhaps explains the paucity and insignificance of innovations that emerged in this context.

From another standpoint, that of the practitioner, medical statistics ceases to look so easy. Apart from the question of experimental design, which in the twentieth century has been answered with the double-blind experiment but which was not addressed in nineteenth-century statistical writings, and leaving aside the vexed issue of whether it was appro-

[10] On Kiaer and the rejection of his ideas by Georg Mayr and Luigi Bodio, see You Poh Seng, "Historical Survey of the Development of Sampling Theory and Practice," in *SHSP2*, 440-457; also William Kruskal and Frederick Mosteller, "Representative Sampling, IV: The History of the Concept in Statistics, 1895-1939," *International Statistical Review*, 48 (1980), 169-195.

priate for physicians to have recourse to statistical thinking at all, the application of probability to therapeutics presented the medical statistician with a serious problem of numbers. Probability, tacitly supposed unnecessary when numbers rose to four digits, was typically seen as ineffectual when they did not reach three. Jules Gavarret, a French disciple of Poisson who in 1840 published the most influential and controversial work on the mathematics of medical statistics, concluded that no reliable judgments were possible unless a minimum of two hundred trials, and preferably three to five hundred trials, had been made.[11] Gavarret's standard of reliability, two units of dispersion (equal to 2.83 times our standard deviation, yielding a probability of .995), was conservative, and even this requirement could be met without hundreds of trials if the experimental probabilities differed widely from one another; but his treatise nonetheless inspired advocates of the numerical method to seek a more practicable criterion by which to assess their statistical results. Jacques Raige-Delorme surmised that rigid adherence to mathematical probability would render statistics useless to medicine, and proposed in effect to supplement mathematical analysis with medical intuition. He argued that the medical trial yields more information than an urn drawing, since the physician observes the rapidity of the cure and the general course of the illness as well as its final outcome. Carl Liebermeister proposed a new statistical test in 1877, which, however, only appeared more powerful than Gavarret's because he chose to illustrate it with comparisons of highly divergent probabilities.[12] The considerable incentive for improved statistical techniques provided by the numerical method yielded virtually no fruit.

Probability was also used increasingly to measure uncertainty in the new experimental psychology of the late nineteenth century. It was introduced by the Leipzig professor Gustav Theodor Fechner, founder of psychophysics, who was given to an energeticist monism and who regarded the law that the perceived force of a given small alteration in a stimulus is inversely proportional to the absolute magnitude of the stim-

[11] Jules Gavarret, *Principes généraux de statistique médicale, ou développement des règles qui doivent presider son emploi* (Paris, 1840).

[12] Jacques Raige-Delorme, "Statistique médicale," in *Dictionnaire de médecine, ou repertoire général des sciences médicales* (30 vols., Paris, 2d ed., 1832 et seq.), vol. 28 (1844), 549-559, pp. 552-554; C. Liebermeister, "Über Wahrscheinlichkeitsrechnung in Anwendung auf therapeutische Statistik," in Richard Volkmann's *Sammlung klinischer Vorträge*, 110 (Leipzig, 1877), 935-962. For references to related works by Adolf Fick, Willers Jessen, and Julius Hirschberg, see his p. 939. On mathematical methods in nineteenth-century medical statistics, see O. B. Sheynin, "On the History of Medical Statistics," *Archive*, 26 (1982), 241-280.

ulus as no less important "for the field of the interrelations of body and soul" than Newton's law of universal gravitation was for physics. Fechner's law could be traced back not only to Ernst Heinrich Weber, who introduced it to psychology, but also to the classification of stellar magnitudes by Karl August Steinheil and Norman Pogson, to the work of Herbart and Drobisch on perceptions of tone intervals, and to Daniel Bernoulli's law of moral expectation, which provided it with a sound ancestry in probability theory.[13] Fechner used the error law, as his admirer Francis Galton pointed out, without incorporating his own geometric law into it,[14] but he was, nevertheless a skilled and enthusiastic user of error analysis. He argued in his *Elemente der Psychophysik* that probabilistic methods based on the law of large numbers "can scarcely be depicted more appropriately than as a Proteus who, instead of answering questions directly and willingly, seems to avoid every answer through the changing forms he assumes; but it suffices to remain undeterred and to hold him constantly to the same point, and a reliable answer will be forced out of him."[15]

Fechner's career, like those of so many contributors to statistical mathematics, was a diverse one, and his methods for analyzing "mass phenomena" came primarily from astronomy. Analysis of the relations between stimuli and internal sensations required great numbers of measurements, which could scarcely be handled without some such techniques. Fechner was principally responsible for the frequent use of error theory by the new generation of quantifying experimental psychologists associated with Wilhelm Wundt's laboratory.[16] He did not simply imitate the astronomers. He developed the use of what Cournot had first termed the "median" as a substitute for the mean, whose computation required, in Fechner's opinion, too much additional labor to justify the slight increase in accuracy it furnished. He also prepared a more general mathematical investigation on minimization of the various powers of errors—least sums, least squares, least cubes, and so on. His last book, published two decades after his death under the editorship

[13] G. T. Fechner, *Elemente der Psychophysik* (2 vols., Leipzig, 2d unaltered ed., 1907), vol. 1, pp. 64, 67.

[14] See Galton and Donald MacAlister, "The Geometric Mean in Vital and Social Statistics," *PRSL*, 29 (1897), 365-376; also Galton letters to a Mrs. Hertz and to G. G. Stokes in *Galton*, vol. 3B, pp. 464, 468.

[15] Fechner, *Psychophysik* (n. 13), p. 78.

[16] See, for example, papers by Friedrich Heerwagen, Julius Merkel, H. Bruns, and G. F. Lipps, respectively, in Wilhelm Wundt's *Philosophische Studien*, 5 (1890), 301-320; 7 (1892), 555-629 and 8 (1893), 97-137; 9 (1894), 1-52; 13 (1898), 579-612.

of G. F. Lipps, was called *Kollectivmasslehre*, a study of "collective objects" and the methods for analyzing them. As Michael Heidelberger has recently shown, Fechner was also perhaps the earliest scientific indeterminist, although it remains unclear just when he began to identify this indeterminism with probability theory.[17]

The academic statisticians of Germany were deeply concerned to work out methods or theories that would establish the worth of their discipline as a dignified scholarly enterprise, and among them a few began during the 1860s to pursue a genuinely mathematical statistics. Some, including Knapp and the physicist and statistician Gustav Anton Zeuner, meant by "mathematical statistics" the formulation of deductive population models with no explicit reliance on probability. This kind of work had long been associated with insurance studies, and eventually came to comprise a large part of the science named by Achille Guillard in 1855—though it was then scarcely distinguishable from statistics—"demography."[18] The application of error theory and of probability models to social statistics was also pursued with growing success in Germany during the last third of the nineteenth century. The most successful and influential of those mathematical writers on statistics was the economist and statistician Wilhelm Lexis.

LEXIS'S INDEX OF DISPERSION

Aside from Quetelet's unwitting reinterpretation of the error law, the most important specific contribution of statistical social science to statistical mathematics was a method for assessing the stability of statistical

[17] Fechner, "Über die Bestimmung des wahrscheinlichen Fehlers eines Beobachtungsmittels durch die Summe der einfachen Abweichungen," in *Annalen der Physik und Chemie*, Jubelband (1874), pp. 66-81; "Über den Ausgangswerth der kleinsten Abweichungssumme, dessen Bestimmung, Verwendung und Verallgemeinerung," *Abhandlungen der mathematisch-physischen Classe der königlich sächsischen Gesellschaft der Wissenschaften*, 11 (1878), 1-76; *Kollectivmasslehre* (Leipzig, 1897). The use of the median as a substitute for the mean had been discussed previously by Laplace in an appendix to the *Théorie analytique des probabilités* (Paris, 2d ed., 1820). On Fechner's indeterminism, see Michael Heidelberger, "Fechner's Indeterminism: From Freedom to Laws of Chance," in *Prob Rev*.

[18] See Gustav Anton Zeuner, *Abhandlungen der mathematischen Statistik* (Leipzig, 1869), and "Zur mathematischen Statistik," *Zeitschrift des k. sächsischen statistischen Bureaus*, 31 (1885), 1-13. The word "demography" was introduced in Achille Guillard, *Eléments de statistique humaine ou démographie comparée, . . .* (Paris, 1855). A distinction between statistics as method and demography as science was adopted by Maurice Block, *Traité théorique et pratique de statistique* (Paris, 1878), pp. 85-86. Ernst Engel introduced the word "demology" with similar intentions.

series. Inevitably, the context of this work was social and ideological as well as mathematical. Perhaps it was not inevitable that these techniques should have been introduced in response to Buckle's exuberant advocacy of social and statistical law, but such was in fact the case. The measurement of dispersion in statistical series was, like Maxwell's discovery of the uncertainty of the second law of thermodynamics, intended as a critique of statistical determinism and a defense of the autonomy of the human will.

In 1859 Robert Campbell published in the *Philosophical Magazine* an investigation of the mathematical import of statistical regularity. Campbell was the brother of Maxwell's biographer, and himself a close friend of the physicist. They had been schoolmates together at the Edinburgh Academy, and one of Maxwell's letters while he was at that institution refers to "R. C." having "written an essay on Probabilities with very grand prop[osition]s in it; everything original, but no signs of reading, I guess."[19] Campbell's 1859 paper, delivered to the very same section of the British Association meetings to which Maxwell read his first paper on the kinetic theory, was in turn neglected by subsequent writers on statistical mathematics, and only rediscovered after similar ideas had been developed more fully by Lexis and others. It was, nonetheless, original, and it is a revealing document.

Campbell's stated purpose was to ascertain whether there was anything "remarkable" in the tables of crime invoked so grandiloquently by Buckle. That author, he explained, had "formed certain moral conclusions from the fact of this [statistical] uniformity, namely, the existence of certain moral laws by which a section of the community, definite in number, is always impelled to such [criminal] acts." The validity of Buckle's conclusions, he thought, was contingent on the supposition that the observed regularity was greater than could reasonably be expected if the individuals in question had acted freely and independently. To test this assumption, Campbell derived a combinatorial formula based on imagined urn drawings, with the ratio of black and white balls set according to the proportion of criminals in the population as revealed by statistics. He concluded that the "observed uniformity" in Buckle's tables was even less than "that which might be expected from events, the occurrence of which to individuals was conceived as purely fortuitous." The militantly positivist historian had at best raised a "metaphysical

[19] *Maxwell*, pp. 127-128.

question, from which it would be hopeless to expect any practical issue," and his statistical argument against free will was worthless.[20]

Wilhelm Lexis, too, intended his analysis of the stability of statistical series as a corrective to an inflated faith in statistical law, but he regarded it further as the basis for a research program in numerical social science. Because of his prominence in the German community of academic statisticians, he attracted the attention of others able and willing to pursue this line of work—which Campbell did not. Lexis's university training, however, was not in statistics; his thesis at the University of Bonn (1859) concerned analytical mechanics, and he spent the following year or so as a researcher in Bunsen's chemical laboratory in Heidelberg. Possibly he learned some statistics while at Bonn. The philosopher Friedrich Albert Lange lectured on social statistics there in 1857-1858, and in 1858 the physicist Gustav Radicke published a rather primitive probabilistic test by which physicians could determine if observed differences in mean values indicated with sufficient certainty the existence of genuine effects. That contribution inspired some controversy, and was discussed sympathetically by Lange in his influential *Geschichte des Materialismus* of 1866.[21] Lexis's real introduction to the social sciences, however, came in 1861, when he traveled to Paris to study economics. His first book, published in 1871, was a study of French trade policy since the revolution.

From the beginning, he displayed the suspicion of abstract theories that marked a convert to the historical school of economics. Although he was less strident in his criticism of the deductive approach than many of his contemporaries, he placed a higher value on empirical study than did any leading economist outside of Germany. Lexis thought utility an imprecise concept, and for that reason believed that the new marginalist approach was unlikely to contribute to the understanding of actual eco-

[20] Robert Campbell, "On a Test for ascertaining whether an observed degree of Uniformity, or the Reverse, in Tables of Statistics is to be looked upon as remarkable," *Phil Mag* [4], 18 (1859), 359-368.

[21] Where it was argued that materialists renounced probability in favor of a "logic of facts." See Friedrich Albert Lange, *Geschichte des Materialismus und Kritik seiner Bedeutung in der Gegenwart* (2 vols., Iserlohn, 1866), vol. 2, pp. 354-357, 474-481. Gustav Radicke's paper is "Die Bedeutung und Werth arithmetischer Mittel . . . ," *Archiv für physiologische Heilkunde*, N.F. 2 (1858), 145-219. The philosopher Friedrich Ueberweg commented also on this debate: "Ueber die sogenannte 'Logik der Tatsachen' in naturwissenschaftlicher und besonders in pharmakodynamischer Forschung," *Archiv für pathologische Anatomie und Physiologie und für klinische Medizin*, 16 (1859), 400-407. See William Coleman, "Experimental Physiology and Statistical Inference: The Therapeutic Trial in Nineteenth Century Germany," in *Prob Rev*.

nomic processes. He readily acknowledged the usefulness of reasoning in economics from the one-sided assumption that individuals pursue exclusively their self-interest, but insisted that these deductions must be checked against empirical facts, and modified as required. He did not reject mathematics from social sciences; on the contrary, he was a champion of it, and his criticism of Carl Menger and the Austrian marginalists was aimed partly at their neglect of exact reasoning.[22]

Given Lexis's epistemological orientation, it is scarcely surprising that he shared much of the outlook of the Verein für Sozialpolitik. He did not take a leading role in that organization, but he was in general agreement with its more liberal members, such as Brentano and Schmoller. He was firmly committed to the establishment of voluntary organizations of workers to defend their interests and to give them a stake in the existing system. In his study of French labor unions and employer associations, published under the auspices of the Verein für Sozialpolitik in 1879, he argued that worker feelings of solidarity could be as potent a force as employer egoism, particularly since the latter tended to undermine the stability of businessmen's agreements. For those rigid economists who denied the possibility of effective labor cooperation on the basis of abstract deductions—the iron law of wages—he had no sympathy. Their blind confidence in these "so-called economic 'natural laws' is no more justified than the sanguine socialists' flights of fantasy."[23]

Lexis's dedication to empiricism, along with his commitment to mathematics, led him to statistics. Like Knapp he was not content with a statistics that consisted of little more than miscellaneous collections of facts. He hoped to raise statistics out of its identity crisis and to bestow on it the dignity prerequisite for a proper academic discipline by infusing it with advanced mathematics. Lexis admired Knapp's mathematical models of population, and contributed to this work, but his main interest was the analysis of empirical statistical data using mathematical probability. His text on the theory of population statistics, written while he was at the University of Strasbourg and published in 1875, embodied the

[22] See Klaus-Peter Heiss, "Lexis, Wilhelm," in W. H. Kruskal and J. M. Tanur, eds., *Encyclopedia of Statistics* (2 vols., New York, 1978), vol. 1, pp. 507-512. For Lexis's skepticism of abstract deduction in economics, see Wilhelm Lexis, "Zur mathematisch-ökonomischen Literatur," *Jbb*, N.F. 3 (1881), 427-434; "Zur Kritik der Rodbertus'schen Theorien," *Jbb*, 9 (1884), 462-463.

[23] Lexis, *Gewerkvereine und Unternehmerverbände in Frankreich: Ein Beitrag zur Kenntnis der socialen Bewegung* (Leipzig, 1879), p. 7.

first significant effort to apply probabilistic error analysis to actual social data since the time of Laplace, or at least Poisson. The mathematics he used was taken from Poisson, and his ideas for its application bore also a close family resemblance to the more prosaic Laplacian efforts. For the limits of random deviation he used the combinatorial result, $3\sqrt{2pq/L}$, where L was the total number of cases, p/L the probability of a positive outcome, and q/L the complementary probability of a negative outcome. The number 3 was his standard of reliability, corresponding to 4.24 standard deviations and yielding a probability of .999978 that the true value lay within the assigned limits. Lexis used his formula to show, for example, that the probabilities of death for twenty-year-olds of both sexes were not demonstrably different from those for men and women of the same age ten years later, but that the probability of death at age twenty for young Bavarian men born in the twelve-year period 1834-35 to 1845-46 was significantly different from that for young women of the same description.

This is not, on its face, an especially inspiring result, but Lexis was already showing a seriousness of concern with his subject matter as something more than an object for the display of mathematical prowess that went far beyond his French predecessors. Alert to the problems of dealing with heterogeneous material, Lexis took seriously the approach to the study of "mass phenomena" discussed by Rümelin. He argued that application of least squares presupposed the existence of a constant, underlying value which the observed measures approximated, and he accordingly viewed his demonstration that the probabilities of death for young Bavarian men and women were reasonably constant over the period in question as a prerequisite to his comparison of males and females based on two mean values taken over the entire period. He insisted that the mere regularity of aggregate figures did not amount to a law of nature, and that populations must be broken down in order that causes might be identified. A population was for him not an undifferentiated mass, but an intricate system of groups, each characterized by different probabilities of death, suicide, and so on. In this regard, he observed:

> Accordingly, the task of social-physiological statistics can be summarized as follows: it should form the most individualized elementary groups possible and characterize through probability relationships the probability schemes (*Chancensysteme*) which condition their important changes; it should further investigate the extent to

which the diversity of probability schemes arises from the diversity of traits that distinguish the elementary groups, and finally, it should determine whether the individual probability schemes remain approximately constant over the course of time or vary in determinate ways.[24]

The next year, Lexis turned his attention to an analysis of the variation within groups. This was by no means unprecedented. An actuary for the Hanover life insurance company named Theodor Wittstein, whose work Lexis knew well, had written extensively on this matter during the 1860s. Wittstein, who in 1863 coined the phrase "mathematical statistics," meaning by it the elevation of the social science statistics to a higher mathematical plane, aspired in the usual manner to make statistics a proper scientific discipline—to advance from the Keplerian to the Newtonian stage by discovering genuine laws. To this end, or so he implied, he presented a study of the fluctuations to be expected in death rates for a given population, based on Poisson's combinatorial formulas. The result, he thought, was both a great truth about nature and a practical aid to insurance institutions, which require some knowledge of the magnitude of these fluctuations in order to calculate the reserves they ought to maintain.[25]

Campbell, of course, had already shown that these combinatorial formulas did not apply, and in 1874 a writer in the *Journal des actuaires français* made the same point, while anticipating a great deal of Lexis's analysis. It had occurred to Emile Dormoy that the variation in statistical series could be compared with that predicted by combinatorial formulas to decide whether the events in question were in fact independent of one another. To do so, Dormoy took the ratio of two values for the mean error of the statistical time series. He calculated his numerator from the actual figures in the series, using the formula for what is now called standard error of the mean that had been given by Laplace and presented with great simplicity by Fourier. For the denominator he used

[24] Lexis, *Einleitung in die Theorie der Bevölkerungsstatistik* (Strasbourg, 1875), p. 121 and *passim*, part 5. See also "Zur mathematischen Statistik," *Jbb*, 25 (1875), 158-163. Probabilistic tests like those in Lexis's *Einleitung* are also to be found in the work of the Austrian civil servant Joseph Hain, *Handbuch der Statistik des Österreichischen Kaiserstaates* (2 vols; Vienna, 1852-53), vol. 1, pp. 8-98.

[25] Th. Wittstein, "Zur Bevölkerungs-Statistik," *Zeitschrift des königl. preussischen statistischen Bureaus*, 2 (1863), 12-16; also *Mathematische Statistik und deren Anwendung auf National-Oekonomie und Versicherungs-Wissenschaft* (Hanover, 1867), translated in *Assurance Magazine and Journal of the Institute of Actuaries* (1873), pp. 178-189, 355-369, 417-435.

Poisson's combinatorial expression, the same one employed by Lexis in 1875 to find limits of variation. If the events under consideration were truly independent, the empirical value should approximately equal the deduced one, and hence this ratio should be close to unity. If the coefficient exceeds or falls short of one by a considerable margin, it follows that these events do react upon one another—positively in the first case, negatively in the second. Dormoy calculated this coefficient of divergence for various statistical measures, and found that only the birth ratio of boys to girls could be interpreted as the result of statistically independent events. The fraction of illegitimate births, the death rate, and the annual rainfall all diverged from their mean with a coefficient substantially greater than one, indicating that deviations in these measures were not independent, but produced by "general causes."[26]

Dormoy was silent on the relation of these discoveries to the statistical argument for determinism, and perhaps thought little of it, since it had long since lost its luster in France. In Germany, however, the philosophical and moral implications of statistical regularity were still too hotly debated for such an issue to be ignored. Oddly, the myth of lawlike regularity had scarcely been challenged in Germany before 1875, even by those critics of statistical law who thought systematic variation far more interesting than undeviating uniformity. That statistical tables did reveal a remarkable regularity was almost always taken for granted.[27] Finally, however, in 1876, the same year that Lexis first published his index of dispersion, there appeared a long essay by the Göttingen philosopher Eduard Rehnisch that was sharply critical of Quetelet, Buckle, and their enshrinement of statistical regularity. This paper, which Lexis had encountered at least by 1877, was intended as an epitaph to the controversy on free will and statistics. That social numbers had been interpreted as proof of philosophical necessity was, according to Rehnisch, preposterous, and could only be explained (with the usual historicist nostrums) as reflecting gallic Enlightenment influences and the materialist *Zeitgeist* of the 1860s.

Rehnisch sought to bring out the absurdity of supporting a deterministic outlook with statistics by demonstrating that the regularity of statis-

[26] Emile Dormoy, *Théorie mathématique des assurances sur la vie* (2 vols., Paris, 1878), vol. 1, pp. 1-47. I have not seen the 1874 *Journal des actuaires française* article of which he claimed this to be a reprint.

[27] The only exception I have seen is the Austrian Leopold Neumann's "Zur Moralstatistik," *Preussische Jahrbücher*, 27 (1871), 223-247, p. 227.

tical series had in fact been grossly exaggerated. Quetelet's so-called budget of crime, he pointed out, showed variation in adjacent years of 3 percent, 21 percent, and 30 percent, while murder statistics revealed variation ranging to 254 percent and 323 percent. That the customary pronouncements of astonishing regularity were based on no more than superficial impressions was most compellingly indicated by a table of murders in France from 1826 to 1844, originally presented by Quetelet in his general essay of 1848 on moral statistics, where it was accompanied by a self-congratulatory remark that his prediction fifteen years earlier of the continued constancy of these numbers had been fully vindicated. In fact, Rehnisch pointed out, the numbers varied from 363 in 1831, to 163 in 1844, and were consistently lower at the end of the period than at the beginning. There was, in fact, good reason; a change in French law in 1833 had placed fully 45 percent of murders into a new category which, if added back into the table, would enhance its regularity markedly. But, wrote Rehnisch, "the penetrating eyes of the 'new Newton,' who fancied himself to be on the track of the laws with which one can succeed in mastering the phenomena of the world of *sciences morales et politiques* in the same way as the world of ponderable matter is ruled by the law of gravitation—. . . these numbers . . . did not startle him in the least."[28]

Lexis, similarly, was concerned about the issue of statistical law and its implications for free will and cultural distinctiveness. For his inaugural lecture at Dorpat, where he taught from 1874 to 1876, he chose the topic "Natural Science and Social Science," and he still felt it necessary to argue at length against the applicability of laws of nature to society in the economics textbook whose second edition appeared in 1913, just one year before his death. Lexis maintained, more explicitly even than Rümelin, that because laws must apply to individuals as well as groups, they are not accessible to statistics, and indeed that the statistical method is defined by its limitation to statistical regularities—to statements of probability. He further held, adumbrating a distinction subsequently developed in detail by Wilhelm Dilthey, that statistics was the "natural-scientific foundation of the social sciences," and as such could no more penetrate to the true underlying cause of social behavior—mo-

[28] Eduard Rehnisch, "Zur Orientierung über die Untersuchungen und Ergebnisse der Moralstatistik," *Zeitschrift für Philosophie und philosophische Kritik*, 68 (1876), 213-264; 69 (1876), 43-115, pp. 70-71.

tives—than physics can get beneath phenomena to forces.[29] In physical science, phenomenological knowledge sufficed to reach genuine laws, but in social science it was confined to empirical regularities. To reach the true causal foundations of social science, the method of natural science must be supplemented with the "unmediated comprehension" of our own consciousness that can "penetrate to the inner causal connection of external phenomena."[30] One such intuition was the psychological assumption of self-interest implicit in theoretical economics which, though one-sided, was at least causal. In any event, "numerical regularities in [ethical] phenomena do not rule, but are conditioned by the moral constitution of human society."[31]

Lexis's first study of the relationship between the deviations from mean values to be expected from a combinatorial model of independent events and those revealed by actual statistical series was carried out in regard to the most time-honored subject of quantitative social research, birth ratios. That his intent in these analyses was not solely to debunk the exaggerated conclusions of the Quetelet school is indicated by the fact that in this study his results were strictly positive. Using Poisson's formula for mean combinatorial, or, as he called it, "statistical" error, and the standard expression for the observed "physical" dispersion, Lexis compared these quantities for birth ratios in the various districts of Prussia and the counties of England to establish that observed deviations were very close to predicted. He also arranged the deviations from the mean ratio (1.065) for each of the 34 Prussian districts and each of the 24 months in the period 1868-1869 into categories by magnitude, and placed them beside the corresponding areas of an error curve whose width, or mean error, was based on a combinatorial prediction. Again, the agreement seemed fully satisfactory. The empirical probability of a male birth, he concluded, belongs among those statistical quantities whose particular values "may be regarded as random modifications of a typical normal value."[32]

Lexis was careful to point out that the match between a model of urn

[29] Lexis, "Naturwissenschaft und Sozialwissenschaft" (1874), in *ATBM*, 233-251, p. 246. See also J. Conrad et al., eds., "Gesetz (im gesellschaftlichen und statistischen Sinne)," in *Handwörterbuch der Staatswissenschaften* (7 vols., Jena, 2d ed., 1898-1901), vol. 4, pp. 234-240; *Allgemeine Volkswirtschaftslehre* (1910; 2d ed., 1913; Leipzig, 3d ed., 1926), pp. 14-25.

[30] Lexis, "Naturwissenschaft," pp. 242-243.

[31] *Ibid.*, pp. 250-251.

[32] Lexis, "Das Geschlechtsverhältnis der Geborenen und die Wahrscheinlichkeitsrechnung" (1876), in *ATBM*, 130-169, pp. 163-164. In this reprint of 1903, Lexis included a footnote acknowledging Dormoy's priority (p. 130). The word "typical" here does not merely mean "ordinary," but "conforming to an underlying type."

drawings and these statistical ratios implied no stabilizing force by which the individual events were regulated, but rather their complete independence. Thus the presence of regularity equal to that predicted by combinatorial formulas was in no way inconsistent with perfect freedom of the will. Even that degree of regularity, however, was by no means universal in statistics. Lexis showed in 1877 that the stability of suicide rates, long regarded as astonishing evidence of the penetration of law into the realm of the most intimate human decisions, was not sufficient to justify the application of a combinatorial model. At the end of the nineteenth century, when he surveyed the results of numerous studies on this matter, the only series from moral statistics that displayed even the degree of stability characteristic of complete independence were the numbers of Danish suicides from 1861-1886, studied by Harald Westergaard. All others showed significantly greater variability, and hence were even less plausibly interpreted as necessary results of rigid statistical laws.[33]

The index of dispersion that Lexis introduced in 1879, possibly after he had encountered Dormoy's identical formulation, provided a coherent mathematical structure for this argument. He defined a quantity Q as the ratio of actual or "physical" dispersion, R, to that predicted from combinatorial formulas, r. The value of Q, he proposed, enabled one to distinguish three types of statistical series. When $Q = 1$, at least approximately, combinatorial and physical dispersions are equal, implying that the binomial model of the urn is more or less applicable. A value of Q significantly greater than one, which is almost universally found in moral and population statistics, implied that the probabilities for sizable subgroups, or even for the whole population, have fluctuated over time—perhaps irregularly, perhaps progressively. The third alternative, Q less than one, would support the mechanical determinism suggested by Quetelet, Buckle, and Wagner, or the mystical unity stressed by Oettingen, for it would imply the existence of some unconscious power which drives individuals to fulfill numerical ratios pertaining to the collective whole. Lexis, bolstered by the absence of any series that revealed so perfect a lawlike regularity, dismissed this "subnormal" dispersion characteristic of a series whose elements are somehow bound together as inconceivable in statistics.[34]

[33] Lexis, *Zur Theorie der Massenerscheinungen in der menschlichen Gesellschaft* (Freiburg, 1877), pp. 19, 87; "Gesetz" (n. 29), p. 239.

[34] Lexis, "Ueber die Theorie der Stabilität statistischer Reihen" (1879), in *ATBM*, 170-212, pp. 175-184. See also "Moralstatistik," *Handwörterbuch* (n. 29), vol. 5, 865-871, p. 868. On mechanical and mystical conceptions of statistics, see *Massenerscheinungen* (n. 33), p. 11.

Although critical of Quetelet's rhetoric about statistical law, Lexis was enthusiastic about his extension of the error law. Indeed, he criticized Quetelet—mistakenly, of course—for failing to apply this curve outside of anthropometry, the measurement of the physical traits of man. Galton's adoption of a wider interpretation, in his "Statistics by Intercomparison" of 1875, seemed to Lexis immensely promising, and he repeated Galton's explanation of how the curve could be used to establish the mean and distribution of a group whose members had only been ordered by rank, and not individually measured. Like Galton, he thought this method especially fruitful for ethnography, where it could be used to portray the racial type (*Stammestypus*) through the identification of the *Mittelmann*. Lexis was likewise hopeful that the curve could be used as suggested by Galton to represent the mean development of the talents of schoolchildren through the selection of an "average type."[35]

Lexis's writings on dispersion and the distribution of human physical attributes were influential within German anthropometry, which began to make increasing use of the analytical techniques associated with the error law during the last quarter of the nineteenth century. Ludwig Stieda, a professor of anatomy at Dorpat, collaborated with Oettingen, professor of statistics there, to train students in the classification of anthropometric data and the use of the error curve to identify and separate racial types.[36] Lexis himself applied the error curve to the distribution of ages of death—nominally to shed light on the mortality curves for the two sexes in the various countries of Europe, but actually to demonstrate its ubiquity. To establish that there exists for each people a normal age of death, from which real individuals diverge as if by accident, it was necessary first to subtract away premature deaths, to which he ascribed the conspicuous asymmetry of real death curves. Infant human mortality seemed unproblematical; it simply reflected defects in the human constitution. His other class of premature deaths extended from puberty to about the age of sixty, and could be ascribed, he thought, to abnormal conditions or external influences rather than to "the natural construc-

[35] Lexis, *Massenerscheinungen* (n. 33), pp. 38-40. See also his subsequent paper, "Die typischen Grössen und das Fehlergesetz," in *ATBM*, pp. 101-129.

[36] See Ludwig Stieda, "Ueber die Anwendung der Wahrscheinlichkeitsrechnung in der anthropologischen Statistik," *Archiv für Anthropologie*, 14 (1883), 167-182. Stieda gives references to dissertations at Dorpat by Hugo Witt, Max Strauch, and A. v. Schrenck, all between 1879 and 1881. See also a paper by Ihering in *ibid.*, 10 (1879), 411-413. Lexis devoted most of his attention to the error curve in his article "Anthropologie und Anthropometrie," in *Handwörterbuch* (n. 29), vol. 1, pp. 388-409.

tion of man." Rather inconveniently, these deaths overlapped that normal mortality expressed by the error law, and for that reason the error curve could be saved only by attributing a (declining) fraction of deaths between ages 40 and 60 to the abnormal circumstances. Lexis did so. Above age 60, the actual mortality curves were in reasonable accord with theoretical predictions based on a normal life span ranging from 67 years for Belgian men to 75 years for Norwegian and Swedish women and a probable error of six or seven years. Only in Norway, however, did the normal group make up more than half the total population.[37]

For Lexis, then, as for Galton and Quetelet, the error function served as a definition of type. Any other probability distribution was, in his view, uninterpretable, and he maintained that an approximate fitting of real data to this symmetric curve was of far more interest than a more exact fitting to an asymmetric curve. This strongly held preference amounted to a sharp limitation on the subjects to which he was willing to apply probabilistic analysis. He could conceive no use for probability in the analysis of numerical data except to assess the significance of an observed discrepancy between related mean values, and he insisted that a test of this sort could only be meaningful if the averages in question were "true means." That is, the underlying value must be stable, or fluctuate around a stable mean, so that each measurement is an approximation to something real. The proper indicator of the existence of such a mean value was the distribution of deviations from it in accordance with the error law. Only then could a series be regarded as "typical," and only then could error analysis be legitimately applied.[38] The class of phenomena to which combinatorial mathematics was suited, those whose index of dispersion equalled one, constituted in turn a subset of these "typical" series.

There were, in accordance with Lexis's typology, three other types of statistical series to which probability could not be applied. There were "undulatory" series, or series whose irregular fluctuations did not conform to the error distribution; there were also "periodic" series and "evolutionary," or "historical" series. Lexis was persuaded that in statistics, typical series were to be found overwhelmingly in the physical side of human life, while in moral statistics, progressive evolution was the

[37] Lexis, *Massenerscheinungen* (n. 33), pp. 41-64.

[38] *Ibid.*, pp. 17, 23-24, 32-34; "Stabilität" (n. 34), p. 175. See also "Naturgesetzlichkeit und statistische Wahrscheinlichkeit," in *ATBM*, 213-232, pp. 230-231, for a concise declaration of the limited role of probability in regard to statistics.

rule.[39] He was further persuaded that, with respect to moral actions, human societies are fundamentally diverse, or heterogeneous, so that even at a specified moment no single value could be assigned as the probability that any given individual will marry or commit a crime. The greater than "normal" dispersion of most statistical series he attributed to the circumstance that events in society are, in effect, drawn not from a single urn, but from a multitude of urns of diverse composition, all subject to independent fluctuations or to evolution at different rates.[40]

This intrinsic diversity was in perfect conformity with the nature of statistics, that science of "mass phenomena" especially designed to deal with it, but it presented an insuperable obstacle to the application of probability mathematics. Lexis agreed with Rümelin that counting and tabular presentation could be applied to any domain of science but that they only provided the basis for an autonomous science when no unified explanation of the phenomena in question was accessible. Statistics could serve as an auxiliary science to check the precision of laws of phenomena whose character was fundamentally uniform, such as economic principles. It was, however, the only available means for studying the "mass phenomena" presented by the ethical life of man, which "consist of single cases, whose similarity we find only in their like end result."[41] Here there was no constant cause, and hence no true underlying value and no basis for applying combinatorial probability or least squares.

Lexis argued consistently that "the transition from the originally purely subjective domain of mathematical probability into that of objective facts is by no means smooth or certain."[42] A subjective conception of probability, he thought, is perfectly appropriate so long as one is dealing with pure mathematics, but probability cannot be applied to nature under the guise of laws of human uncertainty. The use of probability relationships to model nature or society required, in his view, a demonstration that observed frequencies behave like the results of urn drawings, just as the application of error analysis is contingent on the

[39] Lexis, *Massenerscheinungen* (n. 33), pp. 90-92; "Stabilität" (n. 34), pp. 170-171.

[40] Lexis, *Massenerscheinungen* (n. 33), pp. 22-23; *Einleitung* (n. 24), pp. 93-97.

[41] Lexis, "*Massenerscheinungen*" (n. 33), p. 4.

[42] Lexis, "Über die Wahrscheinlichkeitsrechnung und deren Anwendung auf die Statistik," *Jbb*, N.F. 13 (1886), 433-450, p. 433; also *Massenerscheinungen* (n. 33), pp. 14-15. The former is a review of Johannes von Kries, *Die Principien der Wahrscheinlichkeitsrechnung* (Freiburg, 1886).

presence of the error law. Agreement need not be precise, for perfect accuracy is never to be found, but it must be approximate.

His caution in this regard seems perspicacious and admirable, but in conjunction with his narrow definition of the task of probability in statistics and his willingness to admit only close relatives of the binomial and error distribution functions, it imposed sharp limits on the possible range of his accomplishments as a mathematical statistician. Lexis never departed significantly from his conviction that periodic and evolutionary series are not susceptible to formulation in terms of probability. There is a passage in his notable paper of 1879 in which he proposed that evolutionary series could perhaps be modeled by what we would call a regression formula, $v = a + bt$, where v is the dependent variable, perhaps the annual number of suicides, t is time, in years, and a and b are constants, for which the "most probable values" are to be chosen. This equation, he held, is only valid if the residual error is no greater than would be expected in drawings made from an urn whose composition evolved according to the same linear expression. Certainly none of his social statistical series could meet so stringent a requirement and, in any event, Lexis did not present the analysis, though his problem was set up in such a way that he could seemingly have carried it out.[43]

The science of statistics was in serious decline in Germany by 1879, when the last of Lexis's three most important and original works was published, and he had few successors. The most notable of these was his student Ladislaus von Bortkiewicz, who shared Lexis's interest in the analysis of dispersion in statistical series, but not his skepticism about the application of probability to events in the real world. Bortkiewicz wrote abstractly on theoretical statistics, and applied his knowledge of probability to the new problem of radioactive decay,[44] but he is best known, apart from some rather turgid commentary he wrote on Marx, for the so-called "law of small numbers" which he presented in 1898. There he sought to apply mathematical probability to such distinctive subjects as

[43] Lexis, "Stabilität" (n. 34), p. 103. The possibility of using least squares to model change over time in social statistics was also appreciated by the Italian statistician Antonio Gabaglio, *Storia e teoria della statistica* (Milan, 1880), pp. 422-424, whose model was the quadratic $y = a + bx + cx^2$, where a, b, and c were to be set so as to minimize the sum of squares of the residual errors.

[44] See Ladislaus von Bortkiewicz, "Kritische Betrachtungen zur theoretischen Statistik," *Jbb* [3], 8 (1894), 641-680; 10 (1895), 321-360; 11 (1896), 671-705; also "Anwendungen" (n. 9); *Die radioaktive Strahlung als Gegenstand wahrscheinlichkeitstheoretischer Untersuchungen* (Berlin, 1913).

suicides of Prussian children and Prussian soldiers killed by kicks of horses.

Rare events like these, Bortkiewicz pointed out, had long been neglected by mathematical students of statistics, because the fluctuations from year to year are often several times greater than some of the absolute magnitudes. Bortkiewicz argued, however, that on this very account small numbers were all the more appropriate for probability theory. His reasoning was persuasive, and his agreement between predicted and observed variation was indeed good, especially for the kicked soldiers. This was hardly encouraging, however, to anyone who desired actually to learn something about the social processes in question from probabilistic analysis—if indeed anyone cared to know about the social process revealed by the deaths of these soldiers. Lexis's statistical series could not be harmonized with combinatorial probability because the fluctuations of underlying processes were large relative to the error term. Bortkiewicz's solution was to raise the error term, which for one-digit numbers like the annual results of kicked soldiers was overwhelming. Natural fluctuations by comparison paled to insignificance, even though they might be conspicuous if the population were great enough that the tiny fraction in question would represent an absolute number as large, say, as the total of Prussian suicides. Bortkiewicz concluded from this what is doubtless in some sense true, but manifestly does not follow, that probability functions must form the basis of all the numbers of population and moral statistics.[45]

Lexis's analysis of the stability of statistical series was the starting point for a thin but continuing continental tradition of statistical mathematics whose members, increasingly, were interested primarily in mathematical problems rather than social science. Bortkiewicz's probabilistic study of rare events tended already in that direction. The discussion to which he contributed and which also engaged the Russian mathematicians Aleksandr Chuprov and A. A. Markov as to the possibility of a subnormal dispersion was of purely mathematical interest, since the Poisson formulation, which underlay the discovery that subnormal dispersions could be generated, was almost wholly unrealistic from the standpoint of social or even natural science.[46] Lexis's index of dispersion

[45] Bortkewitsch [Bortkiewicz], *Das Gesetz der kleinen Zahlen* (Leipzig, 1898).

[46] See Bortkiewicz, "Anwendungen" (n. 9), pp. 827-829; also Kh. Ondar, ed., *Correspondence between A. A. Markov and A. A. Chuprov on the Theory of Probability and Mathematical Statistics*, Charles M. and Margaret Stein, trans. (New York, 1981).

was widely known in Great Britain and America as well as in Germany, and his applications of error analysis continued to serve as exemplars for statisticians into the twentieth century. In regard to the development of mathematical statistics, his most important influence was exerted through the work of Francis Edgeworth and of Karl Pearson, both of whom were better informed on continental developments than Galton. It was in Britain that the modern mathematical field of statistics was developed—ironically, perhaps, not so much by mathematicians as by biologists and social scientists who were impressed by the power of the statistical method.

EDGEWORTH: MATHEMATICS AND ECONOMICS

Francis Ysidro Edgeworth, the poet of statisticians, was led to probability in the context of his campaign to introduce advanced mathematics into the moral and social sciences. The son of an Irish landlord and eventual inheritor of the impoverished family estate, he early aspired to become a classical scholar, and indeed the lasting effects of his classical education at Trinity College, Dublin, and Oxford University appear in the form of Greek aphorisms and allusions interspersed through even his technical scientific work. During the early 1870s, his career aims changed, and he took up the study of commercial law, but by 1877, when he was called to the bar, his interests had taken yet another turn. In that year he published his first book, *New and Old Methods of Ethics*, an attempt to achieve rigor in that ancient branch of philosophy by applying the calculus of variations to the utilitarian propositions of Henry Sidgwick's recent *Methods of Ethics*. As Stephen Stigler has pointed out, that book revealed the direction of research that Edgeworth would pursue for the rest of his life. Already he had begun in earnest a program of study in the mathematical sciences, which he soon extended to most aspects of mathematics and mathematical physics. He hoped through analogies to bring the same rigor and elegance to economics and ethics.[47]

Edgeworth formulated these aims explicitly and generally in his remarkable book of 1882, *Mathematical Psychics: An Essay on the Appli-*

[47] See Stephen Stigler, "Francis Ysidro Edgeworth, Statistician," *JRSS*, A, 141 (1978), 287-322; J. M. Keynes, "Francis Ysidro Edgeworth" (1926), in *Essays in Biography* (New York, 1963), pp. 218-238.

cation of Mathematics to the Moral Sciences. The particular aim of that work was to show that imperfect competition rendered the outcome of market transactions indeterminate, creating a need for a scientific basis of arbitration which might be furnished by the utilitarian calculus.

> [If] the field of competition is deficient in that continuity of fluid, that multiety of atoms which constitute the foundations of the uniformity of Physics; if competition is found wanting, not only the regularity of law, but even the impartiality of chance—the throw of a die loaded with villainy—economics would indeed be a "dismal science," and the reverence for competition would be no more.
>
> There would arise a general demand for a principle of arbitration.
>
> And this aspiration of the commercial world would be but one breath in the universal sigh for articles of peace. . . .
>
> The whole creation groans and yearns, desiderating a principle of arbitration, an end of strifes.[48]

Yet more specifically, the troubles in Ireland emerged as a special object of his concern, and Edgeworth took some pains to defend the landlord interest by showing how utilitarian ethics could lead not to egalitarianism but to what might be called a hedontocracy, based on the greater capacity of the upper classes for increments of pleasure.[49]

Doubtless Edgeworth's interest in these practical matters was genuine, but he made no attempt to give exact numerical solutions or derive specific policy implications, beyond the general observation that by mathematics we "are biassed to a more conservative caution in reform."[50] Instead he sought to define functions with appropriate mathematical properties and to seek general conditions of maximization. He was not insensitive to the problems of applying abstract scientific knowledge. "The pure theory of Probabilities must be taken *cum grano* when we are treating concrete problems. The relation between the mathematical reasoning and the numerical facts is very much the same as that which holds between the abstract theory of Economics and the actual industrial world—a varying and undefinable degree of consilience, exaggerated by pedants, ignored by the vulgar, and used by the wise."[51]

[48] Francis Ysidro Edgeworth, *Mathematical Psychics: An Essay on the Application of Mathematics to the Moral Sciences* (London, 1881), p. 50.

[49] *Ibid.*, pp. 126-148.

[50] Edgeworth, "The Hedonical Calculus," *Mind*, 4 (1879), 394-408, p. 408.

[51] Edgeworth, "Tests of Accurate Measurement" (1888), in *PPE*, vol. 1, p. 325.

The proper goal is not to demonstrate a precise value, but to keep within a reasonable range of error, "not so much to hit a particular bird, but so to shoot among the most closely clustered covey as to bring down the most game."[52]

John Maynard Keynes contrasted Edgeworth's work with Marshall's by noting that Edgeworth pursued economics and ethics not so much to set forth maxims of behavior as to discover "theorems of intellectual and aesthetic interest."[53] This, surely, is the proper interpretation to be placed on the *mécanique sociale* that Edgeworth, like Quetelet before him, defined on the basis of analogies with the laws of physics. Society, he proposed, may be conceived as a "system of charioteers," the analogue of atoms, governed by hedonic principles as the motion of physical particles is determined by energy constraints.

> "Mécanique Sociale" may one day take her place along with "Mécanique Celeste," throned each upon the double-sided height of one maximum principle, the supreme pinnacle of moral as of physical science. As the movements of each particle, constrained or loose, in a material cosmos are continually subordinated to one maximum sum-total of accumulated energy, so the movements of each soul, whether selfishly isolated or linked sympathetically, may continually be realising the maximum energy of pleasure, the Divine love of the universe.
>
> "Mécanique sociale," in comparison with her elder sister, is less attractive to the vulgar worshipper in that she is discernible by the eye of faith alone. The statuesque beauty of the one is manifest, but the fairy-like features of the other and her fluent form are veiled. But mathematics has long walked by the evidence of things not seen in the world of atoms (the methods whereof, it may incidentally be remarked, statistical and rough, may illustrate the possibility of social mathematics). The invisible energy of electricity is grasped by the marvellous methods of Lagrange; the invisible energy of pleasure may admit of a similar handling.[54]

Notwithstanding the profusion of metaphors in this and other Edgeworthian writings, their author was by no means undiscriminating in his choice of models for social science. As he noted in 1877:

[52] *Ibid.*, p. 331.
[53] Keynes, "Edgeworth" (n. 47), p. 224.
[54] Edgeworth, *Mathematical Psychics* (n. 48), pp. 12-13.

> . . . it has often been triumphantly asked, with what success could mathematical calculation address itself to social phenomena, when it is unable to cope with the problem of three bodies. But perhaps the example of mechanics might suggest another conclusion, namely, that mathematics are capable of advancing victoriously, even while leaving impregnable fortresses in the rear. And so in the class of problems before us, it might be hoped that approximative methods would be attainable, if a sufficiently clear and appropriate conception of the data were obtained.[55]

Here his detour around the impregnable fortress was perturbation theory, but he was soon led to probability as another, and in many respects a more satisfactory, solution to the complexities of mathematical ethics and economics.

By the mid-1880s, Edgeworth had come to see the kinetic gas theory, by then a paradigmatic instance in its own right of the production of mathematical order out of local chaos, as exemplary for economics. On this basis he judged that, despite the intractability of the three-body problem, the analogy of physics "is not altogether discouraging. The difficulty of the problem does not increase indefinitely with its complexity. The case of an immense number of bodies rushing with every degree of speed in all directions, admits of a certain solution. In some respects a crowd of phenomena is more easy to manage than a few individuals. For a certain order is generated by chaos."[56] More concretely, he observed: "There is a certain resemblance between the uniformity of pressure to which the jostling particles of a gas tend, and the unity of price which is apt to result from the play of competition."[57]

Physics was not Edgeworth's only source of analogies. Like Galton, he cultivated an extraordinary range of interests, studied probability and statistics in the full diversity of their applications, and himself applied statistical reasoning and probabilistic mathematics to an impressive range of problems and subject areas. Excepting Cournot and Bienaymé, who lived too early to appreciate in detail the spectrum of potential statistical applications, Edgeworth was the first to conceive statistical think-

[55] Edgeworth, *New and Old Methods of Ethics, or "Physical Ethics" and "Methods of Ethics"* (Oxford, 1877), p. 66.

[56] Edgeworth, "The Element of Chance in Competitive Examinations," *JRSS*, 53 (1890), 460-475, 644-663, p. 471.

[57] Edgeworth, "On the Application of Mathematics to Political Economy" (1889), in *PPE*, vol. 2, p. 280.

ing as a mathematical problem, and to seek methods of analysis that could be applied with great generality. His success, while not total, was considerable, and it flowed from that wide awareness and ability to perceive analogies among diverse subjects that was so characteristic of statistical innovation in the nineteenth century.

Edgeworth first expressed interest in mathematical probability because of its analogies with the hedonic calculus. The success of the "application of mathematics to Belief, the calculus of Probabilities," he wrote, lends plausibility to the mathematization of hedonics, and he added, on the authority of Jevons, that "actions and effective desires can be *numerically* measured by way of statistics."[58] Soon he had become interested in this calculus of belief for its own sake. He was attentive to its philosophical implications, adopting a compromise between subjectivism and frequentism that interpreted statements of probability in terms of "partial incomplete belief" while insisting that the proper standard for allocating such belief was experience of uniformity of statistical ratios. In this vein, he justified the *a posteriori* probability of causes in terms of a presumed general experience common to all men, and perhaps imparted ancestrally, that justified, at least by order of magnitude, an *a priori* assumption of equipossibility. Edgeworth also paid obeisance to the issue of statistics and free will, and he concluded in 1885 that the discoveries of statistics had little value as an argument for either side. The extent of his study is suggested by his sources for this paper, which included Buckle, Quetelet, Mill, Renouvier, and Maxwell.[59]

By the time Edgeworth began writing on statistical subjects, the regularity of social numbers was so well known that George Eliot could treat it in her 1876 novel, *Daniel Deronda*, as a tired cliché.[60] For Edgeworth, who held the stability of aggregate measures to be the "first principle" of all empirical investigations relying on probability, it was simply *pro forma* to invoke Quetelet's great discovery in social science. "It is now a commonplace," he wrote in 1889, "that actions such as suicide and marriage, springing from the most capricious motives, and in respect of which the conduct of individuals most defies prediction, may yet, when taken in the aggregate, be regarded as constant and uniform. The advantage of what has been called the law of large numbers may equally be enjoyed by a theory which deals with markets and combi-

58 Edgeworth, *Mathematical Psychics* (n. 48), p. 1.
59 Edgeworth, "Chance and Law," *Hermathena*, 5 (1885), 154-163.
60 George Eliot, *Daniel Deronda* (New York: Penguin, 1967), pp. 582-583.

nations."[61] The lesson he drew from statistics, and from least squares in astronomy, was hardly novel, but his expression of it was. "It is a useful discipline to walk in a world where, though the objects themselves are fixed, their images are ever vibrating through a large part of their own dimensions. The Calculus of Probabilities . . . conveys a lesson which is required for the study of social science, the power of contemplating general tendencies through the wavering medium of particulars."[62]

The great paradox in probability "is that our reasoning appears to become more accurate as our ignorance becomes more complete; that when we have embarked upon chaos we seem to drop down into a cosmos."[63] Not surprisingly, he was greatly impressed by the wide domain of the error law, which he viewed as "ever tending to be set up throughout all nature—and almost beyond her bound if we may conceive with some metaphysicians a region in which aggregate phenomena show statistical uniformity while the individual events are arbitrary and lawless. Alone or best of propositions outside pure mathematics, the law of error realises the antique ideal of science, as deducible from universal necessary axioms."[64] He was particularly charmed that the order revealed by the error curve derived precisely from the utter irregularity of events when considered individually.

> However we define error, the idea of calculating its extent may appear paradoxical. A science of error seems a contradiction in terms. As the wise slave in the ancient comedy says, when his master begins to reason about love, a thing which is in its nature irregular and irrational, cannot possibly be governed by reason. It would be like going mad according to method.
>
> Natural philosophy has however triumphed over this paradox. Mathematicians have constructed an apparatus for reducing to rule errors.[65]

Edgeworth was also alert to the differences of interpretation required by the error curve in different contexts. He argued that the distinction

[61] Edgeworth, "Application" (n. 57), p. 274; also "The Statistics of Examinations," *JRSS*, 51 (1888), 599-635, p. 627.

[62] Edgeworth, "On Methods of Ascertaining Variations in the Rate of Births, Deaths, and Marriages," *JRSS*, 48 (1885), 628-649, p. 633.

[63] Edgeworth, "The Philosophy of Chance," *Mind*, 9 (1884), 223-235, p. 229.

[64] Edgeworth, "On the Representation of Statistics by Mathematical Formulae," *JRSS*, 61 (1898), 670-700; 62 (1899), 125-140, 373-385, 534-555; 63 (1900), 77-81, p. 551. The metaphysician here referred to is Renouvier.

[65] Edgeworth, "The Element of Chance in Competitive Examinations," *JRSS*, 53 (1890), 460-475, 644-663, p. 462.

between observations and statistics was a crucial one—that deviations in the former are truly errors, since each observation represents an imperfect determination of a true, underlying value, while variation in the latter reflects the diversity of nature itself, and cannot be dismissed as error even if identically distributed. To elaborate, he offered a Galtonian image: "observations are different copies of one original; statistics are different originals affording one 'generic portrait.' "[66] He was also sensitive to the possibility that statistics, at least, might be otherwise distributed. His stance on this issue varied considerably over his career. In his first statistical paper, written in 1883, he announced the end of the "ancient solitary reign" of the error law, and suggested that many or most distributions in the real world depart from it.[67] This nominalist attitude, however, did not last, and by the end of the century he had become critical of Karl Pearson's emphasis on skew variation, arguing that the normal law should be used to represent the frequency distribution unless some other curve was palpably more accurate.[68] Throughout his career, however, Edgeworth regarded the composition of statistics as highly variable so that he was always alert to the modifications of procedure entailed by a rejection of normality.

Mathematics made up an important part of Edgeworth's achievement, but equally significant was his resourcefulness in adapting statistical techniques to apply them to otherwise recalcitrant numerical records. As Stephen Stigler has argued, Edgeworth showed how the lofty ambitions for the use of error theory in the mathematization of the social sciences entertained by Quetelet, Cournot, and Jevons might begin to be fulfilled. Essentially, he accomplished mathematically what social statisticians had increasingly been recommending but almost never doing; he paused only briefly to marvel at the regularity of masses, then began breaking them down into component parts in order to apply appropriate probability tests, analyze causes, and model phenomena.

The problem that gave Edgeworth his start in mathematical statistics was one that for some time had been a matter of much interest to political economists—index numbers. The importance of some measure of

66 Edgeworth, "Observations and Statistics: An Essay on the Theory of Observation and the First Principles of Statistics," *TPSC*, 14 (1889), 138-169, p. 140. See also "On the Application of the Calculus of Probabilities to Statistics," *Bulletin de l'Institut international de statistique*, 18, part 1 (1909), 505-536.

67 Edgeworth, "The Law of Error," *Phil Mag* [5], 16 (1883), 300-309, p. 305. An exclusive reliance on the symmetrical binomial was at the same time criticized by John Venn, "The Law of Error," *Nature*, 36 (1887), 411-412.

68 See Edgeworth, "On the Representation" (n. 64), p. 551.

changes in the value of money had been stressed in 1838 by Cournot, who was a great pioneer of analytical economics. Cournot recommended the use of probability to infer the changing value of precious metals from the prices of numerous commodities, arguing that this mean price was analogous to a *soleil moyen*, or mean sun, which provided a reference frame in astronomy.[69] Measures of changes in the value of money were discussed frequently in Germany during the 1860s and 1870s by such authors as Drobisch, Held, Laspeyres, and Lexis.[70] Edgeworth's principal source, however, was William Stanley Jevons, who, despite his great fondness for probability and his insistence that the "deductive science of Economics must be verified and rendered useful by the purely empirical science of Statistics,"[71] had found no other significant points of contact between the theoretical problems of economics and the numbers of statistics. The same combination of interests, and the same gulf between them, it may be noted, characterized also the work of Cournot and, subsequently, of John Maynard Keynes.

Jevons, whose belief in statistical regularity has already been discussed, argued that changes in the value of currency could be measured only by ignoring the individual circumstances affecting the price of any given commodity and "trusting, that in a wide average . . . all individual discrepancies will be neutralised."[72] Still, the underlying distribution of price fluctuations could not be ignored, since the decision as to how to compute the mean depended on it. "It must be confessed," he wrote, "that the exact mode in which preponderance of rising or falling prices ought to be determined is involved in doubt. Ought we to take all commodities on an equal footing in the determination? Ought we to give additional weight to articles according to their importance, and the quantities bought and sold?"[73] There was also the question of whether to use an arithmetic mean, a geometric mean, or some other kind of average. Jevons chose the geometric mean, which he thought a better

[69] A. A. Cournot, *Recherches sur les principes mathématiques de la théorie des richesses* (Paris, 1838), pp. 22-25.

[70] See E. Laspeyres in *Jbb*, 3 (1864), 81-118, and 16 (1871), 296-314; Adolf Held in *Jbb*, 16 (1871), 315-340; M. W. Drobisch in *Jbb*, 16 (1871), 143-156 and 416-427. See also Drobisch, "Ueber Mittelgrössen und die Anwendbarkeit derselben auf die Berechnung des Steigens und Sinkens des Geldwerths," *Berichte des königl. sächsischen Gesellschaft der Wissenschaften* (1871), Math-Phys. Kl., vol. 1; Lexis, *Erörterungen über die Währungsfrage* (Leipzig, 1881).

[71] W. S. Jevons, *The Theory of Political Economy* (1871; New York, 5th ed., 1957), p. 22.

[72] Jevons, "A Serious Fall in the Value of Gold Ascertained, and its Social Effects Set Forth" (1863), in *Investigations in Currency and Finance* (London, 1884), p. 58.

[73] *Ibid*., p. 21. See also Jevons, "The Variation of Prices and the Value of the Currency since 1782" (1865), and "The Depreciation of Gold" (1869), both in *ibid*.

measure of general price fluctuations than an arithmetic mean because the price of a commodity can double rather more easily than it can vanish. What is more important than his particular solution, however, is the circumstance that the problem of index numbers had given rise to a mathematical discussion of price statistics focused on issues that soon became central to the new mathematical statistics.

Inflation and deflation of the currency, always a matter of great concern to debtors and lenders, were subjects of particularly intense debate in Britain and elsewhere during the deflation of the last decades of the nineteenth century. Debtors agitated for bimetalism, and there were serious proposals, by Alfred Marshall among others, to abandon the use of currency as a standard of value (though not as the medium of exchange) and to substitute a unit of fixed purchasing power determined through the careful and systematic study of price changes.[74] Edgeworth took up the subject in 1883, in one of his very first papers on probability and the methods of statistical analysis. He there discussed the selection of a proper mean value, arguing that it should be based on considerations of utility, and he pointed to the advantageousness of the arithmetic mean in astronomy, where inaccuracies are typically distributed according to the error function. There are, he argued, three good reasons to prefer the arithmetic mean for index numbers as well. First, prices are subject to numerous small causes of deviation, precisely the model according to which the error law was routinely conceived. Edgeworth also thought that the empirical data were in close conformity with the error curve—better, in fact, than the data of physics. Finally, if all commodities are taken into account, the arithmetic mean of price changes, weighted according to the quantities involved, could be said to give by definition the change in the value of money.[75]

Between 1887 and 1888 Edgeworth wrote three long papers for a British Association Committee charged with evaluating the alternative schemes for measuring changes in the value of money. In them, he discussed the implications of the various ways of understanding shifts of currency value. He continued to view respectfully the most straightforward approach, the calculation of an arithmetic mean with commodities weighted according to quantities. If, however, it is supposed that the

[74] Alfred Marshall, "Remedies for Fluctuations of General Prices," *Contemporary Review*, 51 (1887), 355-375.

[75] Edgeworth, "On the Method of Ascertaining a Change in the Value of Gold," *JRSS*, 46 (1883), 714-718.

changing value of money is an objective function which underlies real prices, and hence the constant cause about which diverse commodities are distributed like errors, then these commodity prices should be regarded as indicators, not measures, of monetary change. Since transactions in a given commodity are not independent, it would make no more sense to assign weights in proportion to quantities than to weight barometric readings according to the volume of mercury. Instead, the proper weight to be assigned a price change in a particular commodity as an index of inflation would be inversely proportional to its dispersion—to the magnitude of the random fluctuations that it undergoes.

Edgeworth also discussed the relative merits of mean and median. He thought the median might provide the best measure of long-term price movement, since it is only slightly less accurate than the mean for a normally distributed system, while it has the advantage of ease of calculation and nonresponsiveness to the wild fluctuations sometimes exhibited by individual commodities. His argument, formulated in terms of a rather extravagant meteorological metaphor, was "that the Median is better qualified to fix and measure the mean variation of the monetary barometer, by an average over many observations; at any rate where the commercial weather is extremely variable; where the cyclones of war or anticyclones of speculation produce eccentric deviations from that mean result which may be due to causes operating on all commodities whatever."[76]

During these same years, Edgeworth published a series of abstract mathematical papers on the various possible distributions which a set of quantities can assume, and their bearing on the choice of the most advantageous mean value. Attention to nonnormal distributions played a prominent role in the early development of mathematical statistics; Edgeworth—and perhaps also, indirectly, Pearson—was led to them by problems encountered in his work on index numbers. His aim was to revise the tools and methods of least squares to apply it to the diverse and often irregular distributions of events encountered in economics—to generalize the methods of treating errors of observation developed by astronomers so that they could be applied to the complex data of statistics.[77]

[76] Edgeworth, "Some New Methods of Measuring Variation in General Prices," *JRSS*, 51 (1888), 346-368, p. 361. See Edgeworth's three memoranda in *PPE*, vol. 1.

[77] See Edgeworth, "On Discordant Observations," *Phil Mag* [5], 23 (1887), 364-375; "The Choice of Means," *ibid.*, 24 (1887), 265-271; "The Empirical Proof of the Law of Error," *ibid.*, pp. 330-342; "A New Method of Reducing Observations Relating to Several Quantities," *ibid.*, pp. 222-223; 25 (1888), 184-191.

Edgeworth drew much of his inspiration from particular problems, but he followed from the beginning a general plan of which he never lost sight. That was the application of mathematics, and especially of probability, to the moral sciences. As often as he developed statistical tools to deal with particular issues that had seized his interest, he took up or invented new applications of statistical reasoning because they offered a context for refining his statistical procedures, or demonstrating the power of his methods. He sought continually for new ways to represent social and moral phenomena with mathematical functions or idealized models such as the urn—which, he learned from Lexis, must be conceived in statistics as filled with clusters of varying composition.[78] Like Peirce, he was deeply engaged by the problem of measurement, especially of beliefs and expectations. Accordingly, he followed with interest the work of Fechner and Wundt in exact experimental psychology. Even his writings on the philosophy of probability were inspired in large measure by the problem of finding the relation between objective frequencies and subjective belief.[79]

Much of Edgeworth's most important work involved what is now called significance testing—the use of standard errors and an assumption of normality to determine whether the difference between two mean values is sufficient to rule out the possibility that it had been produced by chance. One of the first fields to which he applied these methods was psychical research, which he recognized as one of the most straightforward applications of probability, since there was good reason to suppose that successive trials were completely independent. Whether a preponderance of correct guesses indicated the existence of special psychical powers, and not merely an experimental imperfection, was outside the competence of the statistician, as he recognized. A test of the significance of observed discrepancies from the expected mean, however, could be performed deductively, using combinatorial formulas, without the need for vast numbers of experiments to attain a reliable approximation to the probable error. There remained the standard obstacle to *a posteriori* calculation, the difficulty of justifying any particular assignment of values to the undermined parameters in the formula. Edgeworth was less skeptical of the "probabilty of causes" than many, but

[78] See Edgeworth, "Methods of Statistics," *JRSS*, Jubilee Volume (1885), 181-217, p. 191.

[79] See especially Edgeworth, "The Physical Basis of Probability," *Phil Mag* [5], 16 (1883), 433-435; also "On the Reduction of Observations," *ibid.*, 17 (1884), 135-141; "A Priori Probabilities," *ibid.*, 18 (1884), 204-210. More generally, see Edgeworth, *Metretike, or the Method of Measuring Probability and Utility* (London, 1887).

here he simply dismissed the apparatus of *a posteriori* analysis and acknowledged that the calculated probability of the observed number of successes and failures arising from chance alone was only an approximation to the probability that no fixed cause had operated.[80]

In 1885, the same year he made his brief foray into psychical research, Edgeworth published a general paper on the methods of statistics in the fiftieth anniversary volume of the Statistical Society. He there applied the ideas and techniques of probability to a set of problems whose most striking attribute was their miscellany. Attendance at London clubs, mortality rates, and Virgilian hexameter—the latter being held up as a counterexample to Lexis's exclusion of subnormal dispersions, thereby inspiring a debate with Bortkiewicz—were selected for discussion principally because of their value as exemplars of statistical methods rather than for their inherent interest or importance. In place of political economy, for example, Edgeworth took up a related problem that was easier, because it was uncomplicated by historical development. He tested his tools on observations of insect economy, examining the statistical significance of variations of "trade" in a wasp's nest based on some observations at his family home in Ireland. He summarized the results of his analysis as follows:

> If in an insect republic there existed theorizers about trade as well as an industrial class, I could imagine some protectionist drone expressing his views about 12 o'clock that 4th day of September, and pointing triumphantly to the decline in trade of 2½ percent as indicated by the latest returns. Nor would it have been easy off-hand to refute him, except by showing that whereas the observed difference between the compared means is only 2, the modulus of comparison is $\sqrt{70/5 + 70/13}$ or 4 at least; and that therefore the difference is insignificant.[81]

Formulas for standard error, of course, had been easily accessible since Laplace. Edgeworth's main achievement in this regard was to find new subjects to which to apply this familiar technique and, concomitantly, new ways to do so. Probability in his hands was a more supple instrument than it had been before, applicable not just to those rare

[80] Edgeworth, "The Calculus of Probabilities Applied to Psychical Research," *Proceedings of the Society for Psychical Research*, 3 (1885), 190-199.

[81] Edgeworth, "Methods of Statistics" (n. 78), p. 209. See also a full paper on wasp economics, "Statistics of Unprogressive Communities," *JRSS*, 59 (1896), 358-386.

problems that were fully analogous to the comparison of mean values in population statistics and barometry, but to a wide range of questions. He estimated the reliability of competitive examinations by considering seriatim the generation of error due to the use of a discrete rather than a continuous scale, to the variation of standards among examiners, and to irregular fluctuations of success in test taking. He proposed only to achieve the limited objective of assessing "the barometrical aspect of examinations," that is, how well they measure whatever it is that they measure directly, under the assumption that intellectual worth is governed by the error function, and he determined that their inaccuracy was appreciable but not overwhelming. The placement of adjacent candidates is so consistently arbitrary, he thought, that numerical rankings should be abandoned—except for the most able individuals who, as Galton had shown, occupied the region of the error curve where differences between candidates became much larger. "The calculus of probabilities," he concluded, "proves to be a leveller of the same sort as the aristocratic reformer in one of Disraeli's stories, who was for abolishing all distinctions of rank, except indeed the order of dukes."[82]

Edgeworth succeeded also in accommodating his methods, via a simplifying model, to values that were in flux. In 1886 he made use of least squares to fit a linear approximation to a series that was evolving over time, and then showed how to calculate the probable error for the estimation of slope.[83] The same year he proposed a statistical contribution to the theory of banking. His problem was to determine the probability that a hypothetical banker will be required in a brief interval to meet more than a certain proportion of his liabilities—that is, enough to exhaust any given level of reserves. Edgeworth did not pretend that he could commend his results as a practical guide to the banker, since there were problems of independence in the behavior of customers, but he was able to model one feature characteristic of banking and inventory theory—the dependence of reserves at any given moment upon those at the time of the previous observation. Records of bank reserves make up an "entangled series," and can be represented by a series of digits, the last of which is dropped at regular intervals and a new one chosen at random. The sum of digits at each succeeding time period forms a se-

[82] Edgeworth, "Element of Chance" (n. 65), p. 656. See also "Problems in Probabilities: No. 2, Competitive Examinations," *Phil Mag* [5], 30 (1890), 171-188; "The Statistics of Examinations," *JRSS*, 51 (1888), 599-635.

[83] Edgeworth, "Progressive Means," *JRSS*, 49 (1886), 469-475.

quence, which Edgeworth took as a model for the banking reserves. Using Galton's method of median and quartile, he calculated the dispersion and estimated its error in order to assign upper and lower limits for the predicted fluctuations.[84]

Edgeworth's most impressive technical achievement was his description of a procedure similar in many ways to the modern analysis of variance. His problem was to take a two-dimensional statistical table giving the death rate, or some other quantity, for several different locations in every year during a given interval, and to separate the fluctuation due to time from that due to location, and both from that irreducible variation which is independent both of time and place. Not uncharacteristically, he prepared the reader and himself for the task by undertaking an analysis of the meter in Virgil's *Aeneid*, asking to what extent the number of dactyls—or metrical feet consisting of one long then two short syllables—varied either from foot to foot or from five-line unit to five-line unit. His discussion was enigmatic in places, but as Stephen Stigler has shown, if read carefully it reveals an impressive understanding of the assumptions and capabilities of his model.[85]

Since this contribution of Edgeworth's seems not to have been read by R. A. Fisher, who introduced the method of analysis of variance more than three decades later, it can be assigned only a place of limited importance in the history of statistical mathematics. These techniques and procedures, however, are primarily illustrative of Edgeworth's work; no particular formula or method can be designated singly as his great statistical discovery. He contributed instead an insight into the ways in which probability could be applied to certain aspects of economics and social science and a vision of a general statistical theory standing above special applications and recipes. The range and quality of his applied mathematical work justifies Stigler's placement of him with Galton and Pearson, in a triumvirate of founders of mathematical statistics.

There are other reasons, however, to regard Edgeworth as *tertium inter pares*. He had only one mathematical follower in statistics, Arthur Bowley.[86] Although economists in succeeding decades knew and admired Edgeworth's statistical work, his ideas were applied mainly to a

[84] Edgeworth, "Problems in Probabilities," *Phil Mag* [5], 22 (1886), 371-384; also "The Mathematical Theory of Banking," *JRSS*, 51 (1888), 113-127.

[85] See Stigler, "Edgeworth" (n. 47), and Edgeworth, "On Methods" (n. 62).

[86] See Bowley's "Francis Ysidro Edgeworth," *Econometrica*, 2 (1934), 113-124; also Bowley, *F. Y. Edgeworth's Contributions to Mathematical Statistics* (Clifton, N.J., 1972).

relatively narrow range of empirical problems, mostly associated with measuring and interpreting changes in the value of money. In this, of course, economists were simply following his lead. But Edgeworth's leadership in statistics did not otherwise extend to issues of recognizable economic importance. In particular, his statistical writings were almost wholly unconnected with his economic theorizing; as Joseph Schumpeter has noted, neither he nor Lexis made much direct contribution to analytical economics through their work on mathematical statistics.[87] Edgeworth tended to define his own problems, often somewhat whimsically, rather than to proffer solutions to statistical questions under study at the time. His failure to demonstrate the pertinence of probability mathematics to the main problems of economics and social statistics surely tended to minimize his influence, and accounts in large measure for the fact that his work gave rise to no statistical movement in social science comparable to the biometric school.

Thus Edgeworth's statistical influence was exerted primarily through the biometric school. His importance here is not inconsiderable, despite Karl Pearson's complaint that he always plowed across the line of the biometricians' furrows and Galton's inability to move him to take up the problems that interested Galton.[88] Stigler argues that Edgeworth kindled Pearson's early interest in probability in 1892. Edgeworth also was first to derive something approximating the modern product moment estimate of correlation, a few years after Galton introduced the idea of mathematical correlation.[89] Pearson was familiar with his work, and used his ideas freely. Still, he used them only insofar as they met the needs of biometrics, and it is to that field we must turn to find the proper origins of mathematical statistics.

[87] Joseph Schumpeter, *History of Economic Analysis* (New York, 1954), p. 961.

[88] See MacKenzie, pp. 97-99, who cites Pearson's comment and notes Galton's attempt to direct Edgeworth's research. Edgeworth's letters to Galton are in file 237, *FGP*. That Edgeworth's failure to concentrate on any particular line of application may account for his limited influence in statistics was also argued by E. S. Pearson, "Some Reflections on Continuity in the Development of Mathematical Statistics," in *SHSP1*, pp. 339-353.

[89] See Stigler; also Edgeworth, "Correlated Averages," *Phil Mag* [5], 34 (1892), 190-204; "The Law of Error and Correlated Averages," *ibid.*, pp. 429-438, 518-526.

Chapter Nine

THE ROOTS OF BIOMETRICAL STATISTICS

That the modern field of mathematical statistics developed out of biometry is not wholly fortuitous. The quantitative study of biological inheritance and evolution provided an outstanding context for statistical thinking, and quantitative genetics remains the best example of an area of science whose very theory is built out of the concepts of statistics—variance-covariance matrices, regression coefficients, and so on. Beyond that, the biometrician-eugenicists were possessed with an intense ecumenical urge and, especially in the case of Karl Pearson, endowed with very respectable talents for academic entrepreneurship. On both counts the contrast with Edgeworth is striking; he wrote only intermittently on statistics after about 1893 (he assumed the economics chair at Oxford in 1891), and from his time to ours, economics has used statistics almost exclusively to analyze data, not as the basis of theory. That these things could not have been otherwise—that economics or sociology could not have provided the chief subject matter for the development of mathematical statistics—would be too brave a claim. In the event, however, it was biometrics. The great stimulus for modern statistics came from Galton's invention of the method of correlation, which, significantly, he first conceived not as an abstract technique of numerical analysis, but as a statistical law of heredity. Here, as throughout the nineteenth century, the special problems of particular fields were of central importance for the development of statistical mathematics.

GALTON'S BIOMETRICAL ANALOGIES

Galton was one of the last universal men of science and a purveyor of analogies scarcely less prolific than Edgeworth. His responsibility for the introduction of the method of correlation constitutes *prima facie* evidence that this landmark in the history of statistics was not entirely the product of close concentration on a single discipline. The examples of

Laplace, Quetelet, Maxwell, Fechner, Lexis, Edgeworth, and Peirce all testify to the generalization that the great nineteenth-century statistical writers were not, and could not have been, narrow-minded specialists. But there can be no better illustration of the bond between diversity of interest and statistical creativity than the career of Francis Galton, who began his scientific work as an African explorer and geographer, became interested in meteorology, ethnology, and anthropology, and was then inspired by his conversion to a creed of eugenic reform to take up biology, psychology, anthropometry, heredity, personal identification, and the new science of sociology. As a result of these diverse pursuits, and of his pronounced statistical bent, Galton encountered statistical reasoning in a variety of forms, and he learned something about it from each of them.

Since Galton routinely associated the stability of aggregate measures with the regularities of social statistics, and since that regularity was regarded as a commonplace by his generation, there is good reason to suppose that he learned of it in his youth.[1] He began to make active use of the methods of statistics, though probably not of error analysis, in his meteorological researches, which commenced during the 1850s. He learned of Quetelet's use of the error curve from a geographer, William Spottiswoode, and he taught himself the fundamental properties of this function using George Airy's textbook of observational astronomy. Galton received assistance in his statistical study of *English Men of Science* from Herbert Spencer and from William Farr of the General Register Office, whose annual reports revealed, according to Galton, a keen appreciation of "what might be called the poetical side of statistics."[2] He collaborated with John Venn, the philosopher of probability, in the collection and analysis of school records to promote the study of heredity.[3] He was an enthusiastic admirer of psychophysics, and studied the statistical methods in Fechner's *Elemente der Psychophysik*.[4] He also received mathematical assistance from, and sought collaboration with, Edgeworth, the kinetic theorists Henry W. Watson and S. H. Burbury, and three other mathematicians, Donald MacAlister, J. D. Hamilton

[1] He wrote in 1844 that he was using a classification of medical records "like the ordinary plan of statistical charts," but I have found no early allusions to statistical regularity. See *Galton*, vol. 1, p. 184.

[2] Francis Galton, *Memories of My Life* (London, 3d ed., 1909), p. 292.

[3] See John Venn, "Cambridge Anthropometry," *JAI*, 18 (1889), 146-154.

[4] See *Galton*, vol. 3B, p. 464; also Galton, "The Just-Perceptible Difference," *PRI*, 14 (1896), 13-26.

Dickson, and W. F. Sheppard.[5] There were, in short, few available sources of useful information or insights on statistics that Galton did not exploit. Still more significantly, he profited greatly in his late work on physical heredity and personal identification from his own experience in the statistical study of meteorology and of hereditary eminence.

Meteorology was most conspicuously involved in these Galtonian interdisciplinary interactions. His use of superimposed composite photographs, for example, was descended from a strategy that he had previously employed to compare "maps and meteorological traces." He refined his techniques for the study of human physiognomy, and then assisted G. M. Whipple, an acquaintance at the Kew Observatory, in importing the improved procedure back into meteorology.[6] More generally, Galton developed in his meteorological studies an extraordinary ability to interpret maps and charts. One of his favorite techniques as a meteorologist was the construction of isobars, isotherms, and other lines on weather maps connecting points of equal magnitude in some variable. During the mid-1880s, he discovered that when the mean parental height is plotted against the height of their offspring, the graph generates lines of constant density whose shape is elliptical, and the slope of whose major axis is determined by the value of the regression. Others might have found this by considering abstractly the properties of the bivariate normal, and indeed his result was subsequently confirmed by mathematical analysis. Galton discovered it simply by totaling the number of points in each square on his graph, and drawing lines by eye connecting all numbers that were approximately equal.[7]

Galton began studying meteorology around 1858, when he was appointed a director of the Kew Observatory, outside of London. Quantitative, empirical studies such as meteorology, geodesy, and observational astronomy, like social statistics, had been flourishing in Europe, and especially in England, for several decades, and the sheer mass of information had increased enormously since 1800. The recent availability

[5] See MacKenzie, pp. 95-97. Galton was pleased to have met Quetelet, and admired his use of the error law (see *Memories*, p. 304, n. 2), but wrote dismissively of his and Buckle's achievements in actual statistical work (see letter to Florence Nightingale, 10 Feb. 1891, in *Galton*, vol. 2, p. 420).

[6] See Galton, "Composite Portraits," *Nature*, 18 (1879), 97-100, p. 98, and G. M. Whipple, "Composite Portraiture Adapted to the Reduction of Meteorological and Other Similar Observations," *Quarterly Journal of the Meteorological Society*, 9 (1883), 189-192.

[7] Galton, "Family Likeness in Stature," *PRSL*, 40 (1886), 42-73, with mathematical appendix by J. D. Hamilton Dickson; see also "Results Obtained from the Natality Table of Körösi, by Employing the Method of Contours or Isogens," *PRSL*, 55 (1894), 18-23.

of telegraphic communication between observatories opened a whole new range of possible techniques for predicting the weather. Even theoretical investigations of weather phenomena depended on extensive records, and the Kew Observatory, with which Galton remained associated until 1901, was probably the leading institution of statistical meteorology in the world during the mid-nineteenth century.[8] Galton, of course, was not a reluctant participant in this movement towards ever greater use of charts and tables; he reveled in it. He was an enthusiastic advocate of the systematic compilation of weather maps, and in 1863 he published a study of methods of mapping the weather.[9]

Galton's ideas about the correlation of variables probably derived in considerable measure from his meteorological work. Most meteorologists, like Galton, worked at observatories, and many divided their time between meteorology and related studies such as geodesy and observational astronomy. For that reason, the science which Galton took up in 1858 had been associated with the techniques of mathematical probability at least since the time of Laplace's work on the diurnal variation of the barometer. Hence meteorologists tended to know about error analysis, and to apply it to their own science with only slight provocation. These familiar mathematical procedures, however, appeared in a new light when incorporated into meteorology.

There was hidden within the method of least squares a mathematical definition of correlation, which appeared in astronomy as the opposite of independence under the title "entanglement of observations."[10] It was almost impossible in astronomy to avoid entanglement, especially when the astronomer sought to carry out measurements that required data from two or more observatories. In that case, any inaccuracy in knowledge of the relative positions of the observatories would reappear in every measurement. Astronomers developed techniques to discover and compensate for these entanglements. In fact, the French astronomer and naval officer Auguste Bravais actually used the term *corrélation* in 1846 to refer to an entanglement of this sort. He even derived a joint error distribution which included a term for this correlation, and he has been credited by Karl Pearson, among others, as the proper discoverer of cor-

[8] See Robert Henry Scott, "The History of the Kew Observatory," *PRSL*, 39 (1885), 37-86, p. 56.

[9] Galton, *Meteorographica*, or *Methods of Mapping the Weather* (London, 1863).

[10] See G. B. Airy, *On the Algebraical and Numerical Theory of Errors of Observation* (London, 1861), sec. 12.

relation. Pearson, an astute historian of science, later changed his mind—correctly, as Donald MacKenzie points out, for Bravais had no interest in studying or even measuring this correlation, and no reason to do so.[11] Astronomers and surveyors quite naturally were mainly concerned to dispose of entanglements, and social or biological statisticians could never have recognized any use for these astronomical concepts in their own work before they were formulated in more congenial terms. To some extent, meteorology served the same function with regard to entanglements as Quetelet's work did with the error function. It gave them some substance, transforming a method for eliminating error into a positive procedure for dealing with natural variation.

The relationship of observational astronomy and statistical meteorology to the idea of correlation is nicely illustrated by a matter that attracted much scientific interest and considerable popular attention at precisely the time that Galton first took up the problem of hereditary regression. At issue was the connection between sunspot cycles, first demonstrated conclusively by Samuel Heinrich Schwabe in 1843, and terrestrial meteorology. That magnetic storms were more frequent and severe during times of highest sunspot density was widely recognized by 1860. Edward Sabine, the dean of "magnetic philosophers" in Victorian Britain, devoted great effort to the investigation of this link during the 1850s and 1860s through the compilation of graphs and tables. Significantly, Sabine was responsible for Galton's appointment to the managing committee of the Kew Observatory, and, as Galton recorded in his memoirs, "exercised great influence in shaping my scientific life."[12] Sabine drew the attention of British astronomers and meteorologists to the importance of sunspot cycles, which were nowhere studied more intensely than at Kew. Sunspots were widely suspected to be convection sinks on the solar surface, and hence indicators of solar activity. It seemed at least inherently plausible that they might be of some importance for terrestrial weather.

In 1873 C. Meldrum, the director of the British meteorological observatory on Mauritius, proposed that these solar cycles ought to have periodic effects on the earth's atmosphere. He assembled records from eighteen observatories around the Indian Ocean to investigate this hy-

[11] See MacKenzie, pp. 68-72. As MacKenzie observes, similar considerations apply to the Dutch military engineer and mathematician Charles Schols, who is sometimes credited with the discovery of correlation for a paper he wrote in 1875.

[12] Galton, *Memories* (n. 2), p. 224.

pothesis, and he concluded that cyclones in the southern part of that ocean were more frequent and more violent in years of maximum sunspot density than at the bottom of the cycle.[13] Within a few years astronomers and meteorologists from throughout the British dominions had begun collecting records and publishing papers on this issue, mostly in support of Meldrum's hypothesis. Already in 1874 J. A. Broun attempted to mathematize the relationship between rainfall and sunspot area using a linear model of the form $\Delta R = f\Delta A$, where ΔR was excess or deficiency of rainfall from the mean, ΔA the deviation of sunspot area from its mean, and f a constant, to be determined empirically. Here was a correlation model which, although not normalized to units of probable error, was clearly expressed in terms of deviations from the mean. The enterprising Broun proposed that such relationships could be confirmed only through the collaboration of meteorologists in observatories throughout the region. He initiated what soon became a chorus of calls for expansion of the imperial meteorological system.[14]

It soon became clear that any number of phenomena could be attributed to sunspot variation. Jevons argued, notoriously, that the regular commercial crises revealed by his study of British monetary history correlated precisely with sunspot fluctuations. He suggested that the cause could be found in the weather cycles of the British south Asian possessions and their influence on trade with the mother country. Later he proposed that these same cycles might have been responsible for the periodic recurrence of the plague during the seventeenth century.[15] A physicist, Arthur Schuster, described the connection, mediated by the weather, between sunspot density and the quality of German wine.[16] The onset of famine in India during 1877 led to a heightening of interest, and soon the director-general of Indian statistics, Hunter, announced his suspicion that sunspots were the culprit.

Balfour Stewart, the director at Kew since 1859, whose research had

[13] C. Meldrum, "On a Periodicity of Rainfall in Connexion with the Sun-spot Periodicity," *PRSL*, 21 (1873), 197-208; also J.H.N. Hennessey, "Note on the Periodicity of Rainfall," *PRSL*, 22 (1874), 286-289.

[14] J. A. Broun, "On the Sun-spot Period and the Rainfall," *PRSL*, 22 (1874), 469-473; see also his articles in *PRSL*, 25 (1876), 24-43, 515-539.

[15] Jevons, "Commercial Crisis and Sun-spots," *Nature*, 19 (1878), 33-37, 588-590; "The Solar Commercial Cycle," *ibid.*, 26 (1882), 226-228; "Sun-spots and the Plague," *ibid.*, 19 (1878), 338. See also a paper he read publicly in 1875, but did not then publish, "The Solar Period and the Price of Corn," in Jevons, *Investigations in Currency and Finance* (London, 1884), no. 6.

[16] Cited in Balfour Stewart, "Suspected Relations between the Sun and the Earth," *Nature*, 16 (1877), 9-11, 26-28, 45-47, p. 45.

been largely devoted to solar physics and spectroscopy, immediately discerned in this situation a splendid opportunity to make a contribution to an important branch of knowledge while promoting his own field of research. Indeed, the urgency of the situation was not diminished by the fact that at this very time a committee on solar physics had begun meeting to establish a policy for the future direction and funding of solar research in Britain. Stewart thus took his case to the public, citing the literature mentioned above in support of his ambitious claims.

> If we now bring together the results of these three papers, we may compare the three problems, solar research, terrestrial magnetism, and meteorology, to three corners of a triangle that are bound together. Of their three relations we are . . . perfectly certain of the relation between solar research and terrestrial magnetism. The connection between solar research and meteorology is perhaps not so well defined, but our evidence is here supplemented by independent traces of a connection between magnetism and meteorology. Thus the three things hang together, and scientific prudence points to the desirability of their being studied as a whole, a consideration which will not, I trust, be overlooked in the contemplated reorganisation of meteorology.
>
> I would desire now to conclude by asking, in all honesty, Have we not here a plea for the establishment of some institution that will keep a daily watch upon that luminary which is thus seen to affect us in such a variety of ways?[17]

Stewart authored and coauthored a series of memoirs, each titled as a report to the Solar Physics Committee, proposing statistical techniques for discovering concurrent periodicities, and explaining a system based on solar observations and magnetic records for predicting both long-term and short-term weather. He suggested that "magnetic weather" is formed simultaneously with atmospheric weather through solar influences, but then travels so much more rapidly that it might be useful for meteorological forecasting.[18] He endorsed the conjecture of the direc-

[17] *Ibid.*, p. 47. Stewart, whose attachments to psychical research may not be wholly irrelevant, saw also in the possible influence of planetary movements and conjunctions on solar activity a plausible scientific basis for astrology.

[18] Papers by Stewart, either alone or in collaboration with Warren De la Rue, Richard Loewy, or William Dodgson, may be found in *Nature*, 16 (1877), 457; 24 (1881), 114-117, 150-153, 260; 26 (1883), 488-489; *PRSL*, 29 (1879), 106-122; 303-324; 32 (1881), 406-407; 33 (1882), 410-420; 34 (1882), 406-409.

tor-general of statistics to the government of India, Hunter, that the Indian famine was due to regular meteorological fluctuations associated with sunspot activity, and proposed that further solar research might enable scientists to predict future famines. Other meteorologists and solar physicists were similarly eager to point out the implications of all these discoveries. Frederick Chambers, for example, explained how solar observations might provide the means "whereby future famines may possibly be foreseen," and drew the inevitable conclusion that since "the whole subject of solar observations is now being investigated by a committee of gentlemen in London, . . . we may therefore hope that the all-important information which solar observations are capable of affording will ere long be at our disposal."[19]

The most significant contribution to the sunspot literature, for present purposes, was a paper by Richard Strachey, whose various activities thicken the plot of this drama immensely. He was a member of the Royal Society Meteorological Committee, served with Galton on the managing committee of the Kew Observatory, and was, as Galton reports, one of his closest friends.[20] He was also a member of the august Solar Physics Committee, a fourth-generation Indian civil servant, and the president of a commission appointed to inquire into the causes of the terrible Indian famine of 1877.[21] Regrettably, he was not persuaded by the argument for a link between sunspot cycles and weather fluctuations. Writing in 1877, in response to Hunter's paper, he proposed to place the entire matter in perspective by applying a general statistical test, which proceeded as follows. First, he computed the mean difference, or error, between the annual values of rainfall and their general mean. Next, he computed the mean difference between these same measures of annual rainfall and the value predicted by the best available model based on a cycle corresponding to sunspot fluctuations. If, he thought, the residual error is not much reduced when a cyclic curve is substituted for the general mean as a prediction of the rainfall, then the evidence does not support the hypothesized cycles in rainfall patterns.

[19] Frederick Chambers, "Abnormal Variations of Barometric Pressure in the Tropics and their Relation to Sun-spots, Rainfall, and Famines," *Nature*, 23 (1880), 88-91, 107-111, p. 110.

[20] See Galton, *Memories* (n. 2), p. 241: "A prominent place ought to be given to him [Strachey] in my 'Memories', for we have been connected in our pursuits very frequently and in very different ways." See also *Galton*, vol. 2, p. 60.

[21] See Robert Hamilton Vetch, "Sir Richard Strachey," in *Dictionary of National Biography*, 23 (Supplement, 1900-1910), part 3, 439-442; also Strachey, "Physical Causes of Indian Famines," *PRI*, 8 (1877), 407-426.

He found, in fact, that the error was little altered when a cycle was substituted for the mean, and he concluded that "the supposed law of variation obtained from the means of the six eleven-year cycles hardly gives a closer approximation to the actual observations than is got by taking the simple arithmetical mean as the most probable value for any year."[22]

Other astronomers hastily pointed out that Strachey's method was not really a test of the existence of regular fluctuations, but a measure of their magnitude relative to the irregular sources of variation.[23] It is, on that account, all the more interesting. Although his mathematics was both simple and unoriginal, it involved a subtle but important difference from the results of least squares, the statistical tests used by Laplace and Lexis, and the linear models that had from time to time been proposed to account for variation in phenomena of one sort or another. Strachey used his statistical methods to measure a real correlation, and not simply to dispose of error, and his measure was of the relation of fluctuations in one variable to those of another rather than merely of the existence of progressive change over time. While it is not clear that Strachey's procedure had no precedent, and uncertain to what extent Galton was personally involved in the sunspot affair, the episode is still illuminating. It shows how the meteorological problems and ideas of Galton's time could have led even an unoriginal mathematician like Strachey—or Galton—to apply methods of error analysis to studies of correlation.

In point of fact, Galton, too, attempted a kind of correlation analysis in reference to meteorological problems. In 1870 he published a paper in which he investigated statistically the connection among barometric pressure, wind velocity, and humidity, and proposed a linear model that Karl Pearson later identified as a multiple regression formula.[24] The similarity of this exercise to the work that eventually led Galton to regression and correlation has doubtless been exaggerated. He set the problem up in terms of deviations from the previously recorded value, rather than from the general mean, and it is not easy to imagine how he could have derived his method of correlation from this work. Hence no

[22] Richard Strachey, "Indian Rainfall and Sun-spots," *Nature*, 16 (1877), 171-172. See also his "On the Alleged Correspondence of the Rainfall at Madras with the Sun-spot Period, and on the True Criterion of Periodicity in a Series of Variable Quantities," *PRSL*, 26 (1878), 249-261. His opinion was supported by G. M. Whipple, "Results of an Inquiry into the Periodicity of Rainfall," *PRSL*, 30 (1880), 70-84.

[23] J. A. Broun, "Rainfall and Sun-spots," *Nature*, 16 (1877), 251-252; W. Clement Ley, "Rainfall and Sun-spots," *ibid.*, pp. 252-254.

[24] *Galton*, vol. 2, p. 54; Galton, "Barometric Predictions of Weather," *BAAS* (1870), pp. 31-33.

specific influence can be assigned to this investigation, or to the sunspot researches of Strachey and Jevons, on Galton's subsequent discovery of the method of correlation. Their general significance, nevertheless, may be considerable. The search for systematic statistical relationships between meteorological variables was evidently standard procedure at the Kew Observatory and elsewhere—there is also an extensive literature, in which Galton's friend Whipple figured prominently, concerning the relation between barometric gradients and wind velocity.[25] The search for correlations in meteorology encouraged Galton to seek correlations in other fields of science, and furnished guidance for his application of related statistical techniques to the study of heredity.

Galton's debt to social statistics was, it would appear, even more significant than that to meteorology. It can, moreover, be identified concretely. The place of social statistics in Galton's work is represented by the simile which constituted not just an illustration, but the substance of his theory of heredity. Here is the most powerful testimony to the role of analogies in the development of statistical thinking, to its dependence on the effective transmission of ideas between disciplines. Galton's social analogy provided the entering wedge by which the statistical approach was introduced to evolutionary biology, and it may be seen in retrospect almost as an involuntary tribute to his statistical sources. He introduced analogies with a frequency approaching that of Quetelet, Maxwell, and Edgeworth, and developed them, at least in the present instance, in yet greater detail. His writing was never dry, and rarely technical, but his discussion of the mechanism of heredity was picturesque even by his own exacting standards.

Galton's starting point in biology was provided by his older cousin's theory of evolution by natural selection, and the inspiration for his work on the mechanism of heredity was provided by Darwin's "provisional hypothesis of Pangenesis." That hypothesis, an attempt to provide a common framework for a much wider range of organic phenomena

[25] See William Ferrel, "The Influence of the Earth's Rotation upon the Relative Motion of Bodies near Its Surface," *Astronomical Journal*, 5 (1858), 99; W. Clement Ley, "Suggestions on Certain Variations, Annual and Diurnal, in the Relation of the Barometric Gradient to the Force of the Wind," *Quarterly Journal of the Meteorological Society*, 3 (1877), 232-237; G. M. Whipple, "Barometric Gradients in Connection with the Wind Velocity and Direction at the Kew Observatory," *ibid.*, 8 (1882), 198-203. Whipple wrote a number of papers giving rough statistical correlations of numerical records; see *ibid.*, 5 (1879), 142-147, 213-217; 7 (1881), 49-52, 53-57; 10 (1884), 45-52. See also James Glaisher in *Proceedings of the British Meteorological Society*, 2 (1865), 350-366; 3 (1867), 359-378; 4 (1869), 1-23, 33-50, 113-132, 327-351.

than interested the nonbiologist Galton, held sexual reproduction to be fundamentally similar to asexual reproduction and to tissue repair. Darwin argued that every part of the body gives off "gemmules" by budding, which circulate through the fluids of the body and are concentrated in the sexual elements. The fertilization of the egg brings together a critical mass of gemmules, which then coalesce and form an embryo. Each element finds its place through "elective attractions" by a process analogous to crystallization. To account for the well-established phenomenon of atavism, Darwin proposed that many more gemmules are transmitted to the offspring than are required to determine its phenotypical structure. He concluded by reflecting on the implications of his model for the understanding of individuality in living things. "An organic being is a microcosm—a little universe, formed of a host of self-propagating organisms, inconceivably minute and numerous as the stars in heaven."[26]

The gemmule theory of heredity was so manifestly supportive of a statistical approach to evolution that it survived in biometry until at least the beginning of the twentieth century, though it was virtually a complete failure among biologists of every other school.[27] Galton adopted it immediately. He was especially impressed by Darwin's idea that the embryo is formed through a complex process of selection involving the affinities and attractions of innumerable distinct elements. Like his cousin, Galton supposed that only a small portion of the gemmules found expression in the physical structure of the offspring, the rest being either discarded or passed on in latent form. Galton, however, cared nothing about budding of gemmules from somatic tissue, or repair of wounded limbs, and since he was doubtful even then of the inheritance of acquired characteristics, he was content to locate all gemmules permanently in the sexual organs. He focused his attention principally on the process by which gemmules are selected, a process that Darwin had only mentioned.

Galton first discussed Pangenesis in the concluding chapter of *Hered-*

[26] This theory makes up chapter 27 of Charles Darwin, *Variation of Animals and Plants under Domestication* (2 vols., London, 1868), quote from vol. 2, p. 399. See Gerald L. Geison, "Darwin and Heredity: The Evolution of His Hypothesis of Pangenesis," *Journal of the History of Medicine and Allied Sciences*, 24 (1969), 375-411.

[27] Galton still favored his variant of Pangenesis when he published *Natural Inheritance* (New York, 1889), see p. 9. Pearson refers to the inheritance of gemmules in Karl Pearson, with the assistance of Alice Lee, "Mathematical Contributions to the Theory of Evolution.—VIII. On the Inheritance of Characters Not Capable of Exact Quantitative Measurement," *Phil Trans*, A, 195 (1900), 79-150, p. 120; also Pearson, *The Grammar of Science* (London, 2d ed., 1900), p. 335.

itary Genius, which appeared just one year after Darwin's *Variation of Animals and Plants under Domestication*, in 1869. He epitomized his interpretation of the theory in two propositions: first, that "germs," or gemmules, are given off in enormous numbers; and second, that "the germs are supposed to be solely governed by their respective natural affinities, in selecting their points of attachment [to the embryo]; and that, consequently, the marvellous structure of the living form is built up under the influence of innumerable blind affinities, and not under that of a central controlling power." Galton continued:

> This theory, propounded by Mr. Darwin as "provisional," and avowedly based, in some degree, on pure hypothesis and very largely on analogy, is—whether it be true or not—of enormous service to those who inquire into heredity. It gives a key that unlocks every one of the hitherto unopened barriers to our comprehension of its nature; it binds within the compass of a singularly simple law, the multifarious forms of reproduction, witnessed in the wide range of organic life, and it brings all these forms of reproduction under the same conditions as govern the ordinary growth of each individual. It is, therefore, very advisable that we should look at the facts of hereditary genius, from the point of view which the theory of Pangenesis affords.[28]

To elucidate Darwin's hypothesis, Galton proposed a set of "similes" to the process of natural inheritance. Since all the gemmules act autonomously, he observed, the characteristics they determine "may be compared to the typical appearance always found in different descriptions of assemblages. . . . The assemblages of which I speak are such as are uncontrolled by any central authority, but have assumed their typical appearance through the free action of the individuals who compose them, each man being bent on his immediate interest, and finding his place under the sole influence of an elective affinity to his neighbours."[29] Just as a "watering-place," or any other distinctive type of human community, can form itself within a society of great diversity purely as a consequence of numerous free decisions by "young artisans, and other floating gemmules of English population," so can distinct types of men be

[28] Galton, *Hereditary Genius* (London, 1869), p. 364.
[29] *Ibid.*, p. 364.

formed from a similar diversity of hereditary elements, simply by the action of their individual affinities and levels of energy.[30]

Galton's model could easily be adapted to explain a great range of hereditary appearances. The tendency exhibited by nature to form certain types of hybrids and to exclude others, for example, could be illustrated in terms of the possible forms of a watering-place, or resort town. Such a community can readily be hybridized with a fishing town, since the seaside delights visitors, the tourists provide a market for the daily catch, and so on. Should a manufacturer seek to establish himself in this location, however, blending would be impossible, for the "dirt and noise and rough artisans engaged in the manufactory are uncongenial to the population of a watering-place." One or the other must give way.[31]

Galton went on to characterize the difference between latent and active (or "patent") gemmules in terms of a political equilibrium. So long as one party has the majority in a region—according to the British and American electoral systems—it completely dominates the political character of that district. "Let, however, by the virtue of the more rapid propagation of one class of electors, say of an Irish population, the numerical strength of the weaker party be supposed to gradually increase, until the minority becomes the majority, then there will be a sudden reversal or revolution of the political equilibrium, and the character of the borough or nation, as evidenced by its corporate acts, will be entirely changed. This corresponds to a so-called 'sport' of nature." And why, asked Galton, are sports and "reversions," or instances of atavism, so often produced by marriages between widely different types? It is for the same reason that two boroughs, one Whig and one Conservative, but each with a substantial Irish minority, might be dominated by the plurality of the Irish party if the boroughs were merged. Galton concluded by defending his mode of expression: "These similes, which are perfectly legitimate according to the theory of Pangenesis, are well worthy of being indulged in, for they give considerable precision to our views on heredity, and compel facts that appear anomalous at first sight, to fall into intelligible order."[32]

[30] *Ibid.*, p. 365. The metaphor of decentralized authority is wonderfully Victorian; compare with the explanations of the action of the heart discussed by Gerald Geison, *Michael Foster and the Cambridge School of Physiology* (Princeton, 1978), pp. 340-355; see also review by Steven Shapin in *Isis*, 71 (1980), 146-149. These decentralizing implications were perhaps already implicit in the cell theory, as is particularly evident in the work of the outstanding German biologist and political liberal, Rudolf Virchow. By Pearson's time even English biometricians found them far less congenial.

[31] Galton, *Hereditary Genius* (n. 28), pp. 366-367.

[32] *Ibid.*, pp. 367-368.

The elaboration of this extended analogy in *Hereditary Genius* might be supposed a mere display of literary imagination, if his other, more technical writings on the subject were cast in a different form. In fact, however, he continued to express himself in terms of the same images, and he insisted that a "strict analogy" obtained between the collective gemmules of a person and the political processes of a parliamentary government.[33] Galton's use of this simile is most striking in his paper "On Blood-relationship," published in the *Proceedings of the Royal Society of London* in 1873, which contained the most detailed exposition of his physiological theory of heredity. Here he explained the processes of gemmule reproduction and transmission over the course of a generation.

Galton began with the "structureless elements," that is, gemmules transmitted by the parents to their future offspring, which have not yet arranged themselves into an embryo. The first process is an "election," whereby a small proportion of these elements are selected by some unknown process to "represent" the genetic material, or "stirp." These "segregated elements" coalesce to determine the overall structure of the embryo, while the great majority of elements, those not selected, remain latent. Both sets of elements replicate themselves, however, and when the offspring reaches maturity, both sets compete to be transmitted to the next generation. This was the second segregation of gemmules, and was called "family representation." Galton deemed it the most fundamental process in heredity, since it determined which elements would be perpetuated in future generations and which would perish. The other division, "class representation," was of secondary importance, for it determined only the characteristics of one offspring in a single generation. Hence the true chain of kinship does not proceed from parent to offspring, as in a pedigree, but from the gemmules of the parent, both latent and patent, to those of the progeny.[34]

Since Galton's model was based on a free political society, it is scarcely surprising that his method for analyzing it was likewise derived from the contemporary numerical science of society. That science, of course, was called statistics. Galton argued, on the basis of the analogy with social science, that the elements which determine the attributes of the offspring may be assumed to represent accurately the totality of gemmules, both latent and patent, subject only to the existence of a "small

[33] Galton, "A Theory of Heredity," *Contemporary Review*, 27 (1875-76), 80-95, p. 86.
[34] Galton, "On Blood-relationship," *PRSL*, 20 (1873), 394-402.

degree of correlation" acting at a very local level. He decided that no special mechanism was required to insure that offspring will receive gemmules representing all bodily parts, any more than formal provision would need to be made in the conscription of an army to guarantee that each town contribute its share of young men. Fair representation occurred spontaneously, in this as in all statistical processes.

Hence, according to this model, hereditary transmission was analogous to a random drawing from an urn, or rather, to a series of such drawings.

> An appropriate notion of the nearest conceivable relationship between a parent and his child may be gained by supposing an urn containing a great number of balls, marked in various ways, and a handful to be drawn out of them at random as a sample: this sample would represent the person of a parent. Let us next suppose the sample to be examined, and a few handfuls of new balls to be marked according to the patterns of those found in the sample, and to be thrown along with them back into the urn. Now let the contents of another urn, representing the influences of the other parent, be mixed with those of the first. Lastly suppose a second sample to be drawn out of the combined contents of the two urns, to represent the offspring. There can be no nearer connexion justly conceived than between the two samples.[35]

Once those processes had been expressed in terms of this classic model, probability relationships followed directly. Galton, significantly, pointed out already in *Hereditary Genius* that "the doctrine of Pangenesis gives excellent materials for mathematical formulae."[36] His review of Darwin's hypothesis of Pangenesis appeared at the end of his book, and given how closely in time it followed Darwin's publication, there is good reason to suppose Galton wrote it under the immediate inspiration of his cousin's hypothesis. Pangenesis, it would appear, gave Galton his idea for a more exact, experimental study of the statistics of hereditary transmission, which he sketched near the conclusion of the volume.

> I do not see why any serious difficulty should stand in the way of mathematicians, in framing a compact formula, based on the the-

[35] *Ibid.*, p. 400.
[36] Galton, *Hereditary Genius* (n. 28), p. 370.

> ory of Pangenesis, to express the composition of organic beings in terms of their inherited and individual peculiarities, and to give us, after certain constants had been determined, the means of foretelling the average distribution of characteristics among a large multitude of offspring whose parentage was known.[37]

The governing principle, he thought, could be foreseen. On the average, children should "inherit the gemmules in the same proportions that they existed in the parents." Superimposed on this average will be a certain amount of deviation, fully predictable in its broad outlines, if not for each individual. "In short, the theory of Pangenesis brings all the influences that bear on heredity into a form, that is appropriate for the grasp of mathematical analysis."[38]

In the end, Galton explained, it ought to be possible to define in advance the proportions of descendants of each level of hereditary talent, from exceptional *F*'s and G's to mediocre A's and *a*'s and incompetent *f*'s and *g*'s that would be produced by any given combination of parental abilities. (See discussion in chapter 5.) But in these preliminary states of investigation, he thought, data on hereditary talents are too fluid and unreliable. Research ought now to be devoted to the averages of "some simple physical characteristic, unmistakeable in its quality, and not subject to the doubts which attend the appraisement of ability." Under the hypothesis of Pangenesis, "there need be no hesitation in accepting averages for this purpose; for the meaning and value of an average are perfectly clear. It would represent the results, supposing the competing 'gemmules' to be equally fertile, and also supposing the proportion of the gemmules affected by individual variation, to be constant in all the cases."[39]

Thus Galton's model of heredity does more than reveal implicitly the lineages of his statistical thought. The supposition that heredity concerned the action of numerous gemmules, subject to processes that may be likened to drawings from an urn or conscription from the young men of a nation, placed his model in conformity with the classic derivation of the normal law of errors. The theory of mutual affinities provided structure to his notion of biological correlation. Most important, the theory of Pangenesis provided the foundation for a model of inheritance

[37] *Ibid.*, pp. 371-373.

[38] *Ibid.*, p. 373.

[39] *Ibid.*, pp. 370-371.

based on the fundamental parameters of statistics—a mean, deriving from the average of parental gemmules or traits, and a certain irregular deviation. That is to say, Pangenesis provided the concepts that led Galton to the statistical investigation of heredity that he began around 1875. Although the results of this investigation were not exactly what he had expected, Galton's model furnished terms of statistical analysis ideally suited to the formulation of regression.

REGRESSION AND CORRELATION

Galton finally gave up theorizing about hereditary models and investigating the inheritance of acquired characteristics about 1873, and turned at last to investigation of the statistical properties of natural inheritance discussed in the conclusion of *Hereditary Genius*. His inability to apply calculation to his survey of men of science completed in 1874 confirmed what he had earlier suspected, that the laws of heredity could best be investigated experimentally, using the simplest possible materials. He was, in any event, confident that these laws were universal and, once discovered, could be applied to inheritance of intellectual and moral traits. Galton resolved to study the inheritance of size in peas. Early in 1876 he sent a few hundred measured seeds to friends, with instructions to plant them in uniform soil and to keep the offspring of each parental size class separate. This appeared to be the simplest experiment possible, for these peas were thought to be self-pollinating, so that only one parent need be taken into account.

Upon receipt of his harvest of peas, Galton "weighed seeds individually, by thousands, and treated them as a census officer would treat a large population."[40] He divided the collection of pea seeds along two dimensions, first according to the weight of their parents, and then according to their own weight. Thus his results took the form of several different weight distributions, each corresponding to parent seeds of a given weight class. From inspection of these distributions, Galton concluded that the dispersion, or probable error, of each batch was independent of the size class of the parent seeds. The measurements also led him to his first statement of the law of regression: the mean of each batch of progeny was displaced from the general mean in an amount propor-

[40] Galton, "Typical Laws of Heredity," *PRI*, 8 (1877), 282-301; reprinted in *Nature*, 15 (1877), 492-495, 512-514, 532-533, p. 512.

tional to the displacement of their parents. The mean displacement of the offspring, however, was always less than that of the parents; they had, on the average, "reverted" part way back to the mean for the entire race.

Given Galton's experience as a meteorologist and his readiness to perceive linear relationships, it is unsurprising that he should have begun by trying a linear formula also for these pea seed results. Here, however, the fit seemed good enough to justify regarding the relationship as a law of nature, whereas the study of barometric changes had indicated little more than a tendency for certain measures to vary in the same direction. Galton was always alert to the danger of relying on the results of intricate analysis when the primary data were coarse, inaccurate, or highly variable, and neither in his study of barometric pressures nor in his survey-based work on *English Men of Science* did he pursue mathematical analysis very far. In his preface to the latter study, Galton commented: "It had been my wish to work up the materials I possess with much minuteness, but some months of careful labor made it clear to me that they were not sufficient to bear a more strict or elaborate treatment than I have now given to them."[41] Galton was, it must be remembered, studying heredity, not seeking new methods of statistics. He had undertaken this work on peas to acquire data that were at least simple and reliable, even if they did not bear directly on the questions of intellectual and moral inheritance that most concerned him. The results seemed to meet his expectations, and Galton was at last able to apply a close statistical analysis to hereditary materials.

Satisfied with the accuracy of the linear "reversion" relationship that he had discovered, the inexorable tendency for each generation to move back towards mediocrity, Galton was confronted with the problem of explaining the evident stability in size distribution for the whole population. "How is it," he asked, "that although each individual does *not* as a rule leave his like behind him, yet successive generations resemble each other with great exactitude in all their general features?"[42] As it has since become clear, this was the question that inspired Galton to develop the basic mathematical apparatus for analyzing variation that he eventually recognized as a general method of correlation. He saw his way to a solution only under one condition—that the distribution of

[41] Galton, *English Men of Science: Their Nature and Nurture* (1874; New York, 1875), p. vii.

[42] Galton, "Typical Laws" (n. 40), p. 492.

traits represented by the gemmules be governed at every point by the law of error, for only the exponential function preserves its form when multiplied by itself.[43]

His explanation of hereditary equilibrium was not difficult. Suppose the size distribution of the parental generation is governed by Quetelet's error law,

$$\frac{1}{c_o\sqrt{\pi}}\, e^{-x^2/c_o^2}$$

where c_o is the measure of dispersion for the curve, and x the deviation from the mean. For purposes of analysis, Galton decomposed the process of reproduction into several parts and evaluated sequentially the effect of each on the dispersion, or width, of the curve. In humans—though not, Galton believed, in peas—the first event in reproduction, mating, causes the gemmules of two different individuals to be mixed together. If it is assumed that choice of mate has no dependence on size, then the mean size of the various pairs, or the the "mid-parental" size, will be distributed according to the same error curve, but with a narrowed dispersion c_1, equal to $c_o/\sqrt{2}$. The next event in the cycle is reversion, which expresses the tendency of the offspring of any given parents to move on average a fixed proportion of the distance back towards the mean. Hence the curve is further narrowed; the dispersion of those mean offspring is reduced to c_2, defined as $c_2 = rc_1$, where r, the measure of reversion, is a constant less than one. Finally, although the offspring of parents of any given size revert, on average, part way to the mean, they are subject to a cause of dispersion which Galton termed "family variability." Representing this factor by v, the effect will be to widen the curve, whose dispersion c_3 will be equal to $\sqrt{v^2 + c_2^2}$. Now, since the offspring population is identical in composition to the parents, this c_3 is equal to c_o. Hence the four unknowns are governed by four equations which, once r and v have been measured, can readily be solved. The whole process can be represented mechanically, and Galton did so using the quincunx lattice model that he made famous.[44]

Galton first presented the results of this study in a lecture at the Royal Institution in 1877. He devoted the next several years to psychological investigations—which, of course, were not without eugenic overtones,

[43] *Ibid.*, pp. 495, 512.

[44] *Ibid.*, *passim*. For purposes of simplicity I have omitted from this model the stage representing natural selection.

since it was inheritance of moral and intellectual characters that most concerned him—and then returned to direct statistical studies of heredity in 1884. Now he wished to confirm his law of heredity for the case of humans, and he devoted enormous effort to the assembly of records giving measures such as height for a family over the course of several generations. The results were largely satisfactory, for it appeared that human heredity was indeed governed by the same statistical relations as peas. Hence his analysis was essentially identical. He did, however, change his terminology from "reversion" to "regression," a shift whose significance is not entirely clear. Possibly he simply felt that the latter term expressed more accurately the fact that offspring returned only part way to the mean. More likely, the change reflected his new conviction, first expressed in the same papers in which he introduced the term "regression," that this return to the mean reflected an inherent stability of type, and not merely the reappearance of remote ancestral gemmules. The type is preserved by regression, he now argued, and the "effective heritage of the child" is "less than the accumulated deviates of his ancestors." Stability resulted from the persistence of gemmule affinities, and could be overcome, as in his political metaphor, only through the creation of a new majority, a "sport," representing a new point of stability. Artificial or natural selection of continuous variation, accordingly, was at best a far less effective agent of evolution than the preservation of sports or saltations.[45]

The study of humans did present one new problem, whose solution was of much importance for Galton's later work on the general problem of correlation. It arose in the form of an apparent paradox—Galton's discovery that the regression from mid-parental height to offspring height was precisely twice as great as the regression from offspring to mid-parental height. What could account for this failure of mathematical reciprocity? The problem weighed heavily on Galton's mind, for a quarter century later he still recalled vividly his sense of triumph when he solved it. As he was walking on the grounds of Naworth Castle, a squall came up, and he sought refuge in the rock beside the pathway. Suddenly it flashed across his mind "that the laws of Heredity were solely concerned with deviations expressed in statistical units."[46] The mid-parental height

[45] Galton, "Family Likeness" (n. 7), p. 62; "Regression towards Mediocrity in Hereditary Stature," *JAI*, 15 (1886), 246-263, p. 258; also *Natural Inheritance* (n. 27), chap. 3, and "Discontinuity in Evolution," *Mind*, 19 (1894), 362-372.

[46] Galton, *Memories* (n. 2), p. 300. Most writers on Galton have thought that this passage must refer to the early 1870s. They could be right. But I can see no problem of heredity upon

is not distributed exactly like that of individuals; it represents the mean height of father and mother (the latter adjusted to conform to the same scale as men's heights). Assuming that mating is not assortative with respect to height, the dispersion of the mid-parental height must be reduced from that of individuals by a factor of $\sqrt{2}$. Hence the effect of a given deviation from the mean in mid-parent is magnified by $\sqrt{2}$ on the individual offspring, while the corresponding regression from offspring to mid-parent is reduced by the same factor.[47]

Despite the identity between the mathematics of regression and correlation, more than a decade passed before Galton perceived this connection. Indeed, it was precisely the mathematical identity that led Galton to recognize his regression problem as a special case of correlation, and not the conceptual similarity which permitted his application of the mathematics. The reason for this should be clear—regression was not originally conceived as a statistical technique at all. It was a law of heredity, albeit one that could be combined with a set of statistical techniques to explain the continuity of the distribution of many traits in a population. The use of a linear model *per se* as a rough approximation to the data was no great advance, for Galton and others had already used such a model in meteorology. Regression was interesting to Galton in a way that his meteorological model was not, because linear regression represented a genuine law of nature, and because it generated an interesting solution to an unsuspected problem—the cause of the stability of natural variation.

Strictly speaking, then, the roots of Galton's qualitative ideas about "correlation" should be traced not to this work on heredity, but to the phenomena of biological inheritance customarily designated with that term.[48] Galton was familiar with these traditional biological uses of this concept, such as Cuvier's notorious belief that a whole skeletal structure could be surmised from the shape of the teeth on account of the correlations of these things. Darwin thought that traits not in themselves ad-

which Galton worked before the 1880s for which he needed the concept of statistical units—however useful they might have been for assigning numerical values to traits that can only be ordered by rank.

[47] Galton, "Family Likeness" (n. 7), pp. 56-57.

[48] And, in Galton's case, perhaps also geography; see Galton, "Address as President of Geography Section," *BAAS*, 42 (1872), 198-203, p. 199: "The configuration of every land, its soil, its vegetable covering, its rivers, its climate, its animal and human inhabitants act and react upon one another. It is the highest problem of geography to analyze their correlations, and to sift the casual from the essential."

vantageous might be passed on because of their favorable correlations.[49] In *Hereditary Genius*, Galton made extensive use of this idea of biological linkage. It provided him, for example, with the reason why children of divines often turn out badly. The piety that is essential to the character of the divine, he explained, involves two distinct attributes: high moral character and a tendency to pronounced oscillations between self-sacrificial adoration and vulgar sensuality. Since, however, "these peculiarities are in no way correlated," it follows that the children of divines will often inherit the instability without the morality, a disastrous combination.[50] Similarly, Galton wrote that in men of science, "energy appears to be correlated with smallness of head," and he undertook a search for physical marks that were correlated with ability in science or in other worthwhile pursuits in the hope that this knowledge would prove advantageous for some future eugenic program to breed great scientists. By 1873, at least, Galton understood this biological form of correlation in terms of the tendency for gemmules to exhibit various affinities and to be transmitted in certain groups.[51]

As is well known, Galton was led to the mathematical idea of correlation by his study of personal identification. The proximate cause of his renewed interest in this problem was furnished by Alphonse Bertillon's new method of identifying criminals. Bertillon, a member of a distinguished French statistical family,[52] proposed to classify all criminals by measuring four bodily parts, height and length of finger, arm, and foot, and classifying each as short, medium, or long, according to some uniform standard. He anticipated that the criminal population would thereby be divided into eighty-one roughly equal classes. Galton had for the previous six years been particularly active in anthropometric work; he had set up a prize system to elicit complete "records of family faculties," had erected an anthropometric laboratory at the International

[49] See Charles Darwin, *On the Origin of Species* (London, 1859), chap. 5.

[50] Galton, *Hereditary Genius* (n. 28), p. 282. Forty years later he gave a similar argument in *Noteworthy Families* (London, 1906), p. xv.

[51] Galton, "Blood-relationship" (n. 34), pp. 395-396; also *Natural Inheritance* (n. 27), p. 9.

[52] Achille Guillard, founder of demography, was his grandfather; Louis-Adolphe Bertillon, author of a well-known article on the "Moyenne" in *Dictionnaire encyclopédique des sciences médicales* (see also "La théorie des moyennes en statistique," *Journal de la Societé de statistique de Paris*, 17 [1876], 265-271, 286-308), was his father; Jacques Bertillon, demographer and crusader for a heightened French birthrate, was his brother. See Bernard-Pierre Lécuyer, "Probability in Vital and Social Statistics: Quetelet, Farr, and the Bertillons," in *Prob Rev*; also Michel Dupaquier, "La famille Bertillon et la naissance d'une nouvelle science sociale: la démographie," *Annales de démographie historique*, 1983, pp. 293-311.

Health Exhibition to measure curious passers-by, and had worked with John Venn to procure and analyze results of measurements from universities and schools. Most of this work was designed to confirm and extend his results on hereditary regression, but he had also given attention to the independence or correlation of human characteristics in the hope that statistically independent features were inherited independently, and might be used to identify the ancestor from which each part was derived. In any event, Galton recognized immediately that Bertillon's measurements were not independent—that, since long feet and long legs would tend to appear together, most of the criminals would end up in a relatively small number of size classes. He resolved to repeat Bertillon's measurements at his anthropometric laboratory in order to "estimate the degree of their interdependence."[53]

Calculation of this "degree of interdependence" had to be performed according to the principles he had uncovered while pondering the reciprocal regression of mean-parents and individual offspring, for there was no other basis for comparing a foot an inch longer than average with an arm three inches longer. In late 1888, after his book *Natural Inheritance* was already in press but before it had appeared, Galton suddenly reached a solution to this problem of criminological entanglement. Following his usual empirical procedure, he drew a graph of stature against left cubit, scaling the axes in units of the respective probable errors, and he began to enter points from his anthropometric data. Faced with Galton's graphical virtuosity, the problem virtually solved itself. "As I proceeded in this way, and as the number of marks upon the paper grew in number, the form of their general disposition became gradually more and more defined. Suddenly it struck me that their form was closely similar to that with which I had become very familiar when engaged in discussing kinships."[54] The problem, he realized, was mathematically identical to regression. "A very little reflection made it clear that family likeness was nothing more than a particular case of the wide subject of correlation, and that the whole of the reasoning already bestowed upon the special case of family likeness was equally applicable to correlation in its most general aspect."[55] The regression of parental height onto off-

[53] Galton, "Personal Identification and Description," *Nature*, 38 (1888), 173-177, 201-202, p. 175; see also p. 202.

[54] Galton, "Kinship and Correlation," *North American Review*, 150 (1890), 419-431, p. 421.

[55] Galton, "President's Address," *JAI*, 18 (1889), 401-419, p. 404.

spring height had been given by an expression of the form $y = rx$; the correlation of height with, say, length of arm could be expressed as $ay = brx$, or $y = rx(b/a)$, where a represents the probable error of stature x, b that of arm length y, and r the coefficient of correlation. Here, he thought, was the solution to the great biological problem of correlation of parts. It seemed to him so obvious, once he had found it, that he rushed his discovery into print, without even correcting the calculations. Galton feared that the same solution would be immediately recognized by readers of *Natural Inheritance*, and that his priority of discovery would be lost.[56]

Linear approximation was not, of course, the "discovery" in the method of correlation. The crucial innovation in Galton's approach was the idea of partitioning variation into classes—one part of which was explained by the correlated variable, and another which remained as error. Calculation of residual error when measurements were compared with some function of the time was common enough, but that function nearly always represented something very concrete—the changing apparent location of a star as the earth rotated. Astronomers were never interested enough in their entanglements to divide the error up as Galton did. Richard Strachey had approached this idea, but he did not reify the connection between sunspots and rainfall by assigning a figure to it. The idea of partitioning variation, a procedure that is absolutely fundamental to mathematical statistics, was not easily conceived even by those who were interested in natural correlations. How was it that Galton, a man of little mathematical skill, reached this subtle conception?

The answer is implicit in his path to its discovery. Galton did not recognize directly from his study of interrelationships between specified human measurements that the total variation in each can be decomposed into two parts. Instead, he discovered empirically that the pattern on his correlation graph was was identical to the one he had already seen many times during his regression studies, and he inferred that the same mathematics apply. Now, although regression and correlation (as Galton defined them) are indeed identical mathematically, regression is far more straightforward conceptually. Once Galton had determined that, on the average, the offspring of each parental size class of peas tend to revert a certain percentage of the distance back to the mean, he was con-

[56] Galton, "Kinship and Correlation" (n. 54), p. 420. His first paper on correlations was "Co-relations and Their Measurement, Chiefly from Anthropometric Data," *PRSL*, 45 (1889), 135-145.

fronted with a well-defined problem formulated not in terms of abstract statistical variation, but concrete biological variation. He had before him two groups of peas, one descended from the other. Both were characterized by the same degree of dispersion, though the mean offspring of each parental size class had reverted to some extent. Here, decomposition of variation into two parts presented no problems of interpretation. Some of the parental variation had been preserved because reversion was incomplete; the remainder was provided by the "family variation" characterizing the offspring of parents in each size class. The general problem of correlation presented nothing that led so naturally to an understanding of correlation in terms of the amount of variation explained. Galton's invention of a method of correlation cannot be ascribed to mathematical acuity—though his intuitive grasp of simple mathematics was indispensable—but to his wide interests and to his ability to use what he learned from one problem as an aid to the solution of others.

Galton's method for partitioning natural variation between that which is explained by the correlation model and that which is not has largely passed into the realm of the obvious, and it is difficult to imagine how readers with any degree of mathematical sophistication could have failed to recognize its validity and usefulness once it had been formulated. Yet it was not obvious to contemporaries, and Galton had difficulty convincing some, especially mathematicians, that his use of probability was legitimate. We find in Galton's *Memories*, for example:

> [H. W. Watson] helped me greatly in my first struggles with certain applications of the Gaussian Law, which, for some reasons that I could never clearly perceive, seemed for a long time to be comprehended with difficulty by mathematicians, including himself. They were unnecessarily alarmed lest the well-known rules of Inverse Probability should be unconsciously violated, which they never were. I could give a striking case of this, but abstain because it would seem depreciatory of a man whose mathematical powers and ability were far in excess of my own. Still, he was quite wrong. The primary objects of the Gaussian Law of Error were exactly opposed, in one sense, to those to which I applied them. They were to get rid of, or to provide a just allowance for errors. But those errors or deviations were the very things I wanted to preserve and to know about.[57]

[57] Galton, *Memories* (n. 2), p. 305.

Although he abstained from divulging this "striking case," Watson himself was not so reticent. The incident illustrates strikingly, as Galton promised it would, the conceptual differences between the law of error and Galton's law of variation. It becomes all the more clear why methods of correlation emerged in biometry rather than in astronomy.

The problem that Galton brought to Watson involved precisely the decomposition of variation. The incident occurred around 1890, at a time when Galton was attempting to work out the expected height of an ancient skeleton after a single thigh bone had been discovered.[58] This differed from the original correlation problem in that the thigh length is itself an element of the height, and Galton was probably inspired by this distinction to feel scruples about his own reasoning, and to consult with a more experienced mathematician. The results of his consultation, however, were not at all satisfactory. Both Watson and some other mathematician, probably J. D. Hamilton Dickson,[59] were unable to view the problem as Galton did. The differences between Galton and Watson on this matter are highly instructive.

Watson narrated the problem in the following terms. Suppose a man takes a hop and a stride; suppose also that both hop and stride are known to cluster around a mean distance in the usual manner with probable errors of a and b respectively. It is obvious, assuming hop and stride length are independent, that the probable error of their sum, h, will be given by $\sqrt{a^2 + b^2}$. Now suppose, instead, that it is not a and b, the probable errors of hop and stride, but a and h, those of the hop and the sum, that are known. Galton wished to know whether the probable error of the stride would be given by $b = \sqrt{h^2 - a^2}$. Watson responded immediately that Galton was "decidedly wrong." The expression $h = \sqrt{a^2 + b^2}$ no longer holds if h instead of b is known, any more than the expected outcome of one hundred coin tosses remains fifty heads and fifty tails after it is known that the first two tosses have come up heads. The errors of the sum and of the hop, Watson maintained, must combine in the usual manner, $b = \sqrt{h^2 + a^2}$.

Watson's reasoning was less than exemplary, even for one who had never used the Gaussian curve for any purpose but to eliminate error—and he had in fact used it in the kinetic theory as well. The hop and the sum are clearly entangled, according to the principles of George Airy, and their errors cannot be combined as if they were independent. Still,

[58] Galton, "Kinship and Correlation" (n. 54), p. 419.
[59] See MacKenzie, p. 236.

Watson was an able mathematician, which Galton was not, and it requires some subtle thinking to understand how he could believe both that $h^2 = a^2 + b^2$ and $b^2 = h^2 + a^2$.

Watson evidently viewed the probable error as nothing more than an astronomer's artifice, an entity so evanescent that it need not even obey the rules of arithmetic. That is, he believed that the value of the probable error is wholly a function of our knowledge, and not an attribute of the object under consideration. Galton took an entirely different view, arguing that the "probable deviation . . . of a scheme is as definite a phrase as the diameter of a circle."[60] Eventually he persuaded Watson that his viewpoint might be justified—though the mathematician was never comfortable with it—by instructing him, in effect, to imagine that all the hops, strides, and sums were carved in stone. The equation $h^2 = a^2 + b^2$ would then be more than a generalization about the imperfections of knowledge; it would be a rigorous truth, which could be confirmed with any measuring stick. Then b^2 would be equal to $h^2 - a^2$, and $b^2 = h^2 + a^2$ would quite simply be wrong.[61] To believe otherwise was to misunderstand the character of statistical mathematics. Galton's ideas could not even be understood until his readers were persuaded that the statistical measure of variation was a real mathematical quantity, not some elusive standard of imperfect knowledge.

PEARSON AND MATHEMATICAL BIOMETRY

The almost simultaneous appearance of Galton's book *Natural Inheritance* and his method of correlation in 1889 marks the beginning of the modern period of statistics. Correlation was a tool that promised to be useful in almost every domain, and especially for those in which it had proved difficult to establish clear lines of causation. It soon began to be applied routinely to anthropometric measurements, and also to a variety of social problems, such as the relations among crime, poverty, ignorance, alcohol, and poor health. *Natural Inheritance* inspired biologists searching for a method of research by which Darwin's theory of evolution could be studied. An especially influential and early advocate of Galtonian methods was Walter Frank Raphael Weldon, who an-

[60] Galton, *Natural Inheritance* (n. 27), p. 58.

[61] See H. W. Watson, "Observations on the Law of Facility of Errors," *Proceedings of the Birmingham Natural Historical and Philosophical Society*, 6 (1889-1891), 289-318.

nounced in no uncertain terms that "the problem of animal evolution is essentially a statistical problem."[62] Weldon, no mathematician, encountered problems of numerical analysis that went beyond his modest competence, and he enlisted the support of his mathematical colleague at University College, London, Karl Pearson. Pearson quickly became impressed by the power of statistics as an agent of mathematization in the biological sciences—which, in his view, encompassed also the social and behavioral sciences. Weldon, he later wrote, "first formulated the view that the method of the Registrar-General is the method by which the fundamental problems of natural selection must be attacked, and this is the essential feature of biometry."[63] Pearson devoted his life from 1893 onwards to working out these methods mathematically, and to demonstrating the extent of their applications. Statistics provided biology with its only alternative to "metaphysical speculation," he argued. "We must proceed from inheritance in the mass to inheritance in narrower and narrower classes, rather than attempt to build up general rules on the observation of individual instances. Shortly, we must proceed by the method of statistics, rather than by the consideration of typical cases."[64]

That statistics could be applied widely to the social and biological sciences and that it would benefit from systematic work on its mathematical foundations were not ideas that had to await Pearson's conversion to statistics in 1893. Francis Galton, long impressed by the wide range of fields to which his cherished numerical approach could be applied, derived new inspiration when the law of regression that he had for more than a decade conceived as a biological principle proved to be a general method of correlation. He had for some years been cultivating acquaintanceships with various mathematicians, whom he hoped to interest in the general problems of statistics by sending them particular mathematical problems which he was unable to solve. In 1892, when one of these mathematicians, W. F. Sheppard, criticized *Natural Inheritance* for its lack of a systematic introduction to the method of statistics, Galton agreed wholeheartedly.

[62] Quoted in E. S. Pearson, *Karl Pearson: An Appreciation of Some Aspects of His Life and Work* (Cambridge, Eng., 1938), p. 26.

[63] Karl Pearson, "Walter Frank Raphael Weldon: 1860-1906, A Memoir," *Biometrika*, 5 (1906), reprinted in *SHSP1*, 264-321, p. 285.

[64] K. Pearson, "Mathematical Contributions to the Theory of Evolution.—III. Regression, Heredity, and Panmixia" (1896), in *ESP*, 113-178, p. 115.

> What is clearly wanted is a clear elegant resumé of all the theoretical work concerned in social and biographical [*sic*, biological] problems to which the exponential law has been applied. I believe the time is ripe for any competent mathematician to do this with much credit to himself. I am *not* competent and know it. Edgeworth has his own work and interests, and fails in sustained clearness of expression. He is moreover somewhat over fond of using higher and more mathematics than is always necessary. Watson is over busy and I think too fastidious and timid. I have often considered what seems wanted and been very desirous of discovering someone who was disposed to throw himself into so useful and such high-class work. He might practically *found* a science, the material for which is now too chaotic.[65]

Tribute to the general power of statistics, in fact, is also abundantly present in *Natural Inheritance*. Galton explained the need to discuss statistical methods before proceeding to his main subject, then added that his obligation to proceed indirectly was no matter for regret, since the subject of statistics "is full of interest of its own."

> It familiarizes us with the measurement of variability, and with various laws of chance that apply to a vast diversity of social subjects. This part of the inquiry may be said to run along a road on a high level, that affords wide views in unexpected directions, and from which easy descents may be made to totally different goals to those we have now to reach. I have a great subject to write upon.[66]

Pearson, in a retrospect on his career, assigned special significance to his first reading of *Natural Inheritance*, and in particular to the spirit of the passage just quoted:

> "Road on a high level", "wide views in unexpected directions", "easy descents to totally different goals"—here was a field for an adventurous roamer. I felt like a buccaneer of Drake's days. . . . I interpreted that sentence of Galton to mean that there was a category broader than causation, namely correlation, of which causation was only the limit, and that this new conception of correlation brought psychology, anthropology, medicine and sociology in large parts into the fields of mathematical treatment. It was Galton who freed me from the prejudice that sound mathematics could

65 Quoted in *Galton*, vol. 3B, pp. 486-487. See also letter, 23 Oct. 1892, in *FGP*, file 315.
66 Galton, *Natural Inheritance* (n. 27), p. 3.

> only be applied to natural phenomena under the category of causation. Here for the first time was a possibility—I will not say a certainty of reaching knowledge—as valid as physical knowledge was then thought to be—in the field of living forms and above all in the field of human conduct.[67]

Pearson was here either forgetful or disingenuous. The method of correlation was not in *Natural Inheritance,* and though Pearson approved in 1889 of Galton's eugenic aims, Galton's book inspired him to remark on the "considerable danger in applying the methods of exact science to problems in descriptive science."[68] Only through Weldon did Pearson come to accept Galton's approach to the study of biological heredity, and he had become skeptical of the category of causation before then. Once he was won to statistics, however, he quickly persuaded himself of the great dictum of Laplace and Quetelet, that mathematics provided the proper approach to every field, and was the unfailing mark of science.

Ironically, it was soon Galton who began to look like the skeptic in regard to the use of advanced mathematics in social and biological statistics. Responding to Pearson's first statistical paper, which provided a technique for separating superimposed normal curves, and which Pearson hoped might provide a criterion for detecting an ongoing process of speciation, Galton wrote critically:

> As Professor [George] Darwin said, it seems to me that observed curves of frequency are never so exact in contour as to lend themselves to exact & minute treatment. I once amused myself by mixing together a large number of measures of male & female adults and was painfully surprised by finding that the result did not deviate anything like so sensibly as I had expected from a normal curve.[69]

The next year, in 1894, Pearson sent Galton a copy of the *Algebra of Animal Evolution* by a former student of W. K. Clifford at University College, Arthur Black, who had recently committed suicide. Pearson

[67] *Speeches Delivered at a Dinner Held in University College, London, in Honour of Professor Karl Pearson, 23 April,* 1934 (Cambridge, Eng., 1934), pp. 22-23.

[68] Pearson, "On the Laws of Inheritance according to Galton," read to the Men's and Women's Club in 1889, quoted in MacKenzie, p. 88. He also complained, in relation to Galton's remarks on stability of biological forms and the dependence of evolution on sports, that "It is merely an analogy without scientific value as to the *how* still less to the why." Penciled note in Pearson's copy of *Natural Inheritance,* p. 30, quoted in Bernard J. Norton, "Karl Pearson and Statistics: The Social Origins of Scientific Innovation," *Social Studies of Science,* 8 (1978), 3-34, pp. 16-17.

[69] Galton to Pearson, 25 Nov. 1893, file 245/18A, *FGP.*

recognized Black to be working along the same lines as he was, but the manuscript was never published despite the favorable opinions of Pearson, Weldon, and Galton. The latter's praise was qualified. He expressed "very great doubt indeed whether any organic laws are so strictly in accordance with that of the error function as to admit of building vast mathematical edifices on that insecure foundation."[70] Early in 1906 Galton indicated regrets that sociology had not undertaken meticulous analysis of individual cases, on the model of Le Play. A few months later, in the context of some circumspect criticisms of Pearson, Galton expressed his viewpoint on the relation of mathematics to statistics as follows:

> You really misread my "heart of hearts" re mathematics. I worship and reverence them, though in their application I have a tendency towards economy in their use, under the ever-haunting fear lest the exactitude of their results may not outrun the trustworthiness of the data. That is all. My fundamental misgiving respects a too free use of the statistical "axiom" "that unspecified influences tend to neutralise one another in a homogeneous series." My doubt always dwells on the questionable assumption of homogeneity, believing that extreme values are liable to be often caused by an heterogeneous admixture, present active, though undiscerned. So I love the rude, but theoretically correct, statistics over much, feeling always *safer* within their moderate limits of one or two decimal places. All this is quite harmless, is it not? It is a purely general statement quite without reference to Biometrika.[71]

Despite Pearson's initial misgivings about statistical biology and the incompleteness of Galton's subsequent accord with Pearson, his reminiscence about his reaction to *Natural Inheritance* captures a deep truth about his conversion to the statistical approach, which is not overturned by his misstatement of fact. If depth of appreciation or consciousness of intellectual debt can be measured in words, then Pearson's four-volume biography of Galton indicates an important bond indeed. Pearson

[70] *Ibid.*, 17 June 1894. On Black, see MacKenzie, p. 88. Galton had long been aware that the fit of the error curve is rarely exact; in 1877 he advised Henry Pickering Bowditch not to trim his anthropometrical results "to the Procrustean bed of the 'Law of Error.' " See Helen M. Walker, *Studies in the History of Statistical Method with Special Reference to Certain Educational Problems* (Baltimore, 1929), p. 101.

[71] Galton to Pearson, 14 Dec. 1906, *FGP* 245/18G. See also letter of 12 Dec. 1906. On Le Play see letter of 29 Aug. 1906.

proved to be the mathematician that Galton had been seeking, and Galton was wise enough to know that a powerful thinker cannot be expected to remain long subservient to another man's ideas. Galton supported Pearson with his money as well as his thoughts, and when he died, without natural issue, his fortune was used to set up a eugenics laboratory at University College, along with a professorship which, in accordance with the recommendation in his will, was offered to Pearson.

Pearson did indeed prove to be a buccaneer. His "decidedly-piratical tendencies"[72] were manifested in his continuing attempts to capture intellectual territory for mathematical statistics—one cause of the fierce disputes in which he was almost continuously engaged. Social meliorists and nonmathematicians of the sort who still dominated the Royal Statistical Society objected to his uncompromising eugenic approach and were subject to his full disdain for their feeble efforts at statistical analysis. Biologists and geneticists such as William Bateson were dismayed by his heavy reliance on advanced mathematics and by his adamant insistence that *Natura non facit saltum* precluded the existence of Mendelian genes and required the exclusive use of smooth, effectively continuous statistical functions in the study of heredity.[73]

Pearson was also a buccaneer in the sense that one gradually learns to take for granted among the most successful early writers on statistical ideas and mathematics—his mind roamed freely over the intellectual globe. Galton and Weldon gave a focus to his intellectual activities that still embraced a sufficiently wide range of subjects to hold his interest. Prior to 1893 he had studied and written on the most astonishing variety of topics. His early works, which are by no means shallow or dilettantish, treat social and religious history, literature, the philosophy of science, passion plays and folk tales, religion and mysticism, socialism, women's rights, and evolution as a framework for history and biology—all this outside of his professional career, which for a time involved law before Pearson shifted to academic life as an applied mathematician. He was first recommended for a professorship at University College, London, under the following remarkable circumstances:

> Then Professor Beesley, just because I had lectured to revolutionary clubs, Professor Croom Robertson, just because I had written

[72] Pearson, *Speeches* (n. 67), p. 22.

[73] On the controversy between biometricians and Mendelians, see William B. Provine, *The Origins of Theoretical Population Genetics* (Chicago, 1971); also MacKenzie, chap. 6, and B. J. Norton, "The Biometric Defense of Darwinism," *JHB*, 6 (1973), 283-316.

on Maimonides in his Journal *Mind*, Professor Alexander Williamson, just because I had published a memoir on Atoms, and Professor Henry Morley, just because I had attended and criticised lectures of his on the Lake Poets, pressed me to become a candidate for the chair of *Mathematics*.[74]

Karl Pearson was born in 1857 in the family of a stern, industrious Yorkshire barrister. Intellectually earnest from the beginning, though not, perhaps, excessively burdened by a spirit of self-criticism, he broke with Christianity about the age of 20. "Personally, I was not born or reared so that rationalism was to me a 'psychological climate'; I spent five years of life in struggling with much bitterness out of the mazes of metaphysics and theology, only to find in agnosticism the peace which arises from understanding."[75] He soon became convinced that the rationalistic metaphysics of Spinoza was the only form of religion that was scientifically respectable—Udny Yule aptly characterized Pearson's Spinozistic faith as "a religion which, one may say, sets an F. R. S. [Fellow of the Royal Society] in the midst and informs you that of such is the Kingdom of Heaven."[76] Despite his considerable disdain for what he regarded as popular superstitions, he made no sustained effort to debunk Christianity. On the contrary, he repeatedly expressed admiration of the medieval Catholic church, though rather for its organization than its dogma, and he castigated Luther for breaking with an admirable institution that hitherto had been able to contain views as heterodox as those of Erasmus.[77] He was sympathetic to the idea of a scientific priesthood, and his political views evoke the spirit of Comte:

> We never think of taking the opinion of the man in the street on the reasons why the moon does not keep her calculated times; we do not ask his opinion on the value of the opsonic index; we recognise that these are problems which require special training and analysis wholly beyond his grasp, but we still think he is quite capable of expressing an opinion on whether the employment of women is good for her infants or not, although he may be in pos-

[74] Pearson, *Speeches* (n. 67), pp. 20-21.

[75] Pearson, "Reaction" (1895), in *The Chances of Death and Other Studies in Evolution* (2 vols., Cambridge, Eng., 1897), vol. 1, pp. 194-195.

[76] G. Udny Yule, "Karl Pearson (1857-1936)," *Obituary Notices of Fellows of the Royal Society*, 2 (1936-1938), 72-104, p. 103.

[77] See Pearson, "Martin Luther" (1884), in *The Ethic of Freethought and Other Addresses and Essays* (1888; London, 2d ed., 1901), p. 199.

session of no data, and although, if he were, he would be quite incapable of interpreting them.[78]

Like Galton before him, Pearson was a man who required a mission in life—if not sacred then secular. In declining Galton's offer to arrange honorary membership for him in the International Statistical Institute—whose clerks Galton thought might arrange data for him—Pearson wrote bluntly, "as one Quaker descended should to a second": "It is perhaps the long line of Quaker yeomen behind me, but I can't help being a fighter, and I don't like to be a member of anything without fighting for what I consider to be the right view. Now nothing wants reorganizing as much as the European government statistical departments," but Pearson needed to devote his time to a higher mission, biometry.[79] Before he was won to biometry and eugenics, his mission was socialism, and among his many activities of the 1880s he found time to lecture on Marx and Lassalle at revolutionary clubs. He was, however, no revolutionary, or at least no advocate of abrupt change; in history, as in biology, he favored continuity with a passion almost equal to Quetelet's. Although he was converted to socialism during his *Wanderjahr* in Germany by a social democratic student, Raphael Wertheimer, his lasting admiration was reserved for a different group of German socialists, who indeed were hostile to social democracy, the *Kathedersozialisten*.[80] The socialism of professors was precisely Pearson's style of socialism. He even was briefly attached at the end of his life to the national variety of German socialism, though he focused, with characteristic self-importance, on Hitler's "proposals to regenerate the German people," and noted that if this German experiment failed, "it will not be for want of enthusiasm, but rather because the Germans are only just starting the study of mathematical statistics in the modern sense!"[81] Donald MacKenzie finds Pearson's ideology in perfect conformity with the aims of the new professional middle classes in Britain, and although his personal ambition must be stressed alongside his class ambition, certainly he identified strongly with those segments of the middle class who derived their livelihood

[78] Pearson, "The Academic Aspect of the Science of National Eugenics," *Eugenics Laboratory Lecture Series*, 7 (London, 1911), p. 20.

[79] First quote from letter of 20 June 1905, second from 18 May 1903; both from Pearson to Galton, file 293 F, *FGP*. See also Galton letter, 18 May 1903, file 245/18F. The exchange was inspired by a meeting between Galton and Luigi Bodio, a leader of the International Statistical Institute.

[80] See Norton, "Pearson" (n. 68), p. 24.

[81] Pearson, *Speeches* (n. 67), p. 23.

from expertise rather than capital. His eugenic society would be a utopia for technocrats, not workers.[82]

Well before 1890, Pearson's socialism had become social Darwinism, or rather, socialist Darwinism. Natural selection at the level of individuals had been eclipsed by the progress of civilization, he thought, and in its place history had arisen as the arena of struggle. There remained a form of individual selection, which he called reproductive selection and which was not based on "social fitness," as he thought it ought to be, but simply on tendency to procreate. Pearson's revered professionals, of course, proved least prolific, while the dregs of society reproduced without restraint. The result was a deterioration of the human stock which education and environmental amelioration could scarcely begin to counteract, and which inevitably resulted in a decline of social fitness. The imperative of survival required maximum social efficiency, which meant, first of all, thorough eugenic measures, for reproduction was the business of the state, not of individuals. Economic competition, similarly, was "not the correct attitude from the standpoint of science,"[83] for victory in the struggle for survival among states will go to the best organized. Individualism must be rooted out for the same reason. Pearson justified *The Grammar of Science*, first published in 1892, by invoking the need to eliminate all judgments based on personal feelings and individual peculiarities. Self-interest must give way to citizenship, which demands a supra-personal standard of judgment, science.[84] Science should assume its rightful place as the nervous system of the social organism—scientific was, for Pearson, synonymous with social. In Pearson's thought, statistics as liberal social science was submerged completely under statistics as the agent of bureaucratic control.

In another sense, Pearson was a true follower of Quetelet. Both agreed on the universality of number and on the absence of discontinuity. Both maintained that the task of science was not to chart a bold

[82] Pearson's desire to identify political and economic power with scientific expertise is clear in a wide range of his popular writings, spanning four or five decades, but see especially *The Function of Science in the Modern State* (1902; Cambridge, 2d ed., 1919). MacKenzie's interpretation of eugenics is more persuasive for Pearson than for Galton. While the latter identified and associated with the nascent professional middle classes, he had of course inherited his fortune, and his ideology was manifestly less self-serving than Pearson's. In any event, this professionalizing self-interest would seem to be equally well served by an environmentalist as by a eugenic reform movement.

[83] Pearson, *National Life from the Standpoint of Science* (London, 1905), p. 59; see also pp. 13-14.

[84] Pearson, *Grammar* (n. 27), p. 6; see also "The Ethic of Freethought" (1883), in *Freethought* (n. 77), p. 8.

new course, but to study the laws of social development so that scientific policy might affirm them and remove all obstacles to their attainment.[85] Although Pearson's hereditarianism contrasted with Quetelet's social ameliorism, both stressed that statistics was fundamentally a practical science. Pearson argued repeatedly that it "will only grow with due sense of proportion, in touch with practical needs, and if it develops in association with anthropometry, medicine, biometry, and the sciences of heredity and psychology."[86] More generally, he wrote that

> . . . the bulk of applied science problems are first formulated, and afterwards the extension of pure science which leads to their solution worked out. . . . Of course this rule is not universal, but it is strikingly illustrated in pure and applied mathematics. Developments of pure analysis are now and then of occasional use, but pure mathematics have over and over again been enriched by analysis directly invented for the solution of special physical or even technical problems.[87]

Egon Pearson, like his father a distinguished historian of statistics as well as an eminent statistician, noted that Karl Pearson was, at least during his most creative period, 1893-1901, "far too much interested in pushing forward with the statistical examination of problems of heredity to question very fully the theoretical possibilities of his results."[88]

Pearson's eugenic convictions provide the principal explanation for the enthusiasm with which he took up the study of statistics in 1893. The inherent compatibility of statistics with his philosophical disposition is also noteworthy, however. Pearson appears in *The Grammar of Science* as a Machian positivist. He argued that there is no knowable "thing in itself" underlying our perceptions of phenomena, that sequences of perceptions provide the whole basis of our knowledge, and that we can know nothing of causation. Scientific principles he thought, are justified by their contribution to economy of thought, which increases our social if not personal fitness. By the time he published his third edition, in 1911, he had become convinced that homogeneity was

[85] Pearson, "Women and Labour" (1894), in *Chances of Death* (n. 75), p. 243; also MacKenzie, p. 80.

[86] K. Pearson (1932), quoted in E. S. Pearson, *Pearson* (n. 62), p. 119.

[87] K. Pearson, *Function of Science* (n. 82), p. 77. The passage following the ellipses is a footnote to the passage preceding it.

[88] E. S. Pearson, *Pearson* (n. 62), p. 29; also pp. 26-27 and "Some Incidents in the Early History of Biometry and Statistics," *SHSP1*, 323-338, p. 323.

a metaphysical conceit of philosophical realists, even with respect to atoms, and that statistics was in the deepest sense inherent in all scientific explanation. His mature view, as Bernard Norton characterizes it, was that "no theory would be an accurate describer of nature if it employed homogeneous classes as explainers."[89]

Pearson found justification for a statistical approach both in his socialism and in his positivism. This science of "mass phenomena," as Pearson quoted "the Germans,"[90] embodied an appropriately small regard for the individual. The journal *Biometrika*, founded by Pearson and Weldon in 1901, launched its campaign for the quantitative study of evolution against the old-school biologists with the following editorial proclamation:

> It is almost impossible to study any type of life without being impressed by the small importance of the individual. In most cases the number of individuals is enormous, they are spread over wide areas, and have existed through long periods. Evolution must depend upon substantial changes in considerable numbers and its theory therefore belongs to that class of phenomena which statisticians have grown accustomed to refer to as mass-phenomena.[91]

The compatibility of statistics with a philosophy that denies knowledge of causation seemed obvious by Pearson's time, and he lectured on the philosophy of chance even before he became involved in Galtonian biology. In fact, his first contact with Galton was a letter of February 1893, in which he asked the well-known biometrician to give a lecture for this course.[92] Incidentally, Pearson borrowed his philosophy of probability, and particularly his defense of *a priori* assumptions of equipossibility within a frequentist framework, from Edgeworth.[93]

[89] Pearson also emphasized successful prediction, but did not explain why a method of organizing observations should permit prediction. See Bernard J. Norton, "Biology and Philosophy: The Methodological Foundations of Biometry," *JHB*, 8 (1975), 85-93, p. 91, and "Metaphysics and Population Genetics: Karl Pearson and the Background to Fisher's Multi-Factorial Theory of Inheritance, *Annals of Science*, 32 (1975), 537-553, pp. 550-551.

[90] Pearson, *Social Problems: Their Treatment, Past, Present, and Future* (London, 1912), p. 19.

[91] Editorial, *Biometrika*, 1 (1901-1902), 3.

[92] Pearson to Galton, 28 Feb. 1893, file 293 A, *FGP*.

[93] See Pearson, *Grammar* (n. 27), pp. 142-146; also excerpts from his 1892 syllabus of lectures on "The Laws of Chance" in E. S. Pearson, *Pearson* (n. 62), p. 23. There is, it may be noted, no reason to think that he believed in irreducible, objective chance. He wrote in "The Ethic of Freethought" (n. 84), p. 19, that regular succession of cause and effect was essential to ordered thought, and that "[t]o conceive of a law as interrupted is to conceive of something in other than the only conceivable way, is impossible." This, however, was in 1883, before he had developed his phenomenalistic philosophy of science.

In the wake of his conversion to statistical biometry, Pearson came to see, in the Laplacian understanding of chance as imperfect knowledge, legitimation for the application of *metron* to *bios*. As a consequence of Galton's investigations, he wrote, "[b]iological phenomena in their numerous phases, economic and social, were seen to be only differentiated from the physical by the intensity of their correlations."[94] He wrote in 1911, in a passage less prophetic than it first appears, that twenty years hence scientists might "recognise that the distinction between the physical and biological sciences is really only quantitative, and the physicists who now see only absolute dependence or perfect independence may then smile over the penurious narrowness of mathematical function as they smile now over the insufficiency of the old laws of motion."[95] The philosophical spirit in which this remark should be interpreted is suggested by Pearson's comment that exactitude even of mathematics is the product of idealization. "Geometry might almost be termed a branch of statistics, and the definition of the circle has much the same character as that of Quetelet's *l'homme moyen*."[96]

Pearson's early technical work virtually defined the new mathematical field of statistics, but it was, as he always insisted, closely allied to certain practical problems of science. His initial papers concerned the form of natural distributions, which he continued to regard throughout his most creative period as a central problem of statistics. He began as a firm believer in what he dubbed in his first statistical paper the "normal curve." "When a series of measurements gives rise to a normal curve, we may probably assume something approaching a stable condition; there is production and destruction impartially around the mean."[97] The absence of normality, he thought, was probably an indication of heterogeneity, and he proposed in this paper to establish a technique for separating these nonnormal distributions into what he presumed were their normal components. His inspiration had in fact come from some measurements by Weldon of Naples crabs, the distribution of whose foreheads seemed to indicate a nascent division into two types. Thus Pearson was seeking a statistical criterion of speciation. He also enter-

[94] *Galton*, vol. 3A, p. 2. See also "Mathematical Contributions to the Theory of Evolution.—XI. On the Influence of Natural Selection on the Variability and Correlation of Organs," *Phil Trans*, A, 200 (1903), 1-66, p. 2.

[95] Pearson, *The Grammar of Science* (3d ed., 1911; repr. New York, 1957), p. xii; see also pp. 157-165.

[96] Pearson, *Grammar*, 2d ed. (n. 27), p. 177n.

[97] Pearson, "Contributions to the Mathematical Theory of Evolution" (1894), in *ESP*, p. 2. Galton had referred to "normal variability" in the title to chapter 5 of *Natural Inheritance*.

tained the possibility, however, that mathematical analysis would reveal a difference rather than a sum of normal curves as the best decomposition. In that case, the positive component would presumably represent the birth curve, and the negative a selective death curve. This, incidentally, was the work Galton thought precarious, because it was based on a too-minute analysis of rough data.

Already when he was preparing this paper, Pearson recognized that distinctly asymmetrical distributions might sometimes appear—if, for example, "the giant tail of a normal curve were being cut off by selection," so that "errors in defect are not equally probable with errors in excess."[98] In the wake of Galton's *Natural Inheritance*, various scientists—among them A. R. Wallace, codiscoverer with Darwin of evolution by natural selection—had come to view the applicability of the error curve more critically.[99] Pearson was impressed by the asymmetry of Edgeworth's economic data and of the results of some experiments he conducted on perception of color, and soon became persuaded that non-normal variability was indeed normal.[100] In his second statistical paper, Pearson announced that skew distributions could not be presumed heterogeneous, and that these were in fact far more common in nature than the symmetrical normal curve. His model for the families of skew curves he presented in this 1895 paper was drawing from an urn without replacement. Others, including Edgeworth and Lexis, found this formulation physically uninterpretable, perhaps unintelligible, but Pearson drew on the deepest reserves of his philosophical positivism to argue that he could see no reason "why nature or economics should, from the standpoint of chance, be more akin to tossing than to teetotum-spinning or card-dealing."[101]

Pearson compared the fit of various distributions to his skew curves with that to the normal curve using material from meteorology, biology, psychology, economics, social research, and even astronomy. In every case the skew was superior. This was hardly surprising, as Pearson himself pointed out, for his family of skews included the normal as a special

[98] Pearson to Galton, 17 Nov. 1893, file 293 A, *FGP*.

[99] See Wallace's letter to Galton, 1 Dec. 1893, in file 336, *FGP*. Wallace evidently did not understand, and certainly did not accept, Galton's statistical approach.

[100] See letter of 11 March 1894 in file 293 A, *FGP*; also E. S. Pearson, "Some Incidents" (n. 88), p. 330.

[101] Pearson, "Contributions to the Mathematical Theory of Evolution.—II. Skew Variation in Homogeneous Material" (1895), in *ESP*, p. 65. See also Edgeworth, "On the Representation of Statistics by Mathematical Formulae," *JRSS*, 61 (1898), p. 551. For Lexis, see n. 104 below.

case, while it contained two additional undefined parameters which could be set to best fit the data. Pearson continued to use the normal where it seemed accurate, or where his complicated skews were inconvenient, and he pointed out quite sensibly that it must be more advantageous and prudent to work with the more accurate and general curve so long as it can be managed. At times, however, he evinced a faith in the reality of his skew curves that belied his militant positivism. This was nowhere more clear than in his essay on "The Chances of Death," which may be viewed as the last contribution to the search by nineteenth-century actuaries for the "law of mortality." Pearson's decomposition of the death curve into five parts bore a strong resemblance to Lexis's division of the same curve into three, although Pearson fit overlapping curves to all five parts, while Lexis had been content to establish the normality of death in old age. Pearson marveled that "from paupers to cricket scores, from school-board classes to ox-eyed daisies, from crustacea to birth rates, we find almost universally the same laws of frequency."[102] He also illustrated his paper with a modern drawing of a medieval image showing marksmen aiming their weapons at mortal man as he passed by the hazards of life's course from birth to old age, then proved the medievals wrong: "Our conception of chance is one of law and order in large numbers; it is not that idea of chaotic incidence which vexed the medieval mind."[103] They also erred in specifying seven ages of man. There are really five, as the empirical fit of Pearson's curves to mortality data confirmed scientifically. Lexis was convinced that this decomposition, however interesting theoretically, "gives us no further information about the causes of the mortality distribution," especially since it could not be interpreted in terms of any model,[104] and it is indeed difficult to see how Pearson could reconcile this empiricist realism with the nominalism to which he nominally adhered. For that reason, this essay illustrates all the more impressively his commitment to these skew distributions.

Pearson was largely successful at least in making his negative point,

[102] Pearson, "The Chances of Death" (1895), in *Chances of Death* (n. 75), p. 20. See also "Reproductive Selection," in *ibid.*, p. 77, where he explains conformity to his skew curves in terms of natural causes and attributes deviations from them to nonnatural causes.

[103] Pearson, "Chances of Death" (n. 102), p. 15.

[104] Lexis, "Die typischen Grössen und das Fehlergesetz," in *ATBM*, 101-129, p. 119; also "Anthropologie und Anthropometrie," in *Handwörterbuch der Staatswissenschaften*, J. Conrad et al., eds. (7 vols., Jena, 2d ed., 1898-1901), vol. 1, 388-409, pp. 394-398, esp. pp. 395-396.

that the reign of the normal law was not quite so universal as had from time to time been imagined. His family of skew curves, however, was less successful,[105] and the particular form of the distribution of variates has come to seem less important than the fact that a wide class of distributions generate mean values which are, to a close approximation, normally distributed. Hence Pearson's work on skew distributions has proven less significant for its own sake than for a particular mathematical technique he developed to demonstrate their superiority. That was the chi-square distribution, which formed the basis of a test of the "goodness of fit" between a set of measures and a curve to which they are supposed to conform, subject to random errors. Pearson developed this analysis in 1900 and used it immediately to show that even the sets of astronomical data presented in textbooks by George Airy and Mansfield Merriman to illustrate conformity to the normal law could in fact be fit to that curve only with very small or even negligible probability.[106] This and an earlier paper written with Pearson's demonstrator L.N.G. Filon formed the basis of a theory of sampling that has had great influence. The latter paper, it may be noted, was written in terms of the theory of evolution and treated sampling error to reach a theory of "random evolution" or, as Sewall Wright later called it, random drift.[107]

Pearson is best known to modern users of mathematical statistics for his method of calculating the correlation coefficient, called the product-moment method, which he derived in 1896. His interest in correlation was not limited to its philosophical implications for causality. Galton had invoked correlation in his argument for discontinuity in evolution, and Pearson worked tirelessly to show that regression could be explained entirely in terms of remote ancestry. His principal incentive for research on biological correlation, however, was the matter of "reproductive selection." Since, as Pearson argued, natural selection scarcely acts on civilized man, the direction of human evolution in advanced societies is determined by the correlation of attributes with what he took to be an inherited capacity for reproduction, and indeed the secondary correla-

[105] See, however, Churchill Eisenhart, "Pearson, Karl," in *DSB*, vol. 10, 447-473, p. 451, who notes that certain sampling distributions are governed by members of Pearson's families of skews.

[106] Pearson, "On the Criterion that a given System of Deviations from the Probable in the Case of a Correlated System of Variables is such that it can be reasonably supposed to have arisen from Random Sampling" (1900), in *ESP*, pp. 339-357.

[107] Karl Pearson and L.N.G. Filon, "Mathematical Contributions to the Theory of Evolution.—IV. On the Probable Errors of Frequency Constants and on the Influence of Random Selection on Variation and Correlation" (1898), in *ESP*, pp. 179-261.

tion of other attributes with these.[108] That it was the less intelligent, less productive, less moral individuals who, according to Pearson, were most fertile seemed to him compelling evidence of the need for eugenic interference on the part of the state. He was not content to make propaganda, however. An understanding of the correlations of organs, with all their complex interactions, was essential if the state was to achieve optimal results from its eugenic measures once it did seize the initiative on human evolution.

One other important matter to which Pearson contributed during his first and most productive decade of statistical research was the analysis of contingency tables. These involve discrete variates, and are commonly used in, for example, medical statistics, where the outcome of competing treatments may be best represented in terms of two alternative outcomes, recovery and death.[109] In 1900 Udny Yule and Karl Pearson proposed rival formulas for measuring contingency correlations. Interestingly, Yule's method was direct and pragmatic, while Pearson's professed nominalism was outweighed by his faith in continuity and his desire to maintain uniformity in the treatment of correlations of all sorts. In the event, Yule's solution has become standard, and his controversy with Pearson on this matter led in effect to his excommunication from the biometric movement. Nevertheless, he was Pearson's student and in 1900 a loyal member of his group, and his result belongs among the central technical innovations in mathematical statistics during the first decade of Pearson's work on it.[110]

These mathematical contributions to statistics form an important part of Pearson's legacy, but they by no means exhaust it. They were only part of his grand vision, the creation of a statistical biology as the basis of effective eugenics and, concomitantly, the development of a mathematical statistics that could be applied to virtually all areas of human knowledge. Pearson was always alert to the general implications of his methods, even though they often arose in a very particular context. His commitment to statistics as a coherent mathematical pursuit is indicated

[108] Pearson, "Math. Cont. III" (n. 64); Karl Pearson, Alice Lee, and Leslie Bramley-Moore, "Mathematical Contributions to the Theory of Evolution.—VI. Genetic (Reproductive) Selection: Inheritance of Fertility in Man, and of Fecundity in Thoroughbred Racehorses," *Phil Trans*, A, 192 (1899), 257-330.

[109] Contingency tables are also used in analyzing continuous magnitudes for which no valid unit of measurement exists.

[110] On the alternative derivations of Pearson and Yule, and on their controversy, see MacKenzie, chap. 7.

by his effort, largely successful, to standardize statistical concepts and terminology. "Normal curve" and "standard deviation," the latter based on an analogy with the "swing radius," or moment of inertia, in mechanics,[111] are only two of the earliest and best-known Pearsonian neologisms which have since become standard.

His adeptness as an applied mathematician was doubtless essential to his achievement as the founder of mathematical statistics, but by itself it would have availed him little. It was no less important that he was able to assemble a research group—to raise funds (a significant portion of which came from Galton), train students, and provide for publication of statistical contributions in *Biometrika*. Most important of all, however, was that Pearson endeavored to show how the mathematical methods of statistics could be usefully applied to problems, in various scientific disciplines, that were of genuine interest to contemporary practitioners. This was no inadvertent byproduct of disinterested mathematical research, but the reward for vast efforts—aided, of course, by the circumstance that scientists in many of these fields were already using crude statistical tools, and were eager to upgrade them. Pearson was no less eager to see his methods prevail. He saw the methods of statistics as unitary, even if the phenomena were diverse, and argued that skill in their use was essential for the competent pursuit of almost every science.

Illustrative of these wide interests and vast ambitions was the memoir on skew variation, the second in Pearson's series of "Contributions to the Mathematical Theory of Evolutions." There he presented his general formula for distributions of variables, then applied it to fourteen different statistical examples: barometric data, Weldon's crab measurements, height of U.S. Army recruits, height of St. Louis school girls, length-to-breadth index of Bavarian skulls, frequency of enteric fever by age, estimations of color, frequency of divorce in the U.S. by duration of marriage, property valuations in England and Wales, number of petals of buttercups, projecting blossoms in clover, percentages of paupers by region in England and Wales, and, finally, the decomposition of English mortality curves.[112] The fifth memoir in the same series was on an anthropological topic, the reconstruction of the stature of primitive races, and was presented not as a narrow contribution to a particular topic, but as an "illustration of a general method."[113] He published in

[111] See "Contributions" (n. 97), pp. 2, 10; on swing radius, see *Grammar* (n. 27), p. 356.

[112] Pearson, "Math. Cont. II" (n. 101).

[113] Pearson, "Mathematical Contributions to the Theory of Evolution.—V. On the Reconstruction of the Stature of Prehistoric Races" (1898), in *ESP*, pp. 263-338.

1897 with Alice Lee, an associate in his laboratory, a major paper on the distribution of barometric heights. Again, their object was "not . . . to make an elaborate investigation of the numerical values of barometric variation or correlation, but rather to indicate to those more directly occupied with meteorological investigations how the mathematical theory of statistics may be applied to barometry with novel and, they [the authors] believe, valuable results."[114]

Pearson, of course, was not the first scientific thinker of eminence to be impressed by the potential power of the mathematical methods of probability and statistics for the social and biological sciences. Laplace, Quetelet, Cournot, Jevons, and Lexis were captivated by probability, but outside of error theory, they used it rather to illustrate its own power than to contribute to natural or social disciplines. They failed on the whole to find particular problems of genuine interest to which probability could be applied. Like many social statisticians, they suffered from the incapacity that Galton shrewdly pointed out in 1891 in characterizing anthropometry.

> So much labor is being applied to Anthropometric observation in various American Colleges, that we may appropriately consider whether it is employed in the best possible direction, and to what really valuable results it is likely to lead. It is a human frailty to which statisticians are eminently liable, to look upon means as ends. They learn to take keen pleasure in the mere accumulation of neatly tabulated figures, carefully added and averaged, quite irrespectively of any use to which those figures can be applied. They are like moneymakers, who spend their lives in piling up wealth for the pure pleasure of doing so, as if wealth were an end in itself, and not a mere instrument for making life more full, more useful, and more bright.[115]

Galton and Pearson knew how to spend their money. Whatever might be thought of their eugenic aims, they at least had aims, whose content was theoretical as well as practical and empirical. Statistics became genuinely fruitful only when wedded to theory.

[114] Karl Pearson and Alice Lee, "On the Distribution of Frequency (Variation and Correlation) of the Barometric Height at Divers Stations," *Phil Trans*, A, 190 (1897), 423-469, p. 467.

[115] Galton, "Useful Anthropometry," *Proceedings of the American Association for the Advancement of Physical Education*, 6 (1891), p. 51, quoted by Victor Hilts in a paper of much interest, as yet unpublished, on the reception of statistical techniques by American social scientists after 1889, and particularly on the early uses of correlation.

The success of Pearson's drive to universalize statistics is indicated by the widespread incorporation of more advanced statistical methods into a variety of sciences during the early decades of the nineteenth century. It is demonstrated no less impressively by the development of new statistical techniques within these disciplines—for example, by the psychologist Charles Spearman, who developed a method of factor analysis in 1904 that he and Cyril Burt used to define statistically the concept of "general intelligence."[116] Spearman, Burt, and James McKeen Cattell relied heavily on Pearson's statistical work. In economics, Warren Persons, R. H. Hooker, and Irving Fisher were soon refining the method of correlation to use it as a test of the quantity theory of money.[117] Biometrics, however, remained the discipline most closely associated with statistical innovation until at least the 1930s. Evolutionary eugenics provided a structure to social and biological statistics that transformed the collection of numbers into an activity of theoretical interest, raising questions of genuine importance that in turn called for more refined statistical techniques. It was as a consequence of their success in establishing a tradition of biometric research that Galton and Pearson must be esteemed as the founders of the mathematical field of statistics.

[116] See S. J. Gould, *The Mismeasure of Man* (New York, 1981), chap. 6. Edward Thorndike, Clark Wissler, and James McKeen Cattell had previously been investigating the structure of abilities through correlations.

[117] See Thomas M. Humphrey, "Empirical Tests of the Quantity Theory of Money in the United States, 1900-1930," *History of Political Economy*, 5 (1973), 285-316.

CONCLUSION

Statistics has assumed the trappings of a modern academic discipline—university departments, professional societies, journals, and the like—primarily during the last half century. Although a department of applied statistics was formed under Pearson at University College in 1911, and Yule was appointed a lecturer in statistics at Cambridge in 1912, the first professorship of mathematical statistics was created only in 1933, upon Karl Pearson's retirement, when his Galton chair of eugenics was split into a eugenics professorship, taken by R. A. Fisher, and a statistical chair, which was offered to Pearson's son Egon. The intellectual character of statistics, however, had been thoroughly transformed by 1900. The period when statistical thinking was allied only to the simplest mathematics gave way to a period of statistical mathematics—which, to be sure, has not been divorced from thinking. In the twentieth century, statistics has at last assumed at least the appearance of conforming to that hierarchical structure of knowledge beloved by philosophers and sociologists in which theory governs practice and in which the "advanced" field of mathematics provides a solid foundation for the "less mature" biological and social sciences. The crystallization of a mathematical statistics out of the wealth of applications developed during the nineteenth century provides the natural culmination to this story.

Most of the standard techniques of modern mathematical statistics, and also most of its formalism, have been developed since 1893, and modern statisticians usually conceive the history of their field as beginning with Galton, if not with Pearson. The correlation coefficient was defined mathematically in 1896, contingency analysis and the chi-square test in 1900, the t-test and its distribution defined by W. S. Gosset (Student) in 1908. Analysis of variance derives from a paper by Fisher in 1918. The representation of each independent datum as a dimension in space and the now-standard terminology of degrees of freedom were conceived also by Fisher, who, not coincidentally, received his Cambridge degree in mathematical physics. He was much impressed by the analogies between population genetics, in which field he began his creative statistical work, and statistical physics, where the representation of gases in a hyperspace of six or more dimensions for each molecule—one for each degree of freedom—had been developed by Maxwell and made standard by Gibbs.[1]

The break that occurred near the end of the nineteenth century in the development of statistics was by no means complete, however. The importance of statistics in this century cannot be identified wholly with the new statistical discipline. Some fields, most notably statistical physics, have remained largely separate from, if not uninfluenced by this field, while population genetics branched off from it during the 1920s and 1930s. Moreover, distinctive statistical techniques and approaches continue to be practiced in psychology, economics, sociology, ecology, agriculture, industrial quality control, medical research, experimental physics, and so on, even if the theory of mathematical statistics has been of great importance for all these fields. The various disciplines frequently teach statistics to their own students, even where departments of statistics are present on the same campus.

More impressively, perhaps, statistical innovation remained closely tied to particular applications throughout what might be called its heroic period, from Pearson to Fisher. Even today, statistics remains predominantly an applied field, studied less often for its abstract mathematical interest than with the aim of learning or developing methods for the analysis of numerical data. Not only Pearson, but Fisher, too, moved to statistics from physics as a result of an infatuation with eugenics. Fisher is as well known for his role in the "evolutionary synthesis" of the biometricians' gradual evolution by natural selection with Mendelian genetics as for his contributions to mathematical statistics. Characteristically, those two areas of activity were for him in some respects inseparable. Analysis of variance, Fisher's most important addition to the techniques of statistical analysis, and one now widely used in a variety of fields, was invented as a method for studying heredity—indeed, one might almost say, as a theory of heredity. Fisher's method permits the partitioning of variance in a given sample or population among the various factors that can be said to explain it. He developed it to resolve the sources of biological variation and particularly to establish, more conclusively than could be done by Pearson's methods, that virtually all variation in human physical measurements—and by implication also in

[1] For the analogy between the kinetic theory and genetics, see R. A. Fisher and C. S. Stock, "Cuénot on Pre-Adaptation: A Criticism," *Eugenics Review*, 7 (1915), 46-61, p. 61; also Fisher, *The Genetical Theory of Natural Selection* (Oxford, 1930), p. 36. On Fisher, see Joan Fisher Box, R. A. *Fisher: The Life of a Scientist* (New York, 1978). It is of interest that the physical concept of entropy has also entered statistics.

mental characteristics—could be attributed to various hereditary Mendelian factors. There remained a negligible five percent to be explained by environment, and Fisher left no doubt as to his view of the relative benefits potentially derivable from environmental improvement and eugenic controls.[2]

The methods of small-sample analysis are also of practical origin. They constitute a decisive break with the science of mass phenomena of the nineteenth century, when inference from small numbers was a sure sign of ignorance of probability, and reflect a shift from observation to experiment. Comparison of mean values in the nineteenth century, when not based on an assumption of perfect independence, required an accurate estimate of the probable error—and hence was scarcely reliable unless observations numbered at least in the hundreds. Small sample statistics were effectively invented by a professional employee of the Guinness brewery, W. S. Gosset, for whom the repetition of trials hundreds of times would have been far more trouble than it was worth. Gosset spent the year 1906-1907 at University College with Pearson, but with the practical needs of the brewery always in mind. In 1908 he derived a sample distribution that incorporated both the error of the mean and that of the standard deviation, what later became known as the t-statistic. Somewhat remarkably, in view of the fate of others who departed from the biometric way, Gosset did not become embroiled in controversy with Pearson, but the ever-industrious biometrician remained firmly in the tradition of mass phenomena. He remarked in a letter of 1912 to Gosset that "only naughty brewers" used numbers so small in their work.[3] Fisher, on the other hand, appreciated the value of Gosset's innovation and incorporated the same strategy in his f-statistic for the analysis of variance.

Perhaps the most significant departure from the tradition of statistical methods extending from Condorcet, Laplace, and Quetelet to Edgeworth, Galton, and Pearson was the integration of statistics with experimental design by R. A. Fisher.[4] Experimental design was closely bound to analysis of variance; Fisher took it up largely in connection with his

[2] See MacKenzie, chap. 8.

[3] See L. McMullen, " 'Student' as a Man," and E. S. Pearson, " 'Student' as Statistician," in *SHSP1*, 355-403, p. 368.

[4] This is to be distinguished from the use of statistics in experimental analysis, which has not been studied systematically. Pearson insisted that computation cannot be purchased at the end of a study, and that statistical expertise must inform the planning as well as the analysis, but he was thinking of the collection of data, not of experiment, and offered little guidance to experimental design. See Pearson to Galton, 20 June 1909, file 293 F, *FGP*.

apointment in 1919 as statistician at the Rothamsted Experimental Station. While there he collaborated closely with agricultural experimenters, and his methods of experimental design were created largely in response to practical needs.

The implication of this work was to change the character of much of statistics, to provide at last a means to move beyond empirical regularities and demonstrate the existence of causal relationships. In this sense, statistics has been brought closer to the classical conception of the scientific method, with its emphasis on analysis to simple underlying laws as the basis for a synthesis of the observed phenomena. The advantageousness of statistics, however, still presupposes an imperfect knowledge of causal structures, if not incomplete causality. Since experimental design does not enable scientists to make exact predictions of the outcome of individual experiments, statistical analysis remains dependent on an assumption that random fluctuations must in the long-run average out—indeed, at a rate and with a probability that has been calculated in the tables. With that assumption, it has become possible since Fisher not only to identify causes, but to estimate their magnitude.

Whether there has been a "probabilistic revolution" associated with the development and diffusion of statistical thinking obviously depends largely on the semantics of "scientific revolution," upon which no two persons seem to be in agreement. The widespread importance of statistical methods and statistical thinking in the natural and social sciences of the twentieth century is, however, clear. The belief in irreducible randomness characteristic of the modern quantum theory and of much philosophical thought is only a secondary consequence of the growth of statistics, and has not become established in the theories or aims of any disciplines not closely based on the quantum theory. The effect of the incorporation of statistical methods into those sciences for which the individuals remain an intractable problem, or at least fail to provide an adequate basis for understanding the populations that form their object, has been no less dramatic and much more widespread.

The development of a procedure by which the study of diverse populations could be effectively pursued was largely the achievement of the nineteenth century. It constitutes the essential precondition for the creation of a useful field of mathematical statistics. That development can be said to have occurred in two large phases.[5] Belief in a regularity of

[5] See Stephen Stigler, "Francis Ysidro Edgeworth, Statistician," *JRSS*, A, 141 (1976), 287-322, pp. 309-310.

masses that does not depend on any strong assumptions about the causes of the behavior of individuals was largely the contribution of social statistics, and more particularly of the drive to make it a science of exact laws. The development of techniques to break down this mass order and to understand the causes of variation was partly also the consequence of work in social statistics, partly of statistical physics, but primarily of the study of heredity and evolution, where the role of variation was most central. Statistical thinking itself evolved in the context of these and other applications, largely as the result of the diffusion of ideas to new subjects with new problems by way of analogy. Its effect has been not just to bring out the chance character of certain individual phenomena, but to establish the regularities and causal relationships that can be shown to prevail nonetheless. Fisher's polemical remark of 1953, interpreted broadly, applies equally to the approach of his opponents and to those other sciences that have incorporated statistical reasoning: "The effects of chance are the most accurately calculable, and therefore the least doubtful, of all the factors of an evolutionary situation."[6]

[6] Fisher, "Croonian Lecture: Population Genetics," *PRSL*, B, 141 (1953), 510-553, p. 515.

INDEX

(Because this book has no bibliography, the index includes secondary as well as primary authors, but not editors and translators.)

Index

Index

LIBRARY OF CONGRESS CATALOGING-IN-PUBLICATION DATA

Porter, Theodore M., 1953-
The rise of statistical thinking, 1820-1900.

Includes index.
1. Mathematical statistics—History. I. Title.
QA276.15.P67 1986 519.5'09 85-43306
ISBN 0-691-08416-5 (alk. paper)